INDUSTRIAL ECOLOGY

INDUSTRIAL ECOLOGY

Second Edition

T. E. Graedel
Professor of Industrial Ecology
Yale University

B. R. Allenby
Research Vice President, Technology and Environment
AT&T

Pearson Education, Inc.
Upper Saddle River, New Jersey

Library of Congress Cataloging-in-Publication Data

Graedel, T. E.

   Industrial ecology / Thomas E. Graedel and Braden R. Allenby.-- 2nd ed.

    p. cm.

Includes index.

   ISBN 0-13-046713-8

  1. Industrial ecology.   2. Product life cycle—Environmental aspects.
  3. Commercial products—Environmental aspects.   I. Allenby, Braden R.
  II. Title.

   TS161 .G74 2003

   658.4'08--dc21

                                    2002011550

Vice President and Editorial Director, ECS: *Marcia J. Horton*
Acquisitions Editor: *Dorothy Marrero*
Editorial Assistant: *Brian Hoehl*
Vice President and Director of Production and Manufacturing, ESM: *David W. Riccardi*
Executive Managing Editor: *Vince O'Brien*
Managing Editor: *David A. George*
Production Editor: *Patty Donovan*

Director of Creative Services: *Paul Belfanti*
Creative Director: *Carole Anson*
Art Director: *Jayne Conte*
Art Editor: *Greg Dulles*
Cover Designer: *Bruce Kenselaar*
Manufacturing Manager: *Trudy Pisciotti*
Manufacturing Buyer: *Lynda Castillo*
Marketing Manager: *Holly Stark*

© 2003, 1995 by AT&T
Published by Pearson Education, Inc.
Pearson Education, Inc.
Upper Saddle River, New Jersey 07458

The author and publisher of this book have used their best efforts in preparing this book. These efforts include the development, research, and testing of the theories and programs to determine their effectiveness. The author and publisher make no warranty of any kind, expressed or implied, with regard to these programs or the documentation contained in this book. The author and publisher shall not be liable in any event for incidental or consequential damages in connection with, or arising out of, the furnishing, performance, or use of these programs.

Printed in the United States of America

10 9 8 7 6 5 4 3 2 1

ISBN 0-13-046713-8

Pearson Education Ltd., *London*
Pearson Education Australia Pty. Ltd., *Sydney*
Pearson Education Singapore, Pte. Ltd.
Pearson Education North Asia Ltd., *Hong Kong*
Pearson Education Canada, Inc., *Toronto*
Pearson Educación de Mexico, S.A. de C.V.
Pearson Education—Japan, *Tokyo*
Pearson Education Malaysia, Pte. Ltd.
Pearson Education, *Upper Saddle River, New Jersey*

In memory of Robert A. Laudise
Enthusiast and supporter of industrial ecology

Contents

Preface

It has been a maxim of many years' standing that the goals of industry are incompatible with the preservation and enhancement of the environment. It is unclear whether that maxim was ever true in the past, but there is certainly no question that it is untrue today. The more forward-looking corporations and the more forward-looking nations recognize that providing a suitable quality of life for Earth's citizens will involve not less industrial activity but more, not less reliance on new technologies but more, not less interaction of technology with society but more, and that providing a sustainable world will require close attention to industry–environment interactions. This awareness of corporations, of citizens, and of governments promises to ensure that corporations that adopt responsible approaches to industrial activities will not only avoid problems but will benefit from their foresight.

Indeed, the involvement of industry is crucial if the world is to achieve sustainable development. Robert Sievers of the University of Colorado points out that governments have many critical short-term issues demanding their attention: achieving economic stability, feeding growing populations, establishing politically viable nations, making the transition from centrally controlled to free market economies, and so forth. The activities of many governments may thus be rather insular for the foreseeable future. At the same time, corporations are increasingly multinational, have longer time horizons, and depend for their survival and prosperity on relatively stable global business conditions and on responding to the desires and concerns of many different cultures and populations. Private firms, not governments, choose, develop, implement, and understand technology. Hence, responsible corporations may turn out to be among the global leaders in the transition between nonsustainable and sustainable development, but they will need the help of governments and nongovernmental organizations in establishing broad, insightful approaches and frameworks to the complex interactions that are involved.

There are three time scales of significance in examining the interactions of industry and environment. The first is that of the past, and concerns itself almost entirely with remedies for dealing with inappropriate disposal of industrial wastes. The second time scale is that of the present, and deals largely with complying with regulations, with preventing the obvious mistakes of the past, and with conducting responsible operations. Hence, it emphasizes waste minimization, avoidance of known toxic chemicals, and "end-of-pipe" control of emissions to air, water, and soil. Corporate environment and safety personnel are often involved, as are manufacturing personnel, in making small to modest changes to processes that have proved their worth over the years. On neither of these time scales do today's industrial process designers and engineers play a significant role.

The third time scale is that of the future. The industrial products, processes, and services that are being designed and developed today will dictate a large fraction of the industry–environment interactions over the next few decades. Thus, process and product design engineers hold much of the future of industry–environment interactions in their hands, and nearly all are favorably disposed toward doing their jobs with the environment in mind.

Their problem is that doing so requires knowledge and perspective never given to them during their college or professional educations and not readily available in their current positions. Remedying this situation for the students and technological professionals of today and tomorrow is the primary focus of this book. In addition, we have crafted the book so that it will be a useful addition to curricula in policy, business, environmental science, law, and other related specialties, since it is equally important that students and professionals in those disciplines understand more about technology's role in mitigating or contributing to environmental problems.

Industrial ecology, which we define in Chapter 2, is a modern approach to thinking about economy–environment interactions. As applied to manufacturing, it requires familiarity with industrial activities, environmental processes, and societal interactions, a combination of specialties that is rare. Accordingly, we have purposely attempted to make this volume useful for those whose primary background is industrial, environmental, or societal, or for those who interact with specialists in those disciplines. Many chapters include discussions of common industrial approaches to the issue being discussed as well as perspective on the environmental impacts produced. For the latter, we emphasize impacts on long time and space scales, as shorter horizons are often better recognized and better regulated either internally or externally. The approaches differ depending on the venue, but all in all, industrial ecology is about ways to transform environmental goals into reality.

The book is divided into five sections. The first provides a brief definition of the topic and outlines the approach of the book. The second, the Physical, Biological, and Societal Framework, describes the playing field on which industrial ecology operates, and the opportunities and constraints that each of these areas provides. In the third section, Design for Environment, topics central to the designers of products, processes, and services are discussed: energy, materials, product delivery and use, end of life. The central concept of life-cycle assessment is introduced here in some detail. The fourth section, Corporate Industrial Ecology, addresses the implementation of industrial ecology within the corporate environment. The final section, Systems-Level Industrial Ecology, deals with issues broader than a design, a factory, or a corporation, as we discuss industrial ecosystems, resource analysis, models and predictions, and Earth system science and engineering.

We have purposely mixed the philosophical with the practical. Avoiding that mix creates a textual product that is easier to write and easier to comprehend, but one that is much less relevant. The essence of industrial ecology is that it is the combination of technology with society, and that combination has many facets and many implications. The industrial ecologist needs to appreciate corporate and societal interactions, and to understand something of the interactions of industrial activity with the environment. Only at that point is there a logical framework in which to place goals and techniques.

We have been gratified at the reception given to the first edition of this book. It was published in 1995, at a time when industrial ecology was a young field searching for definition, approaches, and tools. Our attempt to fill this need was inevitably preliminary. Nonetheless, several thousand students and professionals used the book; Sukehiro Gotoh of Japan's National Institute for Research on the Environment translated it into Japanese; and several thousand students and professionals acquired our subsequent books, *Design for Environment* (1995), *Industrial Ecology and the Automobile* (1998), *Streamlined Life-Cycle Assessment* (1998) (T.E.G. only), and *Industrial Ecology: Policy Framework and Implications* (1999) (B.R.A. only), all published by Prentice Hall. During these few years, the field has begun its own archival publication, the *Journal of Industrial Ecology,* and has formed the International Society of Industrial Ecology. The vigor and lasting nature of industrial ecology is now unmistakable.

As industrial ecology has developed, ideas that were embryonic a decade ago have matured, and new ways of approaching technology–environment interactions have emerged. Accordingly, we spend significant parts of this edition of the book on topics completely unaddressed in the first edition: the relevance of biological ecology, indicators and metrics, the service sector, industrial symbiosis, systems analysis, and scenario development, to name a few. In a textbook for a mature field, these additions would be largely a synthesis of the work of many. Industrial ecology is new enough and its practitioners few enough, however, that parts of these topics represent our efforts to define and describe areas not yet widely addressed, but appearing to hold considerable promise. These new areas seem likely to transform industrial ecology from a topic centered rather narrowly on product design to one defining the parameters and pathways toward sustainable development. In the process, industrial ecology is becoming increasingly interdisciplinary and ever more broadly relevant.

We are grateful to many people for their help during the preparation of this second edition. Detailed reviews were performed by Arpad Horvath, University of California, Berkeley, Timothy Considine, Pennsylvania State University, M. Bertram, R. Gordon, R. Lifset, H. Rechberger, and S. Spatari, and the book is much the better for their comments. For the use of illustrations and examples, we thank I. Horkeby (Volvo Car Corporation), and R. Tierney (Pratt & Whitney). We have appreciated our interactions with the staff at Prentice Hall, especially Marcia Horton and Laura Fischer, who have helped us produce what we feel is a most attractive book. Finally, we thank AT&T, the AT&T and Lucent Foundations, the U.S. National Science Foundation, and the U.S. National Academy of Engineering for their support of industrial ecology initiatives; their help has been essential in the development of this new field.

T. E. Graedel
B. R. Allenby

PART I   Introducing the Field

C H A P T E R   1

Humanity and Environment

1.1   THE TRAGEDY OF THE COMMONS

In 1968, Garrett Hardin of the University of California, Santa Barbara, published an article in *Science* magazine that has become more famous with each passing year. Hardin titled his article "The Tragedy of the Commons"; its principal argument was that a society which permitted perfect freedom of action in activities that adversely influenced common properties was eventually doomed to failure. Hardin cited as an example a community pasture area, used by any local herdsman who chooses to do so. Each herdsman, seeking to maximize his financial well-being, concludes independently that he should add animals to his herd. In doing so, he derives additional income from his larger herd but is only weakly influenced by the effects of overgrazing, at least in the short term. At some point, however, depending on the size and lushness of the common pasture and the increasing population of animals, the overgrazing destroys the pasture and disaster strikes all.

A modern version of the tragedy of the commons has been discussed by Harvey Brooks of Harvard University. Brooks points out that the convenience, privacy, and safety of travel by private automobile encourages each individual to drive to work, school, or stores. At low levels of traffic density, this is a perfectly logical approach to the demands of modern life. At some critical density, however, the road network commons is incapable of dealing with the traffic, and the smallest disruption (a stalled vehicle, a delivery truck, a minor accident) dooms drivers to minutes or hours of idleness, the exact opposite of what they had in mind. Examples of frequent collapse of road network commons systems are now legendary: Los Angeles, Tokyo, Naples, Bangkok, Mexico City.

The common pasture and the common road network are examples of societal systems that are basically local in extent, and can be addressed by local societal action. In some cases, the same is true of portions of the environmental commons: improper trash disposal or soot emissions from a combustion process are basically local problems, for example. Perturbations to water and air do not follow this pattern, however. The hydrosphere and the atmosphere are examples not of a "local commons" but of a "global commons," a system that can be altered by individuals the world over for their own gain, but, if abused, can injure all. Much of society's activities are embodied in industrial activity, and it is the relationships between industry and the environment, especially the global commons, that are the topic of this book.

It is undeniable that modern technology has provided enormous benefits to the world's peoples: a longer life span, increased mobility, decreased manual labor, and widespread literacy, to name a few. Nonetheless, there are growing concerns about the relationships between industrial activity and Earth's environment, nowhere better captured than in the pathbreaking report *Our Common Future*, produced by the World Commission on Environment and Development in 1987. The concerns raised in that report gather credence as we place some of the impacts in perspective. Since 1700, the volume of goods traded internationally has increased some 800 times. In the last 100 years, the world's industrial production has increased more than 100-fold. In the early 1900s, production of synthetic organic chemicals was minimal; today, it is over 225 billion pounds per year in the U.S. alone. Since 1900, the rate of global consumption of fossil fuel has increased by a factor of 50. What is important is not just the numbers themselves, but their magnitude and the relatively short historical time they represent.

Together with these obvious pressures on the system, several underlying trends deserve attention. The first is the diminution of regional and global capacities to deal with anthropogenic emissions. For example, carbon dioxide production associated with human economic activity has grown dramatically (Figure 1.1), largely because of extremely rapid growth in energy consumption. This pattern is in keeping with the evolution of the human economy to a more complex state, increasing growth in materials use and consumption, and an increased use of capital. The societal evolution has been accompanied by a shift in the form of energy consumed, which is increasingly electrical (secondary) as opposed to biomass or direct fossil fuel use (primary), the result being the now familiar exponential increase in atmospheric carbon dioxide since the beginning of the industrial revolution (Figure 1.2). Thus, human activities appear to be rapidly consuming the ability of the atmosphere to act as a sink for the by-products of our economic practices.

Human population growth is, of course, a major factor fueling this explosive industrial growth and expanded use and consumption of materials. Since 1700, human population has grown 10-fold: it is now approximately six billion and is anticipated to peak at between 10 and 12 billion late in the 21st century. It is generally recognized that the human population has exhibited this explosive growth since the Industrial Revolution; what is not frequently realized, but is critical, is how closely human population growth patterns are tied to technological and cultural evolution. As Figure 1.3 shows, the three great jumps in human population have accompanied the initial development of tool use, the agricultural revolution, and the Industrial Revolution. The Industrial Revolution actually consisted of both a technological revolution and a

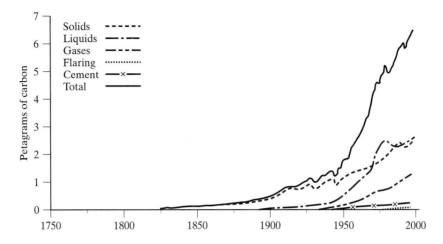

Figure 1.1

Global CO$_2$ emissions from fossil fuel burning, cement production, and gas flaring, 1860–1988. (Adapted from Carbon Dioxide Information Analysis Center, *Trends Online: A Compendium of Data on Global Change*, Oak Ridge, TN: Oak Ridge National Lab., *http://cdiac.esd.ornl.gov/trends*, accessed June 20, 2001.)

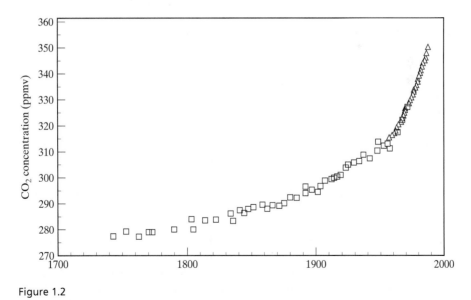

Figure 1.2

The increase in atmospheric carbon dioxide since 1700. (J.T. Houghton, G.J. Jenkins, and J.J. Ephraums, Eds., *Climate Change: The IPCC Scientific Assessment*, Cambridge, U.K.: Cambridge University Press, 1990.)

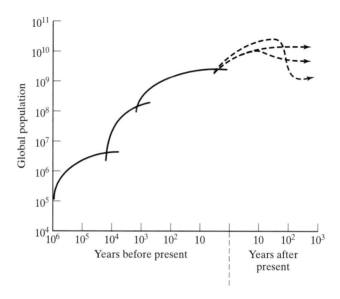

Figure 1.3

Population growth and stages of human cultural evolution. From left to right, the historical stages are tool use, the agricultural revolution, and the industrial revolution. The fourth (present and future) stage, shown in dashed lines, is that of ubiquitous technology and substantial environmental impact. Three possible scenarios for the fourth stage are pictured: one in which population stability is achieved by gradual, ordered approaches; one in which a reduced population stability occurs through a directed program of decreased use of technology; and one in which it is achieved by unmanaged growth followed by an unmanaged crash. (Based on E.S. Deevey, Jr., The human population, *Scientific American, 203* (3), 194–206, 1960, and M.G. Wolman, The impact of man, *EOS-Trans. AGU, 71,* 1884–1886, 1990.)

"neo-agricultural" revolution (the advent of modern agricultural practices), which created what appeared to be unlimited resources for population growth. Our current population levels, patterns of urbanization, economies, and cultures are now inextricably linked to how we use, process, dispose of, and recover or recycle natural and synthetic materials and energy, and the innumerable products made from them.

The above discussion suggests that the planet and its population are far from a steady state and may be on an unsustainable path. Three possible routes toward long-term stability can be postulated: (1) a managed reduction of growth until a long-term sustainable population/technology/cultural dynamic state (which we will call "carrying capacity") is achieved; (2) a managed reduction of population to a lower level sustainable with less technological activity; or (3) an unmanaged crash of one or more of the parameters (population, culture, technology) until stability at some undesirable low level is approached. Figure 1.3 suggests such possibilities.

This perspective has significant implications. When we objectively view the recent past—and 200 years is recent even in terms of human cultural evolution, and certainly in terms of our biological evolution—one fact becomes clear: the Industrial Revolution as we now know it is not sustainable. We cannot keep using materials and resources the way we do now, especially in the more developed countries. But what is the alternative?

1.2   THE MASTER EQUATION

A useful way to focus thinking on the most efficient response that society can make to environmental stresses is to examine the predominant factors involved in generating those stresses. As is obvious, the stresses on many aspects of the Earth system are strongly influenced by the needs of the population that must be provided for, and by the standard of living that population desires. One of the more famous expressions of these driving forces is provided by the "master equation":

$$\text{Environmental impact} = \text{population} \times [\text{GDP/person}] \times$$
$$[(\text{environmental impact})/(\text{unit of GDP})] \quad (1.1)$$

where GDP is a country's gross domestic product, a measure of industrial and economic activity. This equation has traditionally been called the IPAT equation, where A = affluence (GDP/person) and T = technology (impact/unit of GDP). Let us examine the three terms in this equation and their probable change with time.

Earth's population is increasing rapidly, as seen in Figure 1.4 (an expansion of the most recent part of Figure 1.3). For a specific geographical region (city, country, continent), the rate of population change is given by

$$R = [R_b - R_d] + [R_i - R_e] \quad (1.2)$$

where the subscripts refer to birth, death, immigration, and emigration. Different factors can dominate the equation during periods of high birth rates, war, enhanced migration, plague, and the like. For the world as a whole, of course, $R_i = R_e = 0$. Given the rate of change, the population at a future time can be predicted by

$$P = P_0 e^{Rt} \quad (1.3)$$

where P_0 is the present population, t is the number of years in the projection, and R is expressed as a fraction. If R remains constant, the equation predicts an infinite population if one looks far enough into the future. Such a scenario is obviously impossible; at some point in the future R will have to approach zero or go negative and the population growth will thus be adjusted accordingly.

In practice, demographers predict changes in R on the basis of the age structure of populations, cultural evolution, and other factors. Countries differ, of course, and the timing and magnitude of Earth's eventual human population peak remain quite uncertain. Even in the mildest reasonable scenario, however, a global population much larger than the present level is anticipated.

The second term in Equation (1.1), the per capita gross domestic product, varies substantially among different countries and regions, responding to the forces of local

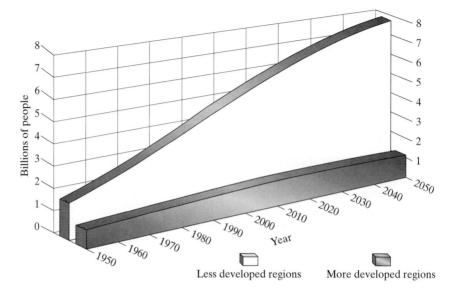

Figure 1.4

Historical and projected human population growth (billions of people) for less developed and more developed regions, 1950–2050. (Reprinted from *World Population Prospects: The 1998 Revision*, New York: United Nations Population Division, 1999.)

and global economic conditions, the stage of historical and technological development, governmental factors, weather, and so forth. The general trend, however, is positive, as seen in Table 1.1. This table is related to the aspiration of humans for a better life. Although GDP and quality of life may not be fully connected, we can expect GDP growth to continue, particularly in developing countries.

The third term in the master equation, the degree of environmental impact per unit of gross domestic product, is an expression of the degree to which technology is available to permit development without serious environmental consequences and the degree to which that available technology is deployed. The typical pattern followed by

TABLE 1.1   Growth of Real Per Capita Income in Developed and Developing Countries, 1960–2000*

Country group	1960–1970	1970–1980	1980–1990	1990–2000
Developed countries	4.1	2.4	2.4	2.1
Sub-Saharan Africa	0.6	0.9	−0.9	0.3
East Asia	3.6	4.6	6.3	5.7
Latin America	2.5	3.1	−0.5	2.2
Eastern Europe	5.2	5.4	0.9	1.6
Developing countries	3.9	3.7	2.2	3.6

*Figures are average annual percentages changes, and for the developing countries entries are weighted by population. 1990–2000 figures are estimated. Data from the World Bank, *World Development Report 1992*, Oxford University Press, Oxford, U.K., 1992.

nations participating in the Industrial Revolution of the eighteenth and nineteenth centuries is shown in Figure 1.5. The abscissa can be divided into three segments: (1) the unconstrained Industrial Revolution, during which the levels of resource use and waste increased very rapidly; (2) the period of immediate remedial action, in which the most egregious examples of excess were addressed; and (3) the period of the longer term vision (not yet fully implemented) in which one hopes that environmental impacts will be reduced to small or even negligible proportions while a reasonably high quality of life is maintained.

Although the master equation should be viewed as conceptual rather than mathematically rigorous, it can be used to suggest goals for technology and society. If our aim is to constrain the environmental impact of humanity to its present level (and one could make arguments that we need to do even better than that), we need to look at the probable trends in the three terms of the equation. The first, as discussed above, will likely increase by a factor of about 1.5 over the next half-century. The second is thought likely to increase over the same time period by a factor of between three and five. Accordingly, just to hold our environmental impact where it is today, the third term must decrease by something between 50 and 90%. This is the inspiration for calls for "Factor Four" or "Factor 10" reductions in environmental impact per unit of economic activity.

Of the trends for the three terms of the master equation, the one which perhaps has the greatest degree of support for its continuation is the second, the gradual improvement of the human standard of living, defined in the broadest of terms. The first term, population growth, is not primarily a technological but a social issue. Although countries and cultures approach the issue differently, the upward trend is clearly strong. The third term, the amount of environmental impact per unit of output, is primarily a technological term, though societal and economic issues provide strong constraints to changing it rapidly and dramatically. It is this third term in the equation that appears to offer the greatest hope for a transition to sustainable development, especially in the short term, and it is modifying this term that is among the central tenets of industrial ecology.

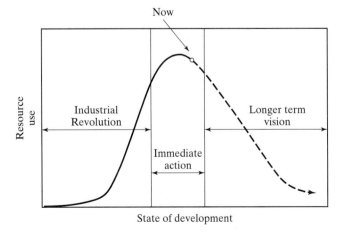

Figure 1.5

A schematic diagram of the typical life cycle of the relationship between the state of technological development of society and its resulting environmental impact.

1.3   THE GRAND OBJECTIVES

Although a number of environmental issues are worth attention, there is indisputable evidence that some environmental concerns are regarded generally or even universally as more important than others. For example, a major global decrease in biodiversity is clearly of more concern than the emission of hydrocarbon molecules from residential heating, and the Montreal Protocol and the Rio Treaty demonstrate that at least most of the countries of the world feel that understanding and minimizing the prospects for ozone depletion and global climate change are issues of universal importance. If one accepts that there are indeed such issues that have general acceptance by human society, one may then postulate the existence of a small number of "Grand Objectives" having to do with life on Earth, its maintenance, and its enjoyment. Determining these objectives requires societal consensus, which may or may not be achievable. For purposes of discussing the concept, a reasonable exposition of the Grand Objectives is the following:

The Ω_1 Objective: Maintaining the existence of the human species
The Ω_2 Objective: Maintaining the capacity for sustainable development and the stability of human systems
The Ω_3 Objective: Maintaining the diversity of life
The Ω_4 Objective: Maintaining the aesthetic richness of the planet

If it is granted that these objectives are universal, there are certain basic societal requirements that must be satisfied if the objectives are to be met. In the case of Ω_1, these are the minimization of environmental toxicity and the provision of basic needs: food, water, shelter. For Ω_2, the requirements are a dependable energy supply, the availability of suitable material resources, and the existence of workable political structures. For Ω_3, it is necessary to maintain a suitable amount of natural areas and to maximize biological diversity in disturbed areas, through, for example, the avoidance of monocultural vegetation. Perturbations due to rapid shifts in fundamental natural systems such as climate or oceanic circulation must also be addressed under this objective. Ω_4 requires control of wastes of various kinds: minimizing emissions that result in smog, discouraging dumping and other activities leading to degradation of the visible world, encouraging farming and agricultural practices that avoid land overuse and erosion, and the preservation of commonly held undeveloped land. The relationships are summarized in Table 1.2.

The Ω framework is an important prerequisite to determining what societal activities would be desirable, but the framework does not ensure progress toward achieving the objectives, especially when social consensus is involved. That progress results when desirable actions encouraged by the framework occur over and over again. In an industrialized society, a number of those actions are decisions made by product designers and manufacturing engineers. Thus, technological recommendations informed by the Grand Objectives are one means by which favorable decisions can be made.

TABLE 1.2   Relating Environmental Concerns to the Grand Objectives*

Grand objective	Environmental concern
Ω_1: Human species existence	
	1. Global climate change
	2. Human organism damage
	3. Water availability and quality
	4. Resource depletion: fossil fuels
	5. Radionuclides
Ω_2: Sustainable development	
	3. Water availability and quality
	4. Resource depletion: fossil fuels
	6. Resource depletion: non-fossil fuels
	7. Landfill exhaustion
Ω_3: Biodiversity	
	3. Water availability and quality
	8. Loss of biodiversity
	9. Stratospheric ozone depletion
	10. Acid deposition
	11. Thermal pollution
	12. Land use patterns
Ω_4: Aesthetic richness	
	13. Smog
	14. Esthetic degradation
	15. Oil spills
	16. Odor

*Numbers in the environmental concern column are for later reference.

1.3.1   Linking the Grand Objectives to Environmental Science

The Grand Objectives are, of course, too general to provide direct guidance to the product designer, who deals with specific actions relating to environmental concerns. The objectives and concerns can readily be related (see Table 1.2), but industrial decisions often require, in addition, a ranking of the relative importance of those concerns. This requirement is, in fact, a throwback to the philosophy that societal actions should be taken so as to produce the maximization of the good, and produces in turn the question "How does society determine the best actions?"

The particular difficulty of identifying the best actions of society in this instance is that societal activities related to the environment inevitably involve tradeoffs: wetland preservation versus job creation, the lack of greenhouse gas emissions of nuclear power reactors versus the chance of nuclear accident, or the preservation and reuse of clothing versus the energy costs required for cleaning, to name but a few. To enable choices to be made, many have proposed that environmental resources (raw materials, plant species, the oceans, and so forth) be assigned economic value so that decisions could be market-driven. The concept, though potentially quite useful, has proven difficult to reduce to practice, and has been further confounded by the fact that the scientific understanding of many of the issues to be valued is itself evolving and thus would require that valuation be continuously performed.

Given this uncertain and shifting foundation for relative ranking, how might specific environmental concerns, several of which are responsive to one or more of the Grand Objectives, be grouped and prioritized in an organized manner? We begin by realizing that sustainability ultimately requires:

- Not using renewable resources faster than they are replenished
- Not using nonrenewable, nonabundant resources faster than renewable substitutes can be found for them
- Not significantly depleting the diversity of life on the planet
- Not releasing pollutants faster than the planet can assimilate them

The relative significance of specific impacts can then be established by consideration of those goals in accordance with the following guidelines for prioritization:

- The spatial scale of the impact (large scales being worse than small)
- The severity and/or persistence of the hazard (highly toxic and/or persistent substances being of more concern than less highly toxic and/or persistent substances)
- The degree of exposure (well sequestered substances being of less concern than readily mobilized substances)
- The degree of irreversibility (easily reversed perturbations being of less concern than permanent impacts)
- The penalty for being wrong (longer remediation times being of more concern than shorter times)

These general criteria are perhaps too anthropocentric as stated, and are, of course, subject to change as scientific knowledge evolves, but are nonetheless a reasonable starting point for distinguishing highly important concerns from those less important. Using the criteria and the Grand Objectives, local, regional, and global environmental concerns can be grouped as shown in Table 1.3. The exact wording and relative positioning of these concerns is not critical for the present purpose; what is important is that most actions of industrial society that have potentially significant environmental implications relate in some way to the list.

Of the seven "crucial environmental concerns," three are global in scope and have very long time scales for amelioration: global climate change, loss of biodiversity, and ozone depletion. The fourth critical concern relates to damage to the human organism by toxic, carcinogenic, or mutagenic agents. The fifth critical concern is the availability and quality of water, a concern that embraces the magnitude of water use as well as discharges of harmful residues to surface or ocean waters. The sixth is the rate of loss of fossil fuel resources, vital to many human activities over the next century, at least. The seventh addresses humanity's use of land, a factor of broad influence on many of the other concerns.

Four additional concerns are regarded as highly important, but not as crucial as the first seven. The first two of these, acid deposition and smog, are regional scale im-

---

TABLE 1.3 Significant Environmental Concerns*

Crucial environmental concerns

1. Global climate change
2. Human organism damage
3. Water availability and quality
4. Depletion of fossil fuel resources
8. Loss of biodiversity
9. Stratospheric ozone depletion
12. Land-use patterns

Highly important environmental concerns

6. Depletion of non-fossil-fuel resources
10. Acid deposition
12. Smog
13. Esthetic degradation

Less important environmental concerns

5. Radionuclides
7. Landfill exhaustion
11. Thermal pollution
15. Oil spills
16. Odor

---

*Numbers are those of Table 1.2. Within groupings, numbers are for reference purposes, and do not indicate order of importance.

pacts occurring in many parts of the world and closely related to fossil fuel combustion and other industrial activities. Esthetic degradation, the third highly important concern, incorporates "quality of life" issues such as visibility, the action of airborne gases on statuary and buildings, and the dispersal of solid and liquid residues. The final concern, depletion of non-fossil-fuel resources, is one of the motivations for current efforts to recycle materials and minimize their use.

Finally, five concerns are rated as less important than those in the first two groupings, but still worthy of being called out for attention: oil spills, radionuclides, odor, thermal pollution, and depletion of landfill space. The justification for this grouping is that the effects, while sometimes quite serious, tend to be local or of short time duration or both, when compared with the concerns in the first two groups.

1.3.2 Targeted Activities of Technological Societies

The mitigation of the environmental impacts of human activities follows, at least in principle, a logical sequence. First is the recognition of an environmental concern related to one or more of the Grand Objectives. Global climate change, for example, is related to two: Ω_1 and Ω_3. Next, once that concern is identified, industrial ecologists study the activities of humanity that are related to it. For global climate change, the activities include (though are not limited to) those that result in emissions of greenhouse gases, especially CO_2, CH_4, N_2O, and CFCs. Figure 1.6 illustrates the relationship

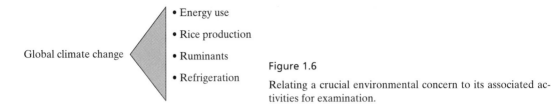

Figure 1.6

Relating a crucial environmental concern to its associated activities for examination.

schematically, and Table 1.4 lists a number of examples. The concept is that societal activities from agriculture to manufacturing to transportation to services can be evaluated with respect to their impacts on the Grand Objectives, and that the link between activities and objectives is the purpose for the environmental evaluation of products, processes, and facilities.

A characteristic of many of the activities of a technological society is that they produce stresses on more than one of the environmental concerns. Similarly, most environmental concerns are related to a spectrum of societal activities. This analytical complexity does not, however, invalidate the framework being developed here.

1.3.3   Actions for an Industrialized Society

The final step in the structured assessment process we are describing is that, given activities for examination, analysts can generate specific design recommendations to improve the environmental responsibility of their products, as shown in Figure 1.7.

Thus, the overall process described here occurs in four stages: (1) the definition by society of its Grand Objectives for life on Earth, (2) the identification by environmental scientists of environmental concerns related to one or more of those objectives, (3) the identification by technologists and social scientists of activities of society related to those concerns, and (4) the appropriate modification of those activities. Note that implementing the fourth step in this sequence depends on accepting the definition of step 1, believing the validity of step 2, and acknowledging the correct attribution in step 3, but not necessarily in knowing the magnitude of the impact of step 4 on improving the environment, i.e., the information that is needed tends to be qualitative, not quantitative. From the standpoint of the industrial manager or the product design engineer, what is important is knowing that if step 4 is taken, the corporation's environmental performance will be improved to at least some degree, and, perhaps at least as important, knowing that customers and policy makers will regard the action as a positive and thoughtful one.

In overview, the four steps of the process are schematically displayed in Figure 1.8, in which it is clear that their influence on the Grand Objectives determines the importance of each of the environmental concerns, and that each of the environmental concerns leads to a group of activities for examination, each of which in turn leads to a set of product design recommendations. Note that each of the Grand Objectives and most environmental concerns relate to a number of recommendations, rather than to only one or two, and, conversely, many recommendations respond to more than one environmental concern and perhaps more than one Grand Objective. As shown at the

TABLE 1.4   Targeted Activities in Connection with Crucial Environmental Concerns

Environmental concern	Targeted activity for examination
1. Global climate change	1.1 Fossil fuel combustion
	1.2 Cement manufacture
	1.3 Rice cultivation
	1.4 Coal mining
	1.5 Ruminant populations
	1.6 Waste treatment
	1.7 Biomass burning
	1.8 Emission of CFCs, HFCs, N_2O
2. Loss of biodiversity	2.1 Loss of habitat
	2.2 Fragmentation of habitat
	2.3 Herbicide, pesticide use
	2.4 Discharge of toxins to surface waters
	2.5 Reduction of dissolved oxygen in surface waters
	2.6 Oil spills
	2.7 Depletion of water resources
	2.8 Industrial development in fragile ecosystems
3. Stratospheric ozone depletion	3.1 Emission of CFCs
	3.2 Emission of HCFCs
	3.3 Emission of halons
	3.4 Emission of nitrous oxide
4. Human organism damage	4.1 Emission of toxins to air
	4.2 Emission of toxins to water
	4.3 Emission of carcinogens to air
	4.4 Emission of carcinogens to water
	4.5 Emission of mutagens to air
	4.6 Emission of mutagens to water
	4.7 Emission of radioactive materials to air
	4.8 Emission of radioactive materials to water
	4.9 Disposition of toxins in landfills
	4.10 Disposition of carcinogens in landfills
	4.11 Disposition of mutagens in landfills
	4.12 Disposition of radioactive materials in landfills
	4.13 Depletion of water resources
5. Water availability and quality	5.1 Use of herbicides and pesticides
	5.2 Use of agricultural fertilizers
	5.3 Discharge of toxins to surface waters
	5.4 Discharge of carcinogens to surface waters
	5.5 Discharge of mutagens to surface waters
	5.6 Discharge of radioactive materials to surface waters
	5.7 Discharge of toxins to groundwaters
	5.8 Discharge of carcinogens to groundwaters
	5.9 Discharge of mutagens to groundwaters
	5.10 Discharge of radioactive materials to groundwaters
	5.11 Depletion of water resources
6. Resource depletion: fossil fuels	6.1 Use of fossil fuels for energy
	6.2 Use of fossil fuels as feedstocks
7. Land-use patterns	7.1 Development of undisturbed land
	7.2 Emissions influencing sensitive ecosystems
	7.3 Restoration of disturbed land

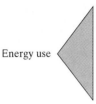

- Practice modular product design
- Develop Energy Star products
- Utilize recycled materials
- Use energy-efficient equipment

Energy use

Figure 1.7

Relating an activity for examination to associated industrial ecology recommendations for ameliorating its impact on a crucial environmental concern.

bottom of the diagram, the relationships among the Grand Objectives and recommendations provide logical interconnections among societal consensus, environmental science, and industrial ecology.

1.4   ADDRESSING THE CHALLENGE

The 20th century was a period of enormous progress, achieved in part by ignoring the possible consequences of the ways in which that progress was made to happen. The conjunction of inadequately thought-out technological approaches with rapidly rising populations and an increasing culture of consumption is now producing stresses obvious to all.

There are roles for many players in addressing the need to transform the technology–society–environment relationship. Social scientists need to understand consumption and how it may evolve and be modified. Environmental scientists and

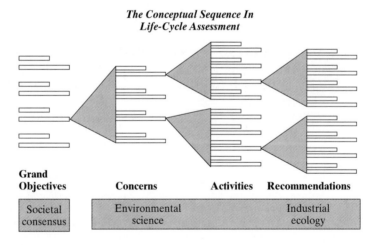

Figure 1.8

A schematic representation of the conceptual sequence in life-cycle assessment. Each of the four grand challenges is related to a number of concerns, such as climate change (each concern suggested by a horizontal bar). Similarly, each of the concerns is related to a number of activities, such as fossil fuel combustion (again, horizontal bars indicate different activities). And each activity is related to a number of recommendations, such as higher efficiency combustion. As noted at the bottom, different specialist fields treat different stages in the sequence.

materials specialists need to understand the limits imposed by a planet with limited resources and limited assimilative capacity for industrial emissions. Technologists need to develop design and manufacturing approaches that are more environmentally sound. Industrialists need to understand all these frameworks for action, and develop ways to integrate the concepts within today's corporate structures. Policy makers need to provide the proper mix of regulations and incentives to promote the long-term health of the planet rather than short-term fixes.

These are great challenges. They are the ones to which this book is addressed. We cannot treat all of them in detail, nor are all of them sufficiently developed to permit doing so even if we wished. Nonetheless, we can see many approaches that will take us in the right direction. It is time to get started.

FURTHER READING

Brooks, H., The typology of surprises in technology, institutions, and development, in *Sustainable Development of the Biosphere*, W.C. Clark and R.E. Munn, eds., Cambridge, U.K.: Cambridge University Press, 325–348, 1986.

Chertow, M., The IPAT equation and its variants: Changing views of technology and environmental impacts, *Journal of Industrial Ecology, 4* (4), 13–29, 2001.

Cohen, J.E., *How Many People Can the Earth Support?* New York: W.W. Norton, 1995.

Graedel, T.E., The grand objectives: A framework for prioritized grouping of environmental concerns in life-cycle assessment, *Journal of Industrial Ecology, 1* (2), 51–64, 1997.

Hardin, G., The tragedy of the commons, *Science, 162,* 1243–1248, 1968.

Robey, B., S.O. Rutstein, and L. Morris, The fertility decline in developing countries, *Scientific American, 269* (6), 60–67, 1993.

Stern, D.I., Progress on the environmental Kuznets curve?, *Environment and Development Economics, 3,* 173–196, 1998.

Von Weizsacker, E., A.B. Lovins, and L.H. Lovins, *Factor Four: Doubling Wealth, Halving Resource Use*, London: Earthscan Publications, 1997.

World Commission on Environment and Development, *Our Common Future*, Oxford, U.K.: Oxford University Press, 1987.

EXERCISES

1.1 In 1983 the birth rate in Ireland was 19.0 per 1000 population per year and the death rate, immigration rate, and emigration rate (same units) were 9.3, 2.7, and 11.5, respectively. Compute the overall rate of population change.

1.2 If the rate of population change for Ireland were to be stable from 1990 to 2005 at the rate computed in problem 1.1, compute the 2020 population. (The 1990 population was 3.72 million.)

1.3 Repeat the previous problem for the situation where the unrest in Northern Ireland is substantially modulated in 2005 and the emigration rate drops by 50%.

1.4 Using the "master equation," the "Units of Measurement" section in Appendix B, and the following data, compute the 1990 GNP/capita and equivalent CO_2 emissions per equivalent U.S. dollar of GNP for each country shown in the table.

1990 Master Equation Data for Five Countries

Country	Population*	GNP**	% of global CO_2[†]
Brazil	150	434,700	3.93
China	1134	419,500	9.12
India	853	262,400	4.18
Nigeria	109	27,500	1.14
United States	250	5,200,800	17.81

* Millions
** Million equivalent U.S. dollars
[†] Percentage of total global greenhouse gas emissions, expressed in equivalent CO_2 (ECO_2) units. ECO_2 is computed by using gas heating effects and lifetimes to adjust emission fluxes for several of the infrared-absorbing gases. The global flux of ECO_2 in 1990 was 13.15 Pg/yr.
Data for this table were drawn primarily from *The 1993 Information Please Environmental Almanac*, compiled by World Resources Institute, Boston, MA: Houghton Mifflin, 1993.

1.5 Trends in population, GNP, and technology are estimated periodically by many institutions. Using the typical trend predictions below, compute the equivalent CO_2 anticipated for the years 2010 and 2025 for the five countries. Graph the answers, together with information from 1990 (problem 1.4), on an ECO_2 versus year plot. Comment on the results.

Master Equation Predicted Data for Five Countries

Country	Population*		GNP growth (%/yr)		Decrease in $ECO_2/GNP(\%/yr)$
	2000	2025	1990–2000	2000–2025	
Brazil	175	240	3.6	2.8	0.5
China	1290	1600	5.5	4.0	1.0
India	990	1425	4.7	3.7	0.2
Nigeria	148	250	3.2	2.4	0.1
United States	270	307	2.4	1.7	0.7

* Millions
Data for this table were drawn primarily from J.T. Houghton, B.A. Callander, and S.K. Varney, *Climate Change 1992*, Cambridge, U.K.: Cambridge University Press, 1992.

C H A P T E R   2

The Industrial Ecology Concept

2.1   FROM CONTEMPORANEOUS THINKING TO FORWARD THINKING

Since the Industrial Revolution, the activities of firms large and small have defined much of the interaction between humanity and the environment. Such interaction has, however, traditionally been outside the topics of major significance for corporate decision makers. Technology's influence on the natural world, and especially the potential magnitude of that influence across the full spectrum of industrial activity, has been underappreciated in the business world.

However, no firm exists in a vacuum. Every industrial activity is linked to thousands of other transactions and activities and to their environmental impacts. A large firm manufacturing high technology products will have tens of thousands of suppliers located around the world, changing on a daily basis. The firm may manufacture and offer for sale hundreds of thousands of individual products to myriad customers, each with her or his own needs and cultural preferences. Each customer, in turn, may treat the product very differently and may live in areas with very different environmental characteristics, considerations of importance when use and maintenance of the product may be a source of potential environmental impact (e.g., used oil from automobiles). When finally disposed of, the product may end up in almost any country, in a high technology landfill, an incinerator, beside a road, or in a river that supplies drinking water to local populations.

In such a complex circumstance, how has industry approached its relationship with the outside world? Satisfying the needs of its customers has always been well done. Industry has, however, been less adept at identifying some of the consequences, especially the long-term consequences, of the ways in which it goes about satisfying needs. Examples of a few of these interactions have been collected by Dr. James Wei of Princeton University; we adapt and display that information here as Table 2.1. The table indicates the difficulties created for society in a world in which industrial operations have been perceived as essentially unrelated to the wider world.

It is important to note that the relationships in Table 2.1 were not the result of disdain for the external world by industry. Several of the solutions were, in fact, great improvements over the practices they replaced, and their eventual consequences could not have been forecast with any precision. What was missing was any structured attempt to relate the techniques for satisfying customer needs to possible environmental consequences. While making such attempts does not ensure that no deleterious impacts will result from industrial activity, such actions have the potential to avoid the most egregious of the impacts and to contribute toward incremental improvements in impacts that are now occurring or can be well forecast. How are such attempts best made?

The approach to industry–environment interactions that is described in this book to aid in evaluating and minimizing impacts is termed "industrial ecology." Industrial ecology (IE) is, in part, technological. As applied in manufacturing, it involves the design of industrial processes, products, and services from the dual perspectives of product competitiveness and environmental concerns. Industrial ecology is also, in part, sociological. In that regard, it recognizes that human culture, individual choice, and societal institutions play major roles in defining the interactions between our technological society and the environment.

In a later chapter we will present an extensive definition of industrial ecology that uses a biological analogy to describe the perspective from which industrial ecology views an industrial system. The essence of industrial ecology can, however, be briefly stated:

> Industrial ecology is the means by which humanity can deliberately and rationally approach and maintain sustainability, given continued economic, cultural, and technological evolution. The concept requires that an industrial system be viewed not in isolation from its surrounding systems, but in concert with them. It is a systems view in which one seeks to optimize the total materials cycle from virgin material, to finished material, to component, to product, to obsolete product, and to ultimate disposal. Factors to be optimized include resources, energy, and capital.

In this definition, the emphasis on *deliberate* and *rational* differentiates the industrial ecology path from unplanned, precipitous, and perhaps quite costly and dis-

TABLE 2.1   Relating Current Environmental Problems to Industrial Responses to Yesterday's Needs

Yesterday's need	Yesterday's solution	Today's problem
Nontoxic, nonflammable refrigerants	Chlorofluorocarbons	Ozone hole
Automobile engine knock	Tetraethyl lead	Lead in air and soil
Locusts, malaria	DDT	Adverse effects on birds, mammals
Fertilizer to aid food production	Nitrogen and phosphorus fertilizer	Lake and estuary eutrophication

ruptive alternatives. By the same token, the definition indicates that industrial ecology practices have the potential to support a sustainable world with a high quality of life for all, as opposed to, for example, an alternative where population levels are controlled by famine.

Practitioners of IE interpret the word industry very broadly: It is intended to represent the sum total of human activity, encompassing mining, manufacturing, agriculture, construction, energy generation and utilization, transportation, product use by customers and service providers, and waste disposal. IE is not limited to the domain within the factory walls, but extends to all the impacts on the planet resulting from the presence and actions of human beings. IE thus encompasses society's use of resources of all kinds.

IE may focus on the study of individual products and their environmental impacts at different stages in their life cycles, but a complementary focus is the study of a facility where products are made. In such a facility, raw materials, processed materials, and perhaps finished components produced by others are the input streams, along with energy. The emergent streams are the product itself, residues to land, water, and air, and transformed energy residues in the form of heat and noise. The IE approach to such a facility treats the budgets and cycles of the input and output streams, and it seeks to devise ways in which smaller portions of the residues are lost and more are retained and recycled within the facility itself or into the facilities of others. Key concepts include conservation of mass (all material must be accounted for), conservation of energy (all energy must be accounted for), and the technological arrow of time—the realization that as society becomes more technologically advanced, it builds on its past technological base and so cannot sustain or improve itself without strong reliance on technology.

One of the most important concepts of industrial ecology is that, like the biological system, it rejects the concept of waste. Dictionaries define waste as useless or worthless material. In nature, however, nothing is eternally discarded; in various ways all materials are reused, generally with great efficiency. Natural systems have evolved these patterns because acquiring materials from their reservoirs is costly in terms of energy and resources, and thus is something to be avoided whenever possible. In our industrial world, discarding materials wrested from the Earth system at great cost is also generally unwise. Hence, materials and products that are obsolete should be termed *residues* rather than *wastes*, and it should be recognized that wastes are merely residues that our economy has not yet learned to use efficiently. We will sometimes use the term wastes in this book where the context refers to material that is or has been discarded, but we encourage the use instead of the term residues, or perhaps the even less pejorative *experienced resources*, thereby calling attention to the engineering characteristics and societal value contained in obsolete products of all sizes and types. In doing so, we acknowledge that the law of entropy prohibits complete reuse without loss, but vision is more useful than scientific rigor in establishing this important perspective.

A full consideration of industrial ecology would include the entire scope of economic activity, such as mining, agriculture, forestry, manufacturing, service sectors, and consumer behavior. It is, however, obviously impossible to cover the full scope of industrial ecology in one volume. Accordingly, we limit the discussion in most of this book to manufacturing activities, although some of the final chapters explore the subject in more general terms.

2.2   LINKING INDUSTRIAL ACTIVITY AND ENVIRONMENTAL AND SOCIAL SCIENCES

The contrast between traditional environmental approaches to industrial activity and those suggested by industrial ecology can be demonstrated by considering several time scales and types of activity, as shown in Table 2.2. The first topic, remediation, deals with such things as removing toxic chemicals from soil. It concerns past mistakes, is very costly, and adds nothing to the productivity of industry. The second topic, treatment, storage, and disposal, deals with the proper handling of residual streams from today's industrial operations. The costs are embedded in the price of doing business, but contribute little or nothing to corporate success except to prevent criminal actions and lawsuits. Neither of these activities is industrial ecology. In contrast, industrial ecology deals with practices that look to the future, and seeks to guide industry to cost-effective methods of operation that will render more nearly benign its interactions with the environment and will optimize the entire manufacturing process for the general good (and, we believe, for the financial good of the corporation). Corporate executives are familiar with the liabilities of past and present industry–environment interactions. A challenge to the industrial ecologist is to demonstrate that viewing this interaction from the perspective of the future is a corporate asset, not a liability.

We began Chapter 1 by discussing the "Tragedy of the Commons," in which a large number of individual actions, clearly beneficial over the short term to those making the decisions, eventually overwhelm a common resource and produce tragedy for all. Originally formulated to describe such local commons resources as public grazing lands, the concept was extended to global resources with discoveries such as the Antarctic ozone hole. Industrial processes and products interact with many different commons regimes, and design engineers should interpret the concept of the commons in a very broad way. The concept certainly includes local venues such as city air, local watersheds, and natural habitats. Regional resources are also included, groundwater and precipitation (and their possible chemical alteration) being examples. The global commons calls other regimes to mind: the deep oceans (can oil spills or ocean dumping significantly degrade this resource, or is humanity's influence modest?), the Antarctic continent (50 years of international scientific activity without environmental controls have left a dubious legacy), and, of course, the atmosphere (and its ozone and climate).

In practice, every product and process interacts with at least one commons regime and probably with several, fragile and robust, monitored and ignored, close and distant. Analyses of industrial designs may produce very different results if the same

TABLE 2.2   Aspects of Industry–Environment Interactions

Activity	Time	Focus	Endpoint	Corporate view
Remediation	Past	Local site	Reduce human risk	Overhead
Treatment, disposal	Present	Local site	Reduce human risk	Overhead
Industrial ecology	Future	Global	Sustainability	Strategic

product or process is to be used under the sea, in the Arctic, or on a spacecraft rather than in a factory. Just as industrial ecology extends to all parts of the life cycle, it extends as well to all commons regimes it may affect. In short, we define *the commons* as everywhere that the human influence may be felt.

2.3   KEY QUESTIONS OF INDUSTRIAL ECOLOGY

As in any field, there are key questions in industrial ecology. Unlike biological ecology, we are interested not in the functioning of the technological system per se, but on the industrial ecosystem's interactions with and implications for the natural systems of the planet. We concentrate on a single species (humans) and its relationship with the environment. From this broad framework, it is possible to propose a set of key research questions for industrial ecology:

1. How do modern technological cycles operate, and what are the environmental implications?
   1.1 How are industrial sectors linked, and what are the corresponding environmental opportunities and threats?
   1.2 Can cycles be established for the technological materials used by our modern society?
   1.3 What elemental cycles, dominated by technology, are constrained by environmental impacts?
   1.4 How might product design and technology's use of resources be moderated?
   1.5 How rapidly can environmentally preferable technological systems be evolved?
2. How do the resource-related aspects of human cultural systems operate, and what are the environmental implications?
   2.1 How do corporations manage their interactions with the environment, and how might corporate environmental management evolve?
   2.2 How can the culture/consumption influence on materials cycles be modulated?
3. What is the future of the technology–environment relationship?
   3.1 What scenarios for development over the next several decades form plausible pictures of the future of technology and its relationship to the environment?
   3.2 What are the implications for technological systems of changes in environmental systems (climate, water, etc.)?
4. How can we operationally define and address sustainability, as contrasted with responsible environmental performance?

These key questions form the intellectual basis for the discussions throughout the rest of this volume.

2.4   AN OVERVIEW

This book, directed toward codifying and explicating ways in which to transform our technological society from what is largely a nonsustainable system to resemble more and more closely a sustainable system, is divided into sections, as shown in Figure 2.1. Part I, Introduction to Technology and Society, is composed of the information in the first several chapters. These descriptions are intended to set the stage by examining trends and patterns of industrial development and environmental impact, particularly where the links between industry and environment are readily apparent.

Part II treats several topics that comprise the framework within which industrial ecology must operate. Separate chapters address the biological system, Earth's resources, and aspects of culture and social science. The relevance of these topics demonstrates that industrial ecology, while springing from a technological base, has strong characteristics of interdisciplinarity.

The concept of industrial ecology and the framework within which it operates lead to two complementary foci. One is quite practical, centered on the ways in which the environmental attributes of products, processes, and services can be evaluated, and on how designs can address environmental issues across product and process life cycles. This is Part III.

Part IV addresses not the designer, but the corporate manager, and provides guidance on "greening" the corporation.

Part V—the final part of the book—is much more sweeping: It deals with corporations and societies as systems, and tries to understand how the human species may address environmental and resource constraints, and how it might begin the transition toward more sustainable development. Thus, the practical segment of the book provides guidance for improving our environmentally related performance, while the systems approaches help define in broad terms where we should be going.

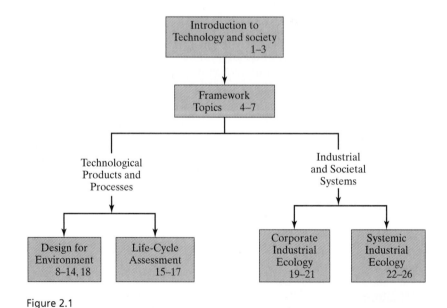

Figure 2.1

The structural outline of *Industrial Ecology*. Each box refers to the numbered book chapters indicated.

TABLE 2.3   The Key Questions of Industrial Ecology, and the Associated Tools
That May Be Brought to Bear Upon Them

Key question	Applicable chapters	Industrial ecology tools
1.1	GIF book*	Industrial sector plots
1.2	5, 23	Material flow analysis, substance flow analysis
1.3	3, 4	Grand nutrient cycles, toxicological assessment
1.4	8–17	Design for environment, life-cycle assessment
1.5	21, 22, 24, 25	Sectoral and infrastructure studies
2.1	18–21	Total environmental management, environ. metrics
2.2	6, 7	Consumption analysis
3.1	24	Scenario development
3.2	24, 25	Global climate models
4	24, 25	Systems analysis

*Greening the Industrial Facility, T.E. Graedel and J. Howard-Grenville, in preparation, 2003.

It is possible also to think of the structure of this book from the perspective of the principal research questions posed earlier in this chapter. We list those questions in Table 2.3, together with the chapters that discuss them and the industrial ecology tools used to address them. It is a rich toolbox, if one that is as yet early in its development. Notwithstanding that limitation, it appears likely that industrial ecology craftspeople will be major participants in the construction of a more sustainable society than that which we now have.

FURTHER READING

Ausubel, J.H., and H.E. Sladovich, eds., *Technology and Environment*, Washington, DC: National Academy Press, 1989.

Ayres, R.U., and L.W. Ayres, *A Handbook of Industrial Ecology*, Cheltenham, U.K.: Edward Elgar, 2002.

Clark, W.C., and R.E. Munn, eds., *Sustainable Development of the Biosphere*, Cambridge, U.K.: Cambridge University Press, 1986.

Lifset, R.J., and T.E. Graedel, Industrial ecology: Goals and definitions, in *A Handbook of Industrial Ecology*, R.U. Ayres and L.W. Ayres, eds., Cheltenham, U.K.: Edward Elgar Publishers, 3–15, 2002.

Scientific American special issues: "Managing Planet Earth," Vol. 261, No. 3, September, 1989; and "Energy for Planet Earth," Vol. 263, No. 3, September 1990.

EXERCISES

2.1 Choose a room of your apartment, dormitory, or house. Conduct an inventory of the physical items or "artifacts" in the room. Divide them into four categories: (1) the artifact is necessary for survival; (2) the function performed by the article is necessary, but the artifact represents unnecessary environmental impact (e.g., clothes may be necessary, but a fur coat or ten pairs of shoes may not be); (3) the artifact is unnecessary for survival but is culturally required; and (4) the artifact is both physically and culturally unnecessary (albeit probably desirable, or it wouldn't be there). Can you extrapolate this result to your general consumption patterns? Based on these results, what percentage of your consumption represents unnecessary environmental impact?

2.2 Technology brings benefits, such as food, home heating, and medications. It also brings potential problems, such as air pollution and ecosystem disruption. Considering your personal interactions with the products of technology (see Exercise 2.1) and those of others, what sort and scope of technology do you think is appropriate for Earth in the 21st century? How should that technology operate?

2.3 Describe a sustainable world in your own words. Include a description of the lifestyle you would expect in such a world, as well as estimates of how large a population could be supported in such a world. What data and analysis would you need to be confident that your vision was, in fact, sustainable?

CHAPTER 3

Technological Change and Evolving Risk

3.1 HISTORICAL PATTERNS IN TECHNOLOGICAL EVOLUTION

Although many people think of technology as physical artifacts, it is a far broader concept than that, particularly in the context of industrial ecology. It cannot be readily separated from the economic, cultural, and social context within which it evolves, nor can technology be separated from the natural systems with which it couples. Technology is the means by which humans and their societies interact with the physical, chemical and biological world.

It was with the advent of agriculture that humans began to exert, through their technology, significant impact on their surroundings. Human migrations spread technology across broad areas of the globe, impacting many local ecosystems. Early civilizations also caused a noticeable increase in atmospheric carbon: for example, one jump resulted from the deforestation of Europe and North Africa in the 11th through the 13th centuries. Greenland ice deposits reflect copper production during the Sung Dynasty in ancient China circa 1000 B.C., and episodes of high lead concentrations in lake sediments in Sweden reflect Greek, Roman, and medieval European production of that metal many centuries earlier.

But the real changes in patterns of human, technological, and environmental interaction date from the Industrial Revolution, and from its concomitant demographic and economic shifts. People moved from agrarian communities to urban centers, and the economy shifted from agrarian activities to manufacturing. Increasingly global transportation and communication infrastructures dramatically increased economic activity, and the Industrial Revolution created the resource base for a significant population increase as well. The results can be seen in the growth in global gross domestic product (GDP); if we take 1500 as a baseline, by 1820 GDP had almost tripled, by 1900 it was up by a factor of 8.2, by 1950 it was up by over 22 times, and by 1998 it had grown over 135 times. The result was a similar accelerating growth in human impacts on natural systems (Figure 3.1).

Regardless of the technology, there is a surprising regularity in technological evolution. At all scales, technology tends to exhibit the familiar logistic growth pattern: It begins in research, invention, and innovation, experiences exponential growth as it is introduced into the market, peaks at market saturation, and is usually replaced by a newer technology as the original becomes obsolete (Figure 3.2). This general pattern, albeit over different periods of time, characterizes electricity, color television, air conditioning, and computers, among many others. (Figure 3.3).

It is also apparent that there are regularities in technological evolution at higher levels, particularly in the way that constellations of core technologies tend to define technology clusters. Table 3.1 presents one example of a technology cluster assessment of the industrial revolution, characterized not only by the major technologies constituting the cluster, but also by the nature of the environmental impacts. Note the shifts in the geographic center of activity, which can be correlated with the movement of technology from the center to the periphery, as indicated in the idealized technology lifecycle of Figure 3.2. Figure 3.4 illustrates an important dynamic of these clusters from the perspective of technology system waves: Even at the overall technology system level, it appears that the rate of innovation and technological change is accelerat-

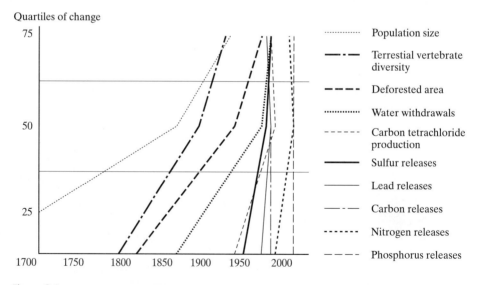

Figure 3.1

Trends in anthropogenic environmental transformation accelerated through the period of the Industrial Revolution. This figure shows the times required to achieve the second and third quartiles of change for a number of parameters: underlying these shifts are a number of technological transformations. (Adapted from R.W. Kates, B.L. Turner II, and W.C. Clark, The great transformation, in *The Earth as Transformed by Human Action*, Cambridge, U.K.: Cambridge University Press, 1–17, 1990.)

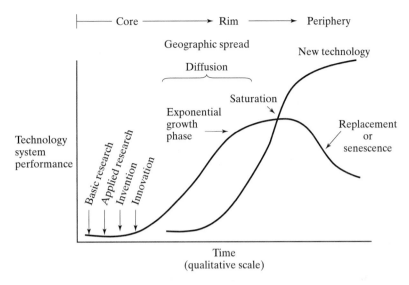

Figure 3.2

The idealized technology lifecycle. Note also the geographic spread of technology, which usually begins in the industrialized core, spreads to rim areas as it becomes more common, and reaches the further peripheries only as it is already becoming obsolete in the core. Technology thus spreads in three dimensions: economically, geographically, and temporally. (Adapted from A. Grübler, *Technology and Global Change*, Cambridge, U.K.: Cambridge University Press, 1998.)

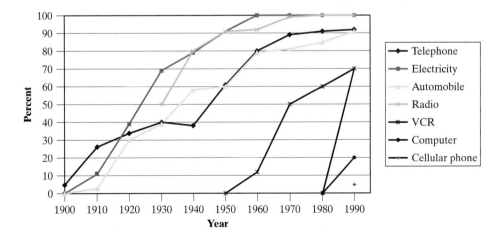

Figure 3.3

Consumer technology penetration rates, based on U.S. data. Such data, and research on the underlying dynamics, lie behind the idealized technology lifecycle model presented in Fig. 3.2.

TABLE 3.1 Technology Clusters Characterizing the Industrial Revolution

Cluster	Major technology	Geographic center of activity	Date	Nature of technology	Nature of environmental impacts
Textiles	Cotton mills/coal and iron production	British midlands/Landshire	1750–1820	Physical infrastructure (materials)	Significant but localized (e.g., British forests)
Steam	Steam engine (pumping from machinery to railroads)	Europe	1800–1870s	Enabling physical infrastructure (energy)	Diffuse; local air impacts
Heavy engineering	Steel and railways	Europe, U.S., Japan	1850–1940	Physical infrastructure (advanced energy and materials sectors)	Significant; less localized (use and disposal throughout developed regions)
Mass production and consumption	Internal combustion engine, automobile	Europe, U.S., Japan	1920s–present	Application of physical infrastructure, mass production	Important contributor to global impacts (scale issues)
Information	Electronics, services and biotechnology	U.S., Pacific Rim	1990s–present	Development of information (nonphysical infrastructure)	Reduction in environmental impact per unit quality of life?

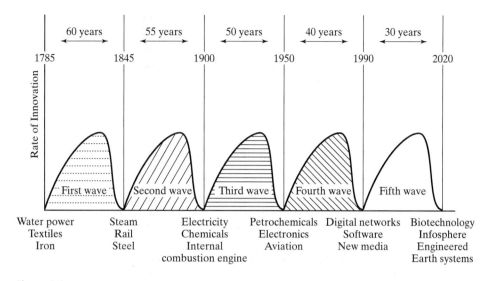

Figure 3.4

The evolution of technology can be mapped as waves of different dominant technology systems. The cycle time for each wave appears to be getting shorter, which means that the rate of technological change is getting faster even at the systems level.

ing, with the information-rich Fifth Wave taking only half the time of the water-powered First Wave.

Is the history of technology a useful guide to the technology of the future? This is a difficult question to answer, because there are important aspects of technological evolution for which we do not as yet have good explanations. Among the most important are:

1. How rapidly can fundamental technologies be evolved, and what are the relevant constraints?

2. Are different paths of technological evolution equivalent, or are there ones that are more likely to achieve desired results? What determines the paths that the evolution of a particular technology is likely to take?

3. What are the differential distributions of risks, costs, and benefits of technological evolution across sectors, social groups, and nations, and who is empowered to decide whether technologies proceed or not?

4. Technology is generally the province of private firms. What should the relationship be among private firms, interested stakeholders, the public, and governmental entities, and which groups are responsible for which elements of technological evolution?

This last question is of considerable interest in such ethically and culturally sensitive areas as biotechnology, drug development and marketing, and the information industry, where the potential for a "digital divide" between the information rich and information poor is a concern. The combination of questions 3 and 4 raise another issue: that of risk, and how to think about it, especially in a complex world, where not acting is frequently as risky as acting, and where risks, as a form of cost, frequently fall on those who are not represented in the political debates. The future of technology is thus inherently bound up with its risks, real or imagined, and with how those risks are assessed and evaluated.

3.2   APPROACHES TO RISK

Risk, that is, the probability of suffering harm from a hazard, is a reflection of a problematic world in which the future cannot be known with certainty. In environmental circles, the hazards are usually defined as impacts to human health or the environment, but in a broader social sense they include economic, cultural, and psychological impacts as well. Risk has two dimensions: objective and subjective. The objective dimension is quantitative, and is frequently captured through a series of algorithmic methodologies. It may involve the engineering of complex systems, evaluating the toxicology of a new substance, or assessing the potential economic implications of a course of action. The subjective cannot be reduced to numbers, but in practice often outweighs more objective approaches.

Many risks can be quantified quite easily. For example, Table 3.2 presents annual and lifetime mortality rates associated with common activities in the Netherlands. Note that the activities involving the highest annual mortality rates, such as smoking or driving various vehicles, are all undertaken voluntarily, an important subjective dimension of risk evaluation by individuals. In connection with the information in Table 3.2, it is of interest that standards for environmental cleanups in the United States are frequently based on a 10^{-6} lifetime risk, or one additional fatality in a million over a

TABLE 3.2   Annual Mortality Rate Associated with Certain Occurrences and Activities in the Netherlands

Activity/occurrence	Annual mortality rate	Lifetime mortality rate
Drowning as a result of dike collapse	1×10^{-7} (1 in 10 million)	1 in 133,000
Bee sting	2×10^{-7} (1 in 5.5 million)	1 in 73,000
Being struck by lightning	5×10^{-7} (1 in 2 million)	1 in 27,000
Flying	1×10^{-6} (1 in 814,000)	1 in 11,000
Walking	2×10^{-5} (1 in 54,000)	1 in 720
Cycling	4×10^{-5} (1 in 26,000)	1 in 350
Driving a car	2×10^{-4} (1 in 5,700)	1 in 76
Riding a Moped	2×10^{-4} (1 in 5,000)	1 in 67
Riding a motorcycle	1×10^{-3} (1 in 1,000)	1 in 13
Smoking cigarettes (one pack per day)	5×10^{-3} (1 in 200)	1 in 3

Source: *Ministry of Housing, Physical Planning, and Environment*, National Environmental Policy Plan: Premises for Risk Management, *7, The Hague, The Netherlands, 1991.*

lifetime. Thus, based on the data in this table, walking, cycling, or driving a car are all far more risky (have a higher probability of resulting in mortality) than the standards the United States chooses to impose on cleanups of contaminated sites. This is a significant efficiency consideration when it is recognized that such standards are the primary determinant of cost for virtually any cleanup.

But with risk evaluations, as with economic analyses, quantitative data must be treated with caution. For example, it is well known that air travel is safer than car travel—or is it? As Table 3.3 shows, this is true on a per-kilometer basis but not on a per-journey basis—and on a per-hour basis, the two are equivalent. Buses, on the other hand, are less risky than any other mode of transportation on either a per-journey or per-hour basis, and virtually as safe as air travel on a per-kilometer basis.

Professionals tend to think of risk as objective, but risks are regarded by most people as intensely subjective. This difference in approach explains studies that have asked experts and informed lay people to rank risks of technology, broadly defined, as shown in Table 3.4. This difference in perception appears to arise because the public integrates a number of subjective factors into its determination of risk, particularly:

1. The extent to which the risk appears to be controllable by the population at risk (note in Table 3.4 that college students, who ride bicycles a lot and thus feel that they are in control on them, rank that risk much lower than the experts);

2. Whether the risk is feared, even dreaded (the focus of much risk assessment on human cancer arises, in part, because of the dreaded nature of the disease);

3. The extent to which the risk is imposed rather than voluntarily assumed (note from Table 3.4 that the League of Women Voters rank contraceptives as much less risky than the experts, probably reflecting not only voluntary assumption of risk, but also familiarity with contraceptives, and quite possibly a personal understanding of the tradeoffs in risk involved in using contraceptives as well);

4. The extent to which the risk is easily observable, especially by the at-risk population, and, if observable, is manageable with current technologies;

5. Whether the victims are especially sympathetic and vulnerable (particularly children);

TABLE 3.3   Using Fatality Rates from Great Britain

Mode of transport	Per 100m passenger		
	Journeys	Hours	km
Motorcycle	100	300	9.7
Air	55	15	0.03
Bicycle	12	60	4.3
Foot	5.1	20	5.3
Car	4.5	15	0.4
Van	2.7	6.6	0.2
Rail	2.7	4.8	0.1
Bus or coach	0.3	0.1	0.04

Based on *The Economist*, January 11, 1997, 57.

TABLE 3.4 Perception of risk from most risky (1) to least risky (30) by three target audiences: educated and politically involved female citizens; college students; and experts.

Activity or technology	League of Women Voters	College students	Experts
Nuclear power	1	1	20
Motor vehicles	2	5	1
Handguns	3	2	4
Smoking	4	3	2
Motorcycles	5	6	6
Alcoholic beverages	6	7	3
General (private aviation)	7	15	12
Police work	8	8	17
Pesticides	9	4	8
Surgery	10	11	5
Fire fighting	11	10	18
Large construction	12	14	13
Hunting	13	18	23
Spray cans	14	13	26
Mountain climbing	15	22	29
Bicycles	16	24	15
Commercial aviation	17	16	26
Electric power (non-nuclear)	18	19	9
Swimming	19	30	10
Contraceptives	20	9	11
Skiing	21	25	30
X-rays	22	17	7
High school and college football	23	26	27
Railroads	24	23	19
Food preservatives	25	12	14
Food coloring	26	20	21
Power mowers	27	28	28
Prescription antibiotics	28	21	24
Home appliances	29	27	22
Vaccinations	30	29	25

From Slovic, P., Perception of risk, *Science*, 236, 280–285, 1987.

6. The extent to which the risk is new and unfamiliar, and unquantifiable or previously unknown to science;

7. The extent to which the victims are identifiable as individuals, as opposed to statistical groupings (media coverage of air crashes, for example, tends to focus on individual victims, particularly children, which may explain why the public perception of the risk associated with them differs significantly from the expert assessment);

8. The extent and type of media attention (sensationalist as opposed to factual reporting); and

9. The perceived equity of the distribution of the risk among different groups: The public is more likely to be concerned where the costs, benefits and risk of a particular activity are disproportionately allocated among groups ("distributed jus-

tice," in other words) than the expert risk assessor, interested primarily in cumulative increases or decreases in absolute risk.

There are three steps in the sequence of risk analysis: assessment, communication, and management. The first is largely objective, and attempts to use statistical and
laboratory data to quantify the risk, as in Table 3.2. The second step involves the ways
in which the results of the assessment are communicated to interested parties. The
third deals with the actions taken by organizations or governments to minimize the
risk. We discuss each of these in turn.

3.3   RISK ASSESSMENT

The type of risk assessment commonly used in environmental regulation is highly
quantitative, typically focuses on health, especially carcinogenic risks to humans, and
generally consists of five stages: (1) hazard identification; (2) delivered dose; (3) probability of an undesirable effect as a result of the delivered dose; (4) determination of
the exposed population; and (5) characterization. The last is the calculation of the total
risk impact: the number of individuals exposed multiplied by the probability that the
delivered dose will cause the undesirable effect. This may be expressed mathematically
as:

$$I = NP(d) \tag{3.1}$$

where I is the total risk impact, N is the number of individuals exposed, and $P(d)$ is the
probability, P, that the indicated dose, d, will cause the effect.

Determining the probability of an undesirable effect at the delivered dose is the
great challenge of risk assessment, particularly where one is dealing with human exposures to low levels of chemicals whose influence may be obvious only years later. Testing for such impacts as carcinogenesis or mutagenesis is not done on humans or at
typical delivered doses, but on laboratory animals and at doses high enough to produce
measurable effects in relatively short times. The results must be evaluated to determine whether they are realistic surrogates for human response, and then extrapolated
to typical delivered dose levels. The methods used to extrapolate the results tend to be
problematic, yet the choice of method can determine the outcome of the risk assessment. Figure 3.5, for example, compares four different methods of extrapolating animal data for trichloroethylene ingestion. If one anticipates drinking water
concentrations of 10 μg/l and wishes to hold lifetime risk below 10^{-6}, the W and L extrapolations indicate a problem while the M and P extrapolations do not. For many
chemicals, the probability determination is more certain than in this example, but in
general quantification becomes less reliable as probability becomes smaller.

Comprehensive risk assessment (CRA) models are based on the recognition that
there are qualitatively different categories of risk associated with environmental concerns. Most models use a taxonomy adopted by the Government of the Netherlands
that establishes three categories of risk. The first concerns damage to biological systems in general and humans in particular. The second category includes risks that aes-
thetically degrade the environment but may or may not damage biological systems.
The final category is risks involving damage to fundamental planetary systems.

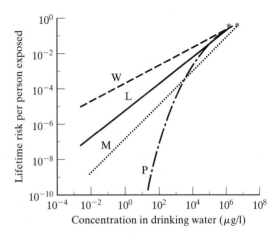

Figure 3.5

Bioassay data for ingestion of trichloroethylene in water (the asterisks in the upper right corner) and extrapolations by the logit (L), multistage (M), log probit (P), and Weibull (W) models. (Adapted with permission from C.R. Cothern, Uncertainties in quantitative risk assessment—Two examples: Trichloroethylene and radon in drinking water, in C.R. Cothern, M.A. Mehlman, and W.L. Marcus, eds., *Risk Assessment and Risk Management of Industrial and Environmental Chemicals*, 159–180. Princeton Scientific Publishing Company, 1988.)

This risk categorization can be used to derive an illustrative CRA methodology. To do so, first consider the generic risk equation (3.1). For the first category, damage to biological systems, the equation can be written as:

$$B = \beta N P(d_i), \tag{3.2}$$

where B is the comprehensive biological risk, i refers to the ith source of impact, and β is a weighting factor, agreed to by social consensus, reflecting both the objective and subjective value placed on biological systems by society. If desired, this term can be broken in two, reflecting different weighting for human and nonhuman systems.

Risk associated with aesthetic degradation can be expressed as follows:

$$A = \alpha N P(d_i) \tag{3.3}$$

where A is the aesthetic risk, N is the number of people affected by aesthetic degradation (including those who may not be physically present, but who value the impacted environment), P is the probability of an effect for the dose d of the ith source of impact, and α is again a weighting factor, reflecting societal consensus. It is likely that α would be less than β, reflecting the fact that aesthetic degradation is felt by most people to be less serious than damage to biological or human systems.

As with other categories of environmental risk, precise weighting factors and dose/response relationships for damage to planetary systems have not been established, though the direction is clear. Weighting factors should be high since the effects in this category potentially constrain the sustainability of the entire planet. It is necessary to integrate global impacts over time since they may extend for several generations. Thus,

$$G = \gamma \int_{t_0}^{t_1} N(t) P(d_{i,t}) dt \tag{3.4}$$

where G is the global risk, γ is the weighting factor, and the integration is performed from the present time t_0 through the lifetime t_1 of the substance or insult in question. Note that the dose and the affected population are time dependent.

From these three equations, the CRA is given by:

$$CRA = B + A + G \tag{3.5}$$

where the comprehensive risk equals the sum of the biological (B), aesthetic (A), and global (G) impacts for any particular subject of assessment.

3.4 RISK COMMUNICATION

Risk communication follows risk assessment, and is the stage during which the risk assessment results are made known to interested communities, organizations, and individuals. Risk assessment, in principle at least, is an objective exercise in the scientific interpretation of data. Risk communication may involve corporations, governments, and the news media. Because of these disparate actors, accurate communication of research results can be challenging, especially in emotionally charged situations.

A classic example of failure in risk communication involved the Brent Spar oil storage platform in the North Sea. When this platform became obsolete in the late 1980s, Royal Dutch Shell (co-owner of the platform with Exxon) hired environmental experts to decide how best to decommission the Brent Spar. The options were:

- Disassembly or disposal on land
- Sinking in its North Sea location
- Disassembly at the North Sea location
- Deep sea disposal

Shell and its experts chose the final option on environmental, employee safety, and economic grounds. The announcement, however, did not effectively discuss the detailed evaluation that had taken place, or the environmental disadvantages of the rejected options. As a result, the choice was widely viewed as corporate disregard for the environment. Much business and good will was lost as a result, and Shell eventually bowed to public pressure and acquiesced to disassembly on land.

In any risk communication, the potential participants include the originator or discoverer of the risk (frequently a corporation), expert analysts, various special interest groups, and the public at large. It is often difficult to find common ground, but experience shows that engaging all likely interested parties as early as possible, coming to agreement on the validity of the risk assessment, and considering everyone's goals and motivations as much as possible has a much higher probability of communicating risks with suitable accuracy and limited acrimony than is the case if these steps are ignored.

3.5 RISK MANAGEMENT

The final step in a structured basis for risk-related regulatory decisions or policy formulation is risk management. At this stage, the risk has been quantitatively evaluated (risk assessment) and interested parties have been informed (risk communication). The task then becomes to decide whether any policy or regulatory action is desirable. These decisions are inherently combinations of the scientific, the economic, and the sociological.

The task is conceptually challenging, but potentially achievable. It may be accomplished, for example, by defining a process similar to the CRA for economic, cultural, and other impacts, assuming one is willing and able to express all the results in comparable units (monetary ones, for example). The quantification of the impacts inherent in the equations can be derived from cost/benefit or other kinds of economic analysis.

For economic impact, for example, one can write:

$$E = \epsilon N P(d_i) \tag{3.6}$$

where E is economic impact, and ϵ is a weighting factor reflecting the fact that the monetary value of economic impacts may be subjectively assessed by the public as more or less than 1 when compared with other values. Also note that economic impacts occur over time; this formulation assumes present discounted value is used, or, in other words, that the integration over time has been performed in the underlying assessment.

A practical approach to risk management has been proposed by Granger Morgan of Carnegie Mellon University. Morgan defines his strategy as follows: "No individual shall be exposed to a lifetime excess probability of death from this hazard to greater than X. Whether additional resources should be spent to reduce the risks from this hazard to people whose lifetime probability of death falls below X should be determined by a careful benefit–cost calculation."

The method is shown in Figure 3.6. It takes the results of the risk assessment, shown as the curve in the figure, and chooses a specific level to be the maximum ac-

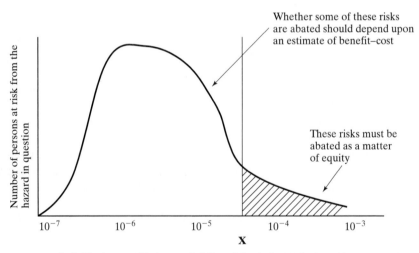

Figure 3.6

A possible risk management strategy. X is the maximum acceptable excess lifetime risk from the hazard for any individual in society. In this example, X is set to 3×10^{-5}. (Adapted from M.G. Morgan, Risk management should be about efficiency and equity, *Environmental Science & Technology, 34*, 32A–34A, 2000.)

ceptable risk (MAR). This example is for human mortality risk, and the MAR is set to that for the probability of death by lightning. Any mortality risk above that level must be abated. Possible risk abatement below that level could be determined by benefit–cost analysis. In Morgan's method, risk management is pursued up to the point where marginal benefits equal marginal costs.

Not all risk management decisions are, or can be, made by this combination of science and economics. Social and cultural values are also commonly included, either implicitly or explicitly. In the latter case, the expression

$$C = \chi N P(d_i) \tag{3.7}$$

can be defined, though no general agreement exists on how to determine the cultural impact C and the appropriate weighting factor χ.

A Comprehensive Policy Support Assessment (CPSA) that integrates the CRA with economic and cultural considerations can in principle be defined for each option under consideration. For the jth policy option, for example, one would obtain:

$$\text{CPSA}_j = B_j + A_j + G_j + E_j + C_j \tag{3.8}$$

or, equivalently:

$$\text{CPSA}_j = \text{CRA}_j + E_j + C_j \tag{3.9}$$

The above methodology implies a quantitative process. In many cases, this may be unrealistic: resources may not permit such detailed analysis, options may be difficult to define with sufficient specificity to support a quantitative assessment, or data may be so sparse or uncertain as to render quantification misleading in itself. The value in such an approach, however, is that it ensures that the full range of implications of the decision under consideration for the system is being considered at some point in the assessment process.

FURTHER READING

Allenby, B.R., *Industrial Ecology: Policy Framework and Implementation*, Upper Saddle River, NJ: Prentice Hall, 1999.

Arthur, B., Positive feedbacks in the economy, *Scientific American*, 92–99, February, 1990.

Cothern, C.R., M.A. Mehlman, and W.L. Marcus, eds., *Risk Assessment and Risk Management of Industrial and Environmental Chemicals*, Princeton; Princeton Scientific Publishing Company, 1988.

Finkel, A.M., and D. Golding, eds., *Worst Things First? The Debate over Risk-Based National Environmental Policies*, Washington, DC: Resources for the Future, 1994.

Gold, L.S., T.H. Slone, B.R. Stern, N.B. Manley, and B.N. Ames, Rodent carcinogens: setting priorities, *Science, 258*, 261–265 (1992).

Grubler, A., Time for a change: On the patterns of diffusion of innovation, in *Technological Trajectories and the Human Environment*, J. Ausubel, ed., Washington, DC: National Academy Press, 14–32, 1997.

Grubler, A., *Technology and Global Change*, Cambridge, U.K.: Cambridge University Press, 1998.

Kleindorfer, P.R., Industrial ecology and risk analysis, in *A Handbook of Industrial Ecology*, R.U. Ayres and L.W. Ayres, eds., Cheltenham, U.K.: Edward Elgar Publishers, 467–475, 2002.

Lofstedt, R.E., and O. Renn, The Brent Spar controversy: An example of risk communication gone wrong, *Risk Analysis, 17*, 131–136, 1997.

Masters, G.M., *Introduction to Environmental Engineering and Science*, Chapter 5, Englewood Cliffs, NJ: Prentice Hall, 1991.

McNeill, J.R., *Something New Under the Sun*, New York: W. W. Norton & Company, 2000.

Zeckhauser, R.J. and W.K. Viscusi, Risk within reason, *Science, 248*, 559–564, 1990.

EXERCISES

3.1 You are an environmental regulator in a developed country faced with a decision as to whether to permit mining in a wetlands wilderness area that probably contains threatened species. You are required to perform an assessment of the desirability of this activity.

  (a) What should your option set be?

  (b) Perform a CPSA for each option (this can be qualitative). What stakeholders should you involve in order to support your CPSAs? What are the major issues that can be resolved by gathering data, and what issues involve value judgments?

  (c) Are there any issues you feel are important, but you cannot fit into a CPSA? Assuming that there are, how would you ensure they are considered as part of the regulatory process?

3.2 You are in charge of the Energy Directorate in a rapidly developing Asian nation. If you do not implement rapid growth in energy production, your analysts tell you that you will reduce your nation's growth rate by approximately 2 percent per year. Your analysts also tell you that nuclear power appears to be the only option that can provide the amount of energy required within the appropriate time frame. On the other hand, Aswame, a neighboring nation with which you have fought two wars in the past decade, has threatened to develop a nuclear weapons capability if you develop nuclear power plants because it fears that your nation will divert nuclear fuel to weapons use. Moreover, you have a small but growing vocal minority that is against anything nuclear.

  (a) Given conditions in your country, develop a set of energy production and consumption options that might be available to you.

  (b) Use the CPSA methodology to organize your thoughts on the costs, benefits, and risks of each option.

3.3 Rice is the staple crop in much of Asia, but its lack of Vitamin A has significant negative health implications for the poor in that region of the world, especially in terms of child mortality. A number of researchers and companies have developed a rice cultivar that has been genetically engineered to contain Vitamin A, and are giving up certain intellectual property rights to allow such rice to be grown in those areas. Some major environmental groups oppose any use of the modified rice regardless of health or mortality benefits, claiming it is an attempt by corporations to make genetically modified organisms (GMOs), which they equate to "playing God," politically acceptable.

  (a) Use the CPSA methodology to identify costs, benefits, and risks of introducing the rice.

  (b) Can this situation be appropriately evaluated using a quantitative methodology like CPSA?

PART II   The Physical, Biological, and Societal Framework

C H A P T E R   4

The Relevance of Biological Ecology to Technology

4.1   CONSIDERING THE ANALOGY

Industrial ecology (IE) consciously incorporates the word *ecology*, a term originated with reference to biological systems. On the face of it, combining a word associated with the natural world with one associated with its exact opposite seems ludicrous, or at least injudicious. However, the idea of conceptualizing human social systems (including industrial systems) from an organismic point of view is at least a century old. The distinction between previous thinking along this line and the evolving IE approach is that the former concentrated on behavior and social structure, whereas the latter focuses especially on physical and chemical parameters: resource flows, energy budgets, and the like. The vision of industry in all its facets engaged in the cycling of resources rather than in one-time resource use was presented in 1989 by Robert Frosch and Nicholas Gallopoulos of the General Motors Research Laboratories in Michigan. Their paper is regarded by many as the first publication in the field of industrial ecology.

A working definition of biological ecology (BE) is the study of the distribution and abundance of organisms and their interactions with the physical world. Along the same lines, IE can be defined as follows:

> Industrial ecology is the study of technological organisms, their use of resources, their potential environmental impacts, and the ways in which their interactions with the natural world could be restructured to enable global sustainability.

IE has proven to be an appealing analogy because it encourages the idea of the cycling (that is, the reuse) of materials. Indeed, that is generally as far as the analogy has been taken. Attempts to view industrial activity with the images and tools of BE are hampered to some degree by BE's being basically an empirical specialty in which systems are complex, data sparse, and perspectives frequently confounded, and by IE's being too early in the process of systemic development for much information of any kind to be available. Nonetheless, parallels between BE and IE in several areas not only exist but seem natural rather than contrived. The purpose of this chapter is to explore the biological analogy in increased depth, with the idea that it holds more value than just the idea of the cyclization of resources. But how much more?

4.2   BIOLOGICAL AND INDUSTRIAL ORGANISMS

The elementary unit of study in BE is the organism, which in the dictionary is defined as "an entity internally organized to maintain vital activities." Organisms share several characteristics, broadly defined, and it is instructive to list and comment on some of them.

1. *A biological organism is capable of independent activity.* Although biological organisms vary greatly in degree of independence, all can take actions on their own behalf.

2. *A biological organism utilizes energy and materials resources.* Biological organisms expend energy to transform materials into new forms suitable for use. They also release waste heat and materials residues. Excess energy is released by biological organisms into the surroundings, as are materials residues (feces, urine, expelled breath, etc). The energy flow through any organism can be diagrammed as shown in Figure 4.1.

3. *A biological organism is capable of reproduction.* Biological organisms are all able to reproduce their own kind, though the lifetimes and number of offspring vary enormously.

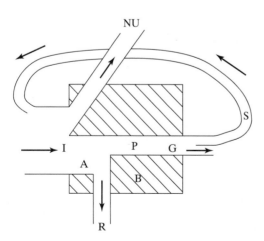

Figure 4.1

A model of an organism's energy flow in biological ecology. I = ingestion, A = assimilation, P = production, NU = not used, R = respiration, G = growth, S = storage (i.e., as fat for future use). B, the biomass in the organism, is stable only if inputs and outputs exactly balance. (Reproduced with permission from R.E. Ricklefs, *The Economy of Nature*, 3rd ed., New York: W. H. Freeman, 1993.)

4. *A biological organism responds to external stimuli.* Biological organisms relate readily to such factors as temperature, humidity, resource availability, potential reproductive partners, and so on.

5. *All multicellular organisms originate as one cell and move through stages of growth.* This characteristic is commonly recognized in every living being from moths to humans (see, for example, Shakespeare's famous "ages of man" speech in *As You Like It*).

6. *A biological organism has a finite lifetime.* Unlike some physical systems such as igneous and metamorphic rocks, which for most purposes can be regarded as having unconstrained existence, biological organisms generally have variable but limited lifetimes.

The word "organism" is not used only to refer to living things. A second definition is "anything analogous in structure and function to a living thing." Hence, we speak of "social organisms" and the like. But what about industrial activity; does it have entities that meet the definition? To make such a determination, let us select a candidate organism—the factory (including its equipment and workers)—and look at its characteristics from the perspective of the biological organism.

1. *Is an industrial organism capable of independent activity?* Factories (and their employees) clearly undertake many essentially independent activities on their own behalf: acquisition of resources, transformation of resources, and so forth.

2. *Do industrial organisms use energy and materials resources and release waste heat and material residues?* Industrial organisms expend energy for the purpose of transforming materials of various kinds into new forms suitable for use. Energy residues are emitted by industrial organisms into the surroundings, as are materials residues (solid waste, liquid waste, gaseous emissions, etc). By analogy with BE, the energy flow of an industrial organism can be diagrammed as shown in Figure 4.2.

3. *Are industrial organisms capable of reproduction?* An industrial organism is designed and constructed not for the purpose of recreating itself, but to create a

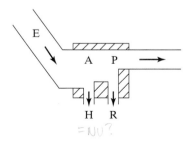

Figure 4.2

A model of an organism's (i.e., a manufacturing facility's) energy flow in industrial ecology. E = energy ingestion, A = assimilation, H = heat loss, R = respiration (i.e., energy used to run motors, etc.), P = production. Unlike the biological analog, there is no storage (at least not by design), and the organism's biomass is static.

nonorganismic product (such as a pencil). Generally speaking, new industrial organisms (factories) are created by contractors whose job is to produce any of a variety of factories to desired specifications rather than to create replicates of existing factories. If reproduction is defined as the generation of essentially exact copies of existing organisms, then industrial organisms do not meet the definition. If substantial modification is allowed for, however, we can recognize that copies or similar organisms are indeed generated. However, industrial organism reproduction is not a function of each individual organism itself, but of specialized external actors.

4. *Do industrial organisms respond to external stimuli?* Industrial organisms relate readily to such external factors as resource availability, potential customers, prices, and so on.

5. *Does an industrial organism move through stages of growth?* The analogy is stretched a bit here. Though few factories are unchanged during their lifetimes, they do not follow the orderly or predictable progression of life stages of the biological organism.

6. *Does an industrial organism have a finite lifetime?* This characteristic is certainly true.

A factory thus seems to be an appropriate candidate as an industrial organism, since it utilizes energy to transform materials just as does a biological organism. A candidate that stores materials but does not transform them is a repository, not an organism, just as soil contains the nitrate that an organism transformed from atmospheric N_2 but does not itself act on the nitrate.

The concept of industrial organisms can be expanded in ways that seem useful, even if all the above conditions are not satisfied. Indeed, the definition of a biological organism—"an animal or plant internally organized to maintain vital activities"— seems to require only two conditions: The candidate organism must not be passive (as is a sedimentary rock or a coffee cup), and the organism must make use of resources during its lifetime (as does a flower or a washing machine factory). Thus, organisms can manufacture other organisms (badgers make little badgers, factories make washing machines) and/or nonorganismal products (badgers make fecal pellets, factories make sludge). The key signature of an organism, biological or industrial, is that it is involved in resource utilization after, as well as during, its manufacture.

Biologists use several measures of resource utilization efficiency to evaluate organisms and their interactions. Three of the most common are

$$\text{Assimilation efficiency} = \frac{\text{Assimilation}}{\text{Ingestion}}$$

$$\text{Gross production efficiency} = \frac{\text{Production}}{\text{Ingestion}}$$

$$\text{Net production efficiency} = \frac{\text{Production}}{\text{Assimilation}}$$

These efficiencies are typically computed for a single organism or a group of similar organisms. The same efficiencies can be computed for industrial organisms, but are complicated in that situation by the fact that industrial organisms are hardly ever identical in capacity, age and type of equipment, or other characteristics. Industrial resource utilization efficiencies are thus most appropriate when comparing only very similar facilities, or the performance of an individual facility over time.

4.3   FOOD CHAINS: NETWORKS OF NUTRIENT AND ENERGY TRANSFER

All organisms must consume resources in order to live and go about their daily functions. The transference from one organism of these resources (nutrients and the embodied energy that they contain) to another and then another forms a food chain. The principal steps in the chain are called *trophic levels* (from the Greek word for food). The first of the traditional trophic levels in BE is that of the primary producers (e.g., plants), who utilize energy and basic nutrients to produce materials usable at the next higher trophic level (seeds, leaves, and the like). The next several BE trophic levels consist of herbivores, carnivores, and detritus decomposers (who receive residues from any of the other trophic levels and from those residues regenerate materials that can again flow to the primary producers). Trophic level types are thus four: (1) extractors, (2) producers, (3) consumers (often at several levels), and (4) decomposers. Within trophic levels it is sometimes useful to distinguish *guilds*, which are groups of species having common methods, location, or foraging practice.

A simplified marine biological food chain is shown in Figure 4.3. In addition to the dominant trophic-level actors the extractor is included, because minerals obtained from inorganic reservoirs are essential to life. Note that the food chain is not completely sequential: Decomposers receive egested material from several trophic levels, for example. Bacteria act as both extractors and decomposers, receiving carbon in one activity, minerals in the other. Omnivory (feeding on more than one trophic level) is common in nature, but it complicates the food chain diagram without adding conceptual insight, so it is not incorporated into the figure.

Similarly, a simplified industrial food chain is pictured in Figure 4.4, again neglecting omnivory. It appears that the biological trophic levels and industrial trophic levels can be described with essentially the same terminology. What is most interesting when comparing the two diagrams is perhaps not the similarities, however, but the differences:

1. *The IE food chain equivalent of the BE decomposer is the recycler.* Unlike the biological decomposer who furnishes reusable materials to primary producers, recyclers can often return resources one trophic level higher, to primary consumers. In practice, of course, the system is seldom this prescribed; many recyclers send intermediate materials to the same smelters used for primary production and are thus more like biological decomposers.

2. *The IE food chain has an additional actor, the disassembler.* The disassembler's goal is to retain resources at high trophic levels, passing as little as possible on to the recycler.

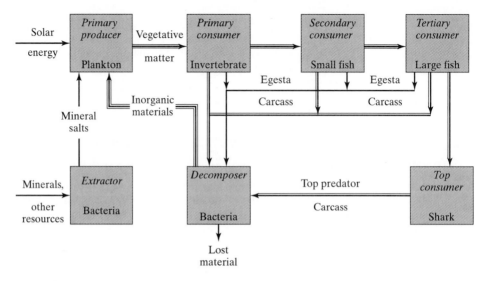

Figure 4.3

The biological food chain. The cycle begins with primary producers, who use energy and materials to generate resources usable to higher trophic levels. The number of consumer trophic levels varies in different ecosystems, and it is in some degree a matter of definition: The final consumer stage is that of the top predator or ultimate consumer. Decomposers return materials to the primary producers, thus completing the cycle. At the bottom of each box is an example of an organism in an aquatic ecosystem that plays the trophic level role. Types of resources are indicated along the flow arrows. The widths of the arrows are very rough indications of typical relative quantities of materials flow.

3. *Little in the way of resources is lost in the overall biological food chain, but much is lost in the industrial food chain.* As a consequence, the industrial ecosystem must extract a substantial portion of its resources from outside the system, the biological ecosystem only a small amount.

A bit of reflection on these diagrams point up some of the other distinctions between biological and industrial ecosystems:

1. *Speed in adapting to change.* Consider what happens if, for reason of disease in BE or regulatory inhibition in IE, the nutrient supply from an intermediate trophic level is restricted or disrupted. The industrial organism that is affected can, in all probability, quickly develop an alternative nutrient supply, perhaps by process or product redesign or by negotiation with new suppliers. The biological species can do the same, but generally over time scales of decades to millennia, rather than days to weeks.

2. *Response to increased need for resources.* Suppose a species in an intermediate trophic level finds conditions favorable for multiplication (due to a lack of competition, for example), provided suitable nutrients are available. In the IE case, increased activity on the part of extractors and primary producers can generally

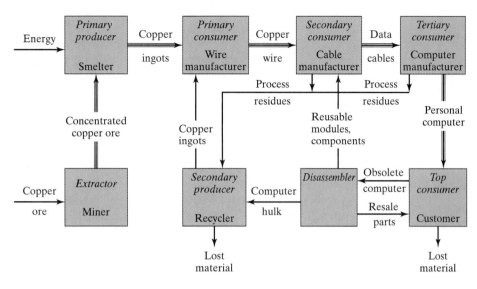

Figure 4.4

The industrial food chain, drawn by analogy with the biological food chain of Figure 4.3. Examples of organisms that play specific trophic-level roles for the example of the use of copper in personal computers appear at the bottom of each box. As in Figure 4.3, the widths of the arrows are very rough indicators of typical relative quantities of materials flow.

supply the needed materials, perhaps with lag times of days to a few years. In the BE case, little systemic capacity exists for substantially increasing the flow of nutrients, and the likely result is that mean individual growth and/or population increase does not take place, or does so at the expense of another species at the same trophic level.

3. *Approach to initiative.* Biological organisms are expert at working within the environment in which they find themselves. In contrast, industrial organisms strive to define the environment for themselves. BE systems are responders, IE systems are initiators.

The concept of the ecosystem (an ecological unit and its physical environment) can be generalized by drawing on the trophic-level discussions. Strictly speaking, only the initial user of materials from an extractor is an unambiguous primary producer, but organisms from many trophic levels are consumers, and many organisms operate at multiple trophic levels. In biological systems, for example, trees use nutrients to produce nuts, which are eaten by squirrels, and the nut resources are (among other things) used to produce baby squirrels. Some of those babies become food for foraging mammals and birds. The squirrel is thus both prey and predator, secondary producer and consumer. A similar situation exists in industry, where a factory acting as a consumer may receive disk drives, housings, and keyboards as a consumer and assemble computers as a secondary producer.

As with individual organisms, food chains can be studied by energy flow models. The classical BE version is shown in Figure 4.5. Energy is supplied to the chain from an external source (the sun) only at the level of the *autotrophs*, those organisms that are capable of assimilating energy from inorganic compounds. The remaining levels are made up of *heterotrophs*, which use organic material from lower trophic levels as sources of energy and nutrients. As energy flows from one trophic level to the next, it is diminished by respiration and nonutilized foodstuffs. Studies of many natural systems have shown that assimilation efficiency increases at higher trophic levels (i.e., less incoming energy is discarded without use), but that both net and gross production efficiency decrease as a consequence of the respiratory requirements of organisms operating at increasingly higher metabolic rates.

The IE version of a food chain is shown in Figure 4.6. There is one important difference between this diagram and Figure 4.5: In the former, external energy is an input at all trophic levels, not just the earliest. This energy, mostly from fossil fuel combustion but also from nuclear power and various renewable sources, reverses the energy flow trend of BE so that in IE it is generally the case that $P_n \ldots > P_3 > P_2 > P_1$. In BE, the autotrophs provide energy and nutrients for the entire food chain. In IE, we generally find three types of *acquirotrophs*: (1) those such as mining companies that extract metal ores and other useful materials to inaugurate the food chain; (2) those such as oil drillers that provide energy-producing resources that can be furnished at any point in the chain; and (3) recyclers such as scrap dealers who inject previously used materials at various points in the food chain. This situation means that the actors at higher trophic levels—the *transformotrophs*—operate as the recipients of resources from three complementary food chains rather than one. In addition, the diversity of their nutrient requirements is far greater than is the case for biological heterotrophs.

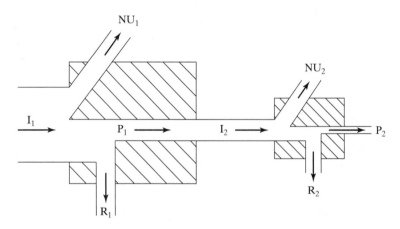

Figure 4.5

A model of energy flow in a biological food chain. The notation is the same as in Figure 4.1. The key characteristic of the chain is that the net production of one trophic level becomes the ingested energy of the next higher level. For simplicity, energy storage is not shown. (After R.E. Ricklefs, *The Economy of Nature*, 3rd ed., New York: W. H. Freeman, 1993.)

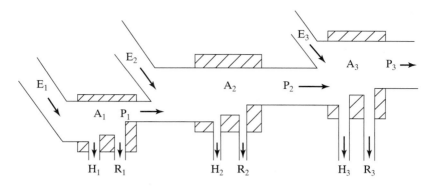

Figure 4.6

A model of energy flow in an industrial food chain. The notation is the same as in Figure 4.2. The key characteristic is that the net production of one trophic level plus added energy becomes the ingested energy of the next higher level.

At the top of the food chain is the "top predator," known alternatively as the human being, who employs products for various purposes but does not transform them; we refer to this actor as the *employotroph*.

4.4   POPULATION ECOLOGY

Population ecology is the study of individuals of a given species or several closely related species—their spatial distribution, their demography, their temporal dynamics, and their evolution. From an industrial ecology perspective, it is not necessarily very helpful to slavishly follow the BE analogue and attempt to study the characteristics of groups of industrial factories. Rather, it turns out to be of interest to look at population ecology with a focus on products. While products are not appropriately considered as organisms, since in general they have no transformative function, it appears that some of the environmentally related characteristics of products are nicely addressed by the concepts and tools of population ecology.

Products, or families thereof, exist because they fill a niche; that is, they provide a service not previously available or less well served. The numbers of products and the breadth of product families are indicators of the scale and importance of the niche that is filled. As in BE, a product dies if its niche disappears (e.g., buggy whips) or if it is subsumed by an organism whose ability to fill or expand the niche is in some way superior (e.g., the replacement of vinyl records by compact discs).

While the absolute numbers of a product or its skill in economic niche-filling may be important areas of study for economists, in IE the topics we wish to study relate more to potential resource consumption and potential environmental impacts as a result of industrial product development, dynamics, and evolution: that is, the consequences of a product rather than merely its existence. The resource consumption attributable to a product or product family depends upon the size of the family and the quantity of resources required by each individual. Mathematically, the absolute resource consumption rate α of resource i in a specific geographical area may be expressed as

$$\alpha_i = Q U_i \qquad (4.1)$$

where Q is the number of individual products annually manufactured in that area (the product production rate), and U_i is the average resource use of an individual product.

For products made from local resources, the local area is the appropriate geographical region. For many products, made from resources acquired through international trade, the global scale should be chosen.

For example, Figure 4.7 shows the use of different resources in automobiles over a 15-year period. (The data are U.S., but typical of world automobile designs.) During that same period, the number of automobiles manufactured grew dramatically. Table 4.1 shows resource utilization rates over time for the aluminum use shown in the figure.

Perhaps of more interest, at least for products that are abundant enough to have global implications for resources, are the relative resource consumption rates ρ_i, given by

$$\rho_i = \frac{Q U_i}{R_i} \qquad (4.2)$$

where R_i is the annual use rate of resource i for all purposes. For aluminum, R_i and ρ_i values are also shown in Table 4.1. If the product in question is entirely responsible for consumption of the resource, $\rho_i = 1$. Automobile manufacturers accounted for 11% of aluminum use in 1980. By 1995, the proportion had increased to 15%.

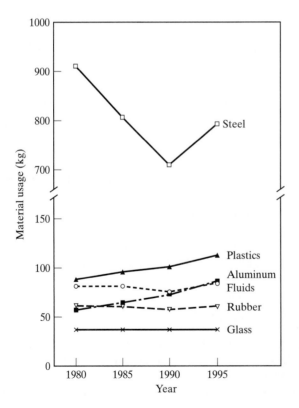

Figure 4.7

The use of different materials in the average automobile built in the United States, 1980–1995. (Data from the American Automobile Manufacturers Association.)

TABLE 4.1   Rates of Aluminum Resource Consumption by the World's Automotive Fleet

	1980	1985	1990	1995
Q (million vehicles/yr)	28.6	32.4	36.3	36.1
U (kg Al/vehicle)	60	65	70	80
α (Tg/yr)	1.7	2.1	2.5	2.9
R (Tg/yr)	15.4	15.4	19.4	19.7
ρ (unitless)	0.11	0.14	0.13	0.15

U_i values are from Figure 4.7; aluminum use rates are from the 1999 edition of the *Materials Yearbook,* Washington, DC: U.S. Geological Survey; vehicle manufacture rates are from *World Motor Vehicle Data,* Washington, DC: American Auto Manufacturers Association, 1998.

The equation demonstrates that there are two possible ways to reduce ρ should that be desired. One can reduce the product density Q, perhaps by substituting a product that performs an equivalent function while utilizing a different resource. Alternatively, one can reduce U_i while maintaining product functionality; this is termed dematerialization. In principal, one could also decrease ρ by increasing R, but this constitutes an increase in overall use of resource i, so is unhelpful.

Populations of products can also be related to the potential environmental impact that they generate during manufacture, use, or at the end of life. This potential impact is a function of the population (or population change) of a product or product family and its environmentally related characteristics. Thus, the absolute environmental impact rate ε of impact type j in a specific geographical area is

$$\varepsilon_j = P\,I_j \tag{4.3}$$

where I_j is the average type j impact of an individual product and P is the number of products.

In the case of environmental impacts, it is important to determine the relative importance of specific products so as to determine the degree to which changes might be desirable. The relative environmental impact rate is given by

$$\varphi_j = \frac{P I_j}{\Phi_j} \tag{4.4}$$

where Φ_j is the overall environmental impact of type j. If the product in question is entirely responsible for impact j, $\varphi_j = 1$.

4.5   CLASSIFICATION OF SPECIFIC LINKAGES

It is instructive to think of the materials cycles associated with a postulated primitive biological system such as might have existed early in Earth's history. At that time, the potentially usable resources were so large and the amount of life so small that the existence of life forms had essentially no impact on available resources. This process was one in which the flow of material from one stage to the next was essentially independent of any other flows. Schematically, what we might term a Type I system takes the form of Figure 4.8a.

As the early life forms multiplied, external constraints on the unlimited sources and sinks of the Type I system began to develop. These constraints led in turn to the development of resource cycling as an alternative to linear materials flows. Feedback and cycling loops for resources were developed because scarcity drove the process of change. In such systems, the flows of material within the proximal domain could have been quite large, but the flows into and out of that domain (i.e., from resources and to waste) eventually were quite small. Schematically, such a Type II system might be expressed as in Figure 4.8b.

The above discussion of Type I and Type II systems refers to the planet as a whole and regards the ecosystem types as sequential. Individual ecosystems, however, may be of either type in any epoch. A Type I system, known ecologically as an open system, is one in which resource flows into and out of the system are large compared with the flows within it. A Type II system, known as a closed system, is the reverse. Complicating the issue, especially on the small scale, is the fact that an ecosystem can be open with respect to one resource (water, for example) and closed with respect to another (nitrogen).

A Type II system is much more efficient than a Type I system, but on a planetary scale it clearly is not sustainable over the long term because the flows are all in one direction, that is, the system is "running down." To be ultimately sustainable, the global biological ecosystem has evolved over the long term to the point where resources and waste are undefined, because waste to one component of the system represents resources to another. Such a Type III system, in which complete cyclicity has been achieved (except for solar energy), may be pictured as in Figure 4.8c.

As suggested above, recycle loops have, as inherent properties, temporal and spatial scales. The ideal temporal scale for a resource recycle loop in a biological ecosystem is short, for two reasons. First, material viewed by a specific organism as a resource may degrade if left unused for long periods and thus become less useful to that organism. Second, resources not utilized promptly must be retained in some sort of storage facility, which must first be found or constructed, then defended. Still, examples of resource storage in biological systems are relatively common: Squirrels store nuts, birds store seeds, and so forth; the organisms trade off the costs of storage against the benefits to themselves or their offspring.

Analogously, the ideal spatial scale for most resource recycle loops in a biological ecosystem is small. One reason is that procuring resources from far away has a high energy cost. A second is that it is much more difficult to monitor and thus ensure the availability of spatially distant resources. As with temporal scales, counterexamples come readily to mind, such as eagles that hunt over wide areas. However, the eagles would probably have evolved to hunt close to home if such a strategy provided adequate resources.

The ideal anthropogenic use of the materials and resources available for industrial processes (broadly defined to include agriculture, the urban infrastructure, etc.) would be one similar to the cyclic biological model. Historically, however, human resource use has mimicked the Type I unconstrained resource diagram above (Figure

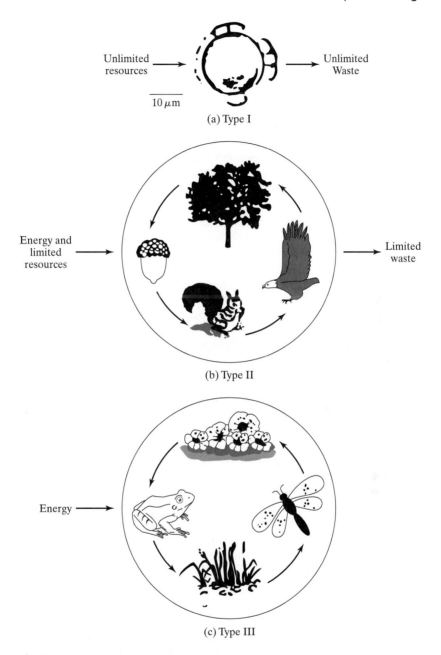

Figure 4.8

(a) Linear materials flows in Type I biological ecology. (b) Quasicyclic materials flows in Type II biological ecology. (c) Cyclic materials flows in Type III biological ecology.

4.8a). Such a mode of operation is essentially unplanned and imposes significant economic costs as a result. IE, in its implementation, is intended to accomplish the evolution of technological systems from Type I to Type II, and perhaps ultimately to Type III, by optimizing in ensemble all the factors involved, as suggested in Figure 4.9.

Temporal and spatial scales are also important considerations in IE. As with biological systems, temporal resource loops should, in general, be short, lest corrosion or other processes degrade reusable materials. Society's long-term resource storage facilities, many of which are termed landfills, cannot be recommended from an ecosystem viewpoint: They are expensive to maintain, and they mix materials, which makes recovery and reuse difficult. The ideal spatial scale in an industrial ecosystem also mimics its biological analog: Small is best, again because of the energy requirements of long-range resource procurement and the uncertainty of supply continuity in a world where political issues as well as resource stocks may provide constraints. However, industrial organisms can view resources on a global basis (whereas biological systems generally cannot), and can acquire resources on very large spatial scales if the combination of resource attributes and resource cost is satisfactory.

4.6   THE UTILITY OF THE ECOLOGICAL APPROACH

This chapter has demonstrated that the concepts and tools of BE are important to our perspective of industry and its relationship to the natural world. The concept of scale is a central idea, enabling us to consider industrial facilities as organisms acting both individually and collectively. Quantitative measures of efficiency in resource use provide useful comparative benchmarking. Also important is the concept of the food chain, which provides a framework for and an analytical approach to the flows of resources in the technological society.

Perhaps just as important is the symbolic appreciation of both nature and industry as interconnected systems, systems that ideally act to conserve and reuse resources, to be resilient under stress, and to evolve in response to need. Industrial ecology includes such highly focused topics as detailed product design and methods for choosing materials, but at heart it is a systems science.

While this chapter illustrates the value of the analogy between industrial and biological systems, an analogy that will be explored in more detail in Chapter 22, one should also recognize that industrial (human) systems differ from biological systems in fundamental ways. Human-centered systems exhibit three separate modes of evolution: biological, cultural, and technological. Biological systems are essentially limited to only the first two of these. Human systems are reflexive in that the learning that characterizes them changes them in the process: To study a salt marsh does not itself change the marsh, but to study a human culture tends to change that culture. Human systems are also less predictable and more transformable than biological systems in the short term. Overall, the biological analogy is obviously useful in framing industrial ecology, and its tools promise to be usefully employed, but the analogy should not be given more stature than it deserves.

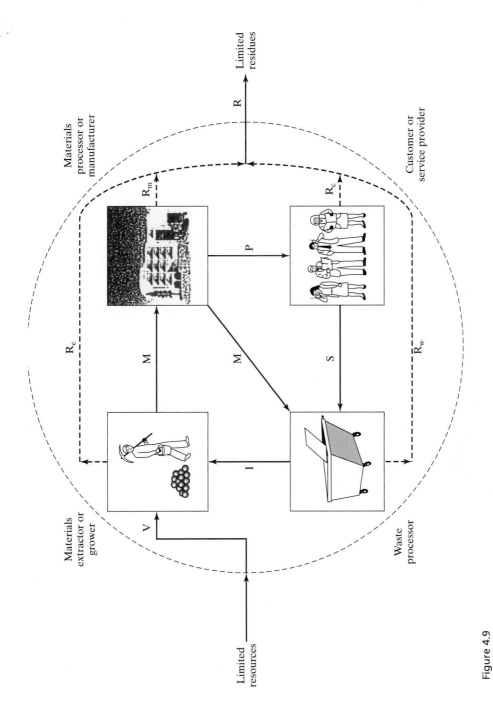

Figure 4.9

The Type II materials flow model of industrial ecology. The letters refer to the following mass flows: V = virgin material; M = processed material; P = product; S = salvaged material; I = impure material; R = uncaptured residues; e = extractor; m = manufacturer; c = customer; w = waste processor.

FURTHER READING

Archer, K., Regions as social organisms: The Lamarckian characteristics of Vidal de la Blache's regional geography, *Annals of the Association of American Geographers, 83*, 498–514, 1993.

Frosch, R.A., and N.E. Gallopoulos, Strategies for manufacturing, *Scientific American, 261* (3), 144–152, 1989.

Graedel, T. E., On the concept of industrial ecology, *Annual Reviews of Energy and the Environment, 21*, 69–98, 1996.

Levine, S. H., Products and ecological models: A population ecology perspective, *Journal of Industrial Ecology, 3* (2, 3), 47–62, 2000.

Pimm, S. L., *Food Webs*. London: Chapman Hall. 220 pp., 1982.

Ricklefs, R. E., *Ecology*, 3rd ed., New York: Freeman. 896 pp., 1990.

EXERCISES

4.1 The process line in a factory draws 6 J/day of electrical power. 15% is lost to heat, and another 25% to respiration. Compute the assimilation efficiency and production efficiency.

4.2 The three-stage process line of Figure 4.6 has the following characteristics: E_1 = 6 J/day, E_2 = 4 J/day, E_3 = 3 J/day, H_1 = 0.9 J/day, R_1 = 1.5 J/day, P_2 = 6 J/day, H_3 = 1 J/day, R_3 = 1.2 J/day. Compute the overall net production efficiency and comment on the results.

4.3 A factory is proposed in this chapter as a candidate industrial organism, and some of its characteristics evaluated from that perspective. There are other possible candidate organisms, however. Evaluate the following as industrial organisms, compare their characteristics to those of a factory, and determine the most appropriate organism analogue: (a) a multinational corporation, (b) a city of one million people.

4.4 A package delivery company has been in business in the same city since 1974. Over that period, its number of vehicles and distance driven have increased, the city has grown, and exhaust emissions have been increasingly controlled. Using the data given below, compute ϵ and φ for emissions of smog-forming hydrocarbons in each of the three years.

Year	P (vehicles)	VOC emission rate (g/km)	Distance driven (km/y)	Overall city VOC emissions (Tg/y)
1974	100	0.30	20,000	1.4
1985	150	0.15	25,000	2.6
1995	200	0.02	30,000	4.3

C H A P T E R   5

The Status of Resources

5.1 INTRODUCTION

One impediment to sustainable development that is frequently mentioned is pending shortages of some of the resources on which our technological society depends. Observing these concerns, corporations seeking to improve their environmental performance typically list resource conservation as an item for attention, and resource use is a commonly included component in sets of environmental performance metrics. Clearly there is the impression that resource availability is a legitimate concern.

Conversely, there are voices in opposition to this position. These "cornucopians" are of the opinion that supplies of resources are ample. If supplies should somehow fall short anyway, the cornucopians predict that substitutes will rapidly be developed to fill the need.

Is one of these positions true and the other false? Or, like so many contentious issues, is the correct answer somewhere in between? Or, is the question incomplete, so that resource availability should be considered in a different framework?

5.2 DEPLETION TIMES AND UNDERABUNDANT RESOURCES

The mineral and energy resources used for millenia, and some part of water resources as well, have come from rich deposits located near Earth's surface, and thus mineable with relative ease. Consider the case of mineral resources. As the use of poorer deposits is contemplated, it is useful to know the distribution of occurrence of elements of interest. Each element is present to some degree in each kind of rock, and the occurrence spectrum of at least the common chemical elements appears to display lognormal distributions in common rocks, as suggested in Figure 5.1a. The lognormal

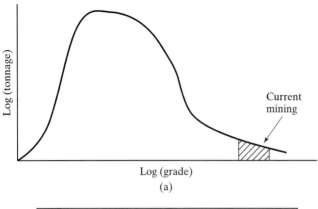

(a)

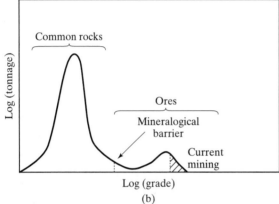

(b)

Figure 5.1

(a) Lognormal distribution of a technologically useful material in Earth's crust. (b) A bimodal distribution of a a technologically useful material in Earth's crust. In each diagram, the area represented by current mining is shaded. (Reproduced with permission from R.B. Gordon, T.C. Koopmans, W.D. Nordhaus, and B.J. Skinner, *Toward a New Iron Age?*, Cambridge, MA: Harvard University Press, 1987.)

distribution has not been well established, however, especially for trace elements, and there is some weight of evidence that the abundance spectrum of most technologically desirable minerals forms a bimodal ore grade distribution, as shown in Figure 5.1b. A mineralogical barrier is indicated on the figure; this is the ore grade at which atoms are present in sufficient abundance to form distinct minerals. Below that concentration, the individual metal atoms substitute for other atoms in minerals of the most abundant elements in common rocks. Regardless of which abundance spectrum is correct, humans have thus far utilized only the highest ore grades (Figure 5.2). With increased ef-

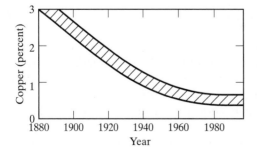

Figure 5.2

The minimum grade of copper ore mined economically in the United States over the period 1880–1985. (Reproduced with permission from R.B. Gordon, T.C. Koopmans, W.D. Nordhaus, and B.J. Skinner, *Toward a New Iron Age?*, Cambridge, MA: Harvard University Press, 1987.)

TABLE 5.1   Classes of Abundance Relative to Rate of Use

Abundant ($t_D > 100$ yr)	Al, B, C, Ca, coal, Cr, Fe, I, K, Li, Mg, Na, Nb, Os, Pt, rare earths, Ru, Si, Ti, V, Yt
Plentiful ($t_D = 50–100$ yr)	Co, Hf, natural gas, Ni, P, Pd, Rh, Sb, Ta, W, Zr
Constrained ($t_D = 25–50$ yr)	Ba, Bi, Cd, Cs, Cu, Mn, Mo, oil, Se, Sn, Sr, U
Rare ($t_D < 25$ yr)	Ag, Au, Hg, In, Pb, S, Th, Zn

This list is limited to major and selected trace elements and energy sources, and does not include gases or chalcocides.

Based on S. Kesler, *Mineral Resources, Economics, and the Environment*, New York: Macmillan, 1994.

fort, poorer ore grades can be mined. The quantity of the resource residing in deposits above the mineralogical barrier (assuming Figure 5.1b is correct) is termed the *reserve base*, and it is estimated by geological techniques.

The scarcity issue can first be addressed by examining the basis for resource limitations. The concerns relate to the total amount of resources that are thought to be present and to the rates at which they are being utilized. If the reserve base numbers are divided by annual consumption rates, the result is an estimate of the depletion time, i.e., the number of years left until the resource is exhausted under a constant use rate. For a number of materials, the depletion times are well under an average human lifetime (Table 5.1). That is, unless more reserves are located or rates of use fall, we and/or our children will surely see the effective disappearance of some resources now regarded as common. It is true that the resource bases are constantly being enhanced by new discoveries, the development of improved extractive techniques, and, in some cases, higher rates of recycling. However, it is equally true that the global rates of increase in population and standard of living, and the additional resource use thereby demanded, are counterweights to resource base enhancement. Economic factors obviously play a major role in both the pace of discovery of new concentrations of resources and the rate at which virgin resource reservoirs are exploited, and higher prices will support increased exploration, extraction, and recycling. All things considered, the depletion times of Table 5.1 are probably reasonably good estimates of the time periods during which virgin materials will be available at more or less today's relative prices, and will thus be employed in the current spectrum of high value and lower value uses.

5.3   HITCHHIKER RESOURCES

Many of the materials in common use tend not to occur in nature in concentrated form, but as coproducts or "hitchhikers" in ores or other reservoirs containing more abundant constituents. A summary of the principal hitchhiker materials appears in Table 5.2. For some elements, acquisition as hitchhikers is essentially the only source. For others, concentrated deposits are occasionally found, but most of the bulk material is acquired from hitchhiker sources.

The challenge to the availability of hitchhiker resources is thus not their absolute abundance, but the abundance and rate of use of their parent materials. Consider, for example, the situation of cadmium. It occurs almost entirely as a minor constituent in

TABLE 5.2   Hitchhiker Materials (Parent Materials in Parentheses)

Materials acquired solely as hitchhikers	Cd (Zn), Ga (Al, Zn, Pb), Ge (Zn), In (Zn), Rh (Cu), Se (Cu), Te (Cu)
Materials acquired largely as hitchhikers	Au (Cu, Ag), Ag (Au, Pb, Cu), Bi (Pb, Cu), Mo (Cu), F (PO_4^{2-}), Co (Ni, Cu), Pt group (Ni, Cu), V (Fe, petroleum ash), Hg (Au, Zn), Tl (Au), As (Cu, Pb, Au, An, Fe), Th (Ti, Zr, rare earths), U (PO_4^{2-}, Cu, rare earths), S (Cu, other sulfide ores)

Adapted from P. Barton, Hitchhiker or co-occurring resources, *Workshop on Material Flow Accounting*, Washington, DC: U.S. National Research Council, Jan. 26, 1998.

zinc ores, and is isolated from zinc as part of the smelting and refining process. Cadmium's abundance and rate of use give it a depletion time of 27 years. However, that of zinc is shorter—about 20 years as of 1999. As productive zinc deposits are depleted and substitutes for zinc are developed, as they inevitably must be, the availability of cadmium will decline unless poor grade zinc ore is mined principally for its cadmium content. This will raise the price of cadmium substantially, which, combined with cadmium's biotoxic nature, will further contribute to its decline as an industrial material. Thus, we are likely to see the effective disappearance of virgin cadmium as an industrial material within the next few decades, not because of its absolute abundance but because of the forms in which it occurs and because of its toxicity.

5.4   ENERGY RESOURCES

5.4.1   Trading Energy for Mineral Resources

Modern technology has two basic requirements: the acquisition of potentially beneficial materials and the transformation of those materials into forms suitable for use. In accomplishing these requirements, energy is consumed. Ultimately, all technology is involved in this trade: To have available the materials needed to provide the products of modern industry, we must invest energy to acquire resources, and use energy to put those resources in suitable form. If we desire the materials, we must pay the energy price.

A substantial fraction of Earth's resources is acquired by mining. In this process, many tons of material covering the desired resource are moved aside (and perhaps eventually back again): as much as 1000 tons of "overburden" for every ton of material eventually recovered. The ore that is the goal of this activity may be moved again, depending on the location of the smelter. The acquisition of other resources, such as wood products, requires less gigantic but still extensive energy consumption. Most of the energy needed to accomplish these tasks comes from gasoline and diesel fuel, and a great attribute of these fuels is that they can readily be transported to where needed. It is this "mobile energy" that has made the extraction of Earth's resources so geographically widespread and so efficient.

Once extracted, resources must generally be transformed if they are to be useful. The simplest transformations are physical in nature, an example being the crushing of stone to create material suitable for highway foundations or as an ingredient in concrete. More commonly, a resource undergoes chemical transformation, an example being the production of a pure metal from the metal oxide or sulfide found in Earth's crust. The actual process involved is the use of energy to break the chemical forces bonding the material in its original form, and to generate chemical bonds that create new forms useful to technology.

5.4.2   Energy Sources

The use of fossil fuels dominates humanity's energy budget. As seen in Table 5.3, oil is the largest current energy source, accounting for about a third of the world's primary energy consumption in recent years. (*Primary energy* is the energy embodied in resources as they exist in nature; it is not possible for a variety of technical and thermodynamic reasons to recover all of it.) Extracted coal and natural gas contribute much of the rest. Biomass, mostly used locally for heating and cooling, generates 10–15% of the total. Nuclear power and hydropower account for a few percent each. The table shows that *final energy* (the energy actually supplied to the point of final use) was slightly less than three-fourths of the primary energy in the extracted fuels in 1990. Of the final energy amount, nearly a third is used industrially.

Table 5.3 also includes primary energy source data for 1998. Several items are of particular interest. First, the global energy consumption increased by more than 6% over that period. Second, the proportion provided by oil increased from 34 to 37%, that for natural gas from 19 to 20%. Third, coal and biomass percentages decreased slightly. It is clear that it is the highly transportable fuels that are enabling the increased use of energy in the early 21st century.

5.4.3   Energy Resource Status

Fossil fuels (coal, petroleum, natural gas) are formed, as their name suggests, from the decay and alteration of deeply buried organic matter, mostly woody plants. These processes occur over millions of years, so Earth's fossil fuel resources are, for all practical purposes, nonrenewable.

The depletion times for oil and natural gas are each near 50 years. Exploration for these resources has been intense, and there is now general agreement among pe-

TABLE 5.3   Global Energy Consumption by Energy Source and by Sector, in EJ/yr

	Coal	Oil	Gas	Nuclear	Hydro	Electricity	Heat	Biomass	Total
Primary (1990)	91	128	71	19	21	—	—	55	385
Final (1990)	36	106	41	—	—	35	8	53	279
Industry	25	15	22	—	—	17	4	3	86
Transport	1	59	0	—	—	1	0	0	61
Others	10	18	18	—	—	17	4	50	117
Feedstocks	0	14	1	—	—	—	—	0	15
Primary (1998)	92	151	80	27	9	—	—	44	403

Data sources: N. Nakićenović et al., Energy primer, in *Climate Change 1995: Impacts, Adaptations and Mitigation of Climate Change*, R.T. Watson, M.C. Zinyowera, and R.H. Moss, eds., 75–92, Cambridge, U.K.: Cambridge University Press, 1996, and International Energy Association, *Energy Statistics, http://www.iea.org/statist*, accessed June 18, 2001.

troleum geologists that production of conventional oil and gas will reach its all-time peak somewhere in the 2010–2020 period, depending on rate of use and speed in bringing known deposits into production. From that point on, the world will need to begin to shift to a different spectrum of energy sources. The unconventional sources of tar sands and oil shale may help with this situation, but total supplies, environmental damage, and water requirements associated with these sources make it unlikely that they can fill the role now played by conventional oil and gas.

Coal is abundant, with a depletion time greater than 200 years. Strip-mining for coal, a common practice, is problematic, however, as are the environmental challenges of coal's combustion. Coal will doubtless continue to be a major component of the energy supply, but will need to be developed with care.

Uranium, the fuel for nuclear power, has a depletion time similar to that for oil. Nuclear reactors do not have wide public acceptance, however, and long-term storage of spent fuel is a continuing concern. As a result, nuclear power use is not expanding, and uranium supplies appear adequate for the 21st century unless major changes occur.

Other sources of energy—biomass, hydro, geothermal, solar, wind—appear unlikely to provide as much as 25% of energy needs in the next few decades even under the most optimistic scenarios. Broad implementation is limited by supplies (biomass), nature (hydro, geothermal, wind), technology and nature (solar), and environmental concerns (hydro).

As a result of these supply constraints, it is likely that the middle of the 21st century will see a major shift in the sources of energy. Coal gasification may be a significant contributor, although it is technologically limited at present and the economics are unfavorable. The possible alternative, especially if global climate change accelerates the shift away from fossil fuels, is a new generation of fail-safe nuclear reactors. Such a transition would require the development of reliable international approaches to nuclear waste disposal, and considerable public education. It may prove, however, to be the best of the options available within a few decades.

5.5 ENERGETICALLY LIMITED MINERAL RESOURCES

The extraction and processing of minerals requires large amounts of energy, the energy needed depending on the metal itself and the ore grade being processed. When an element is sufficiently abundant to be above the mineralogical barrier; i.e., when the matrix can be called an ore and not a rock, the minerals are freed from the surrounding matrix by crushing and grinding, and concentrated by selective processes such as flotation. The resulting concentrate of minerals can then be purified to acquire the target metal.

The situation for copper presents clearly the relationship between ore grade and energy. Extraction and processing are relatively efficient for copper ore (the rock matrix containing copper sulfide minerals), though considerations of energy are certainly present. At mineral concentrations below 0.1 percent, however, the copper is thought to be dispersed in solid solution within silicate minerals rather than concentrated into mineral form. To recover the dispersed copper, the silicate minerals themselves must be separated and processed. The energy costs of so doing are very large, because the

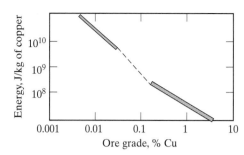

Figure 5.3

The energy necessary to recover copper from its ores and by atomic substitution from common silicate minerals. (Reproduced with permission from R.B. Gordon, T.C. Koopmans, W.D. Nordhaus, and B.J. Skinner, *Toward a New Iron Age?*, Cambridge, MA: Harvard University Press, 1987.)

chemical bonds holding the atoms together in a silicate mineral are much stronger than those holding the atoms of a copper sulfide mineral together. Metallurgical experience suggests that the overall energy needs will be about 10 times as great per recovered copper atom if recovery from silicate minerals is required. Figure 5.3 shows the ore grade–energy relationship for copper. This is a log–log plot, indicating that recovery from ore of grade 0.01 % will require roughly 100 times the energy needed for ore of grade 0.1%.

An energy increase of this magnitude is not likely to be a major factor in overall energy supply, since the nonferrous extractive industries use only a small part of the global energy supply. Nonetheless, higher energy use will inevitably cause higher prices for the extracted metal. Recycling rates will increase in such a circumstance, but are unlikely to keep up with the growing demand that is anticipated for low-cost materials. The result seems likely to be the reduction or cessation of low-value uses for materials whose extraction or processing imposes high energy costs, and an impetus to substitute more abundant materials for the less abundant where such substitutions make technological sense.

5.6   GEOGRAPHICALLY INFLUENCED RESOURCE AVAILABILITY

Resources are not distributed equally around the planet, but reflect the specialized concentrating processes of geochemistry and hydrology. Some resources are much more spatially homogeneous in occurrence than others. Limestone, iron, and sodium chloride are available nearly everywhere, while gold, diamonds, the platinum group metals, and petroleum are found in reasonable quantities only in a few locations. From the standpoint of availability, it is easy to purchase the latter type of materials in a world where global shipments of commodities occur freely. Such a free-trade world is not always the case, however. Wars, political unrest, cartels, natural disasters, and other difficulties sometimes render reasonably abundant materials difficult to come by. During World War II, for example, the disruption of global shipping forced countries on both sides of the conflict to develop substitutes—often inadequate ones—for a number of industrial materials.

Few countries can ignore the potential for politically induced resource unavailability, because modern technology uses such a large fraction of the periodic table, and everyone uses petroleum. Even in a resource-rich country like Australia or Canada, imports provide virtually the entire supply of a number of materials. For countries like

TABLE 5.4   Geographical Supply Constraints for Resources

Sources on several continents	Ag, Al, Au, B, Ba, Bi, Br, Ca, Cd, Cl, Co, coal, Cr, Cs, Cu, F, Fe, Ga, Hf, I, In, K, Li, Mg, Na, natural gas, Ni, oil, P, Pb, Pt group, rare earths, Rb, S, Sb, Sc, Se, Si, Sn, Sr, Ta, Te, Ti, U, V, W, Y, Zn, Zr
Sources on only two continents	As, Ge, Hg, Mn, Mo, Nb, Tl
Sources on only one continent	Be

Derived from information in S. Kesler, *Mineral Resources, Economics, and the Environment*, New York: Macmillan, 1994.

Japan, with vigorous industrial activity but limited natural resources, a very large fraction of the resources that are employed are imported.

A guide to the geographical limitations to resource availability is given in Table 5.4, where metals, major minerals, and energy resources have been divided into three groups: (1) those widely available (defined as those with significant deposits occurring on at least three continents); (2) those with modest geographical constraints (those with significant deposits occurring on two continents); and (3) those with substantial geographical constraints (those with significant deposits occurring on only one continent). Although a single continent is often the prime supplier, the table shows that few resources are severely constrained by geography. It is clear that two possible positions can be adopted concerning resources that may be limited because of geographical occurrence. One is to assume that world trade will carry on regardless, which has historically been the case more often than not. The second is to be more conservative, and to recognize that even short-term (e.g., a few years') disruptions strain technology severely. Designers of products and processes may therefore wish to emphasize those resources whose availability is unlikely to be dependent on factors beyond the potential user's control.

5.7   ENVIRONMENTALLY LIMITED RESOURCES

It has become increasingly apparent that the extraction and processing of resources carries with it the potential for severe environmental impacts. As a consequence, we commonly see situations in which the location of a resource is known and its extraction and processing are perfectly feasible from a technological standpoint, but the potential environmental liabilities dictate that it not be utilized. As Earth's population grows so that urban expansion invades traditional resource recovery areas, as pristine natural areas become fewer and fewer, and as environmental sensitivity becomes ever more heightened, environmentally related limitations on resource availability can only grow.

The simplest constraint, but one of the most important, is the disturbance of surface and subsurface ecosystems as resource extraction occurs. Strip or open-pit mining is the most extreme example, but the effects of opening mine shafts or drilling for oil from off-shore platforms can also cause major disruptions in local ecosystems.

A second constraint, sometimes overlooked, is the use of water in resource extraction. Resources are not necessarily located in regions where water is abundant, as the copper mines in the southern Arizona desert of the United States demonstrate. The separation by flotation from unwanted gangue of copper minerals following milling typically requires about one liter per kilogram of metal. The use for other metals is similar. In locations where water is scarce, the processing of ores can be a heavy load on the regional water supply, either because of sheer volume of use or because of contamination.

The chemicals used in extraction or processing may be a further environmental hazard. Gold is particularly problematic in this regard. A common extraction technique, now largely abandoned, involved bathing the ore in mercury, the amalgam of mercury and gold being more easily extracted than the metal itself. A second technique involves leaching the gold with cyanide to dissolve it; since the leaching occurs in open pits that may become breached, the hazard for wildlife and ecosystems is obvious. Somewhat less hazardous leaching techniques for other metals may also be potential environmental problems.

An awkward but unavoidable situation is that many metal resources occur in nature largely as sulfide compounds. This presents two environmental challenges. The first is that smelting liberates the sulfur in the process of recovering the target metal. Unless the sulfur is captured, it is emitted as toxic, acid-rain–producing sulfur dioxide. The second challenge relates largely to abandoned mines and unworked mine tailings containing metal sulfides. Once exposed to the air, the sulfides are slowly oxidized, forming the sulfuric acid that renders the environmental impacts of drainage from many metal mines quite a severe problem.

Many of these challenges come together as lower mineral ore grades are mined— more energy is required, more water is needed, and more land area undergoes displacement, even with the most modern technology. Robert Gordon and colleagues at Yale University deduced what would be required if the United States copper needs in the mid-21st century were to be supplied not from the reserve base, but from the "backstop copper" in the highly dispersed left-hand segment of the abundance diagram of Figure 5.1b. The requirements that they derived were:

- The opening of 300–500 large open pit mines
- The removal and processing of 13 cubic kilometers of igneous rock per year
- The use of water equivalent to 20% of the flow of the Mississippi River
- The use of half the current world energy supply

These requirements appear insurmountable. They seem to indicate that it is completely unrealistic to anticipate using materials in ores leaner than the mineralogical barrier, and that we must be protective of our reserve base supplies if we wish to continue using the present suite of technological materials, at least in the present spectrum of uses.

5.8   CUMULATIVE SUPPLY CURVES

A helpful perspective on the relationships of total supply of nonrenewable materials to their price over all time is provided by cumulative supply curves, developed by John Tilton of the Colorado School of Mines, and Brian Skinner of Yale University. There are three scenarios of interest. That of Figure 5.4a assumes that moderate increases in demand will result over time in gradual increases in price. Figures 5.4b and 5.4c, how-

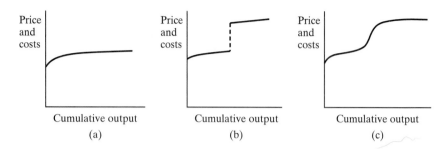

Figure 5.4

Cumulative supply curves for nonrenewable resources. (a) Slowly rising slope due to gradual increase in costs. (b) Discontinuity in slope due to jump in costs. (c) Sharply rising slope due to rapid increase in costs. (Reproduced with permission from Tilton, J.E., *On Borrowed Time? Civilization and the Threat of Mineral Depletion,* publication forthcoming.)

ever, represent situations in which high grade ore deposits are exhausted and much lower grade deposits are utilized.

Several groups of factors relate to the cumulative supply curves. The first group is geological, and determines the slope of the curve. As described earlier in this chapter and illustrated in Figure 5.1, if ore grade distributions are lognormal, they will lead to a cumulative supply curve of the form of Figure 5.4a. If they are bimodal, the result is likely to be a cumulative supply curve of the form of Figure 5.4b or 5.4c. There is, at present, insufficient data to determine which alternative is correct.

The second group of factors is related to demand, and determines how rapidly the world moves along the supply curve. Here are included global population, per capita income, and intensity of material use (the consumption of a mineral commodity per unit of income). As discussed in Chapter 1, the first two are clearly and strongly increasing. Intensity of use appears to be trending downward for most resources as a consequence of new technologies and cultural preferences for "upscale" products. Also included in the demand equation for minerals (but not for energy resources) is the availability of recycled material as a substitute for virgin material. While a helpful supplement to extracted stock, recycled material cannot be expected to play a dominant role in material supply in a time of rapidly increasing consumption.

The third group of factors has the potential to cause the cumulative supply curve to shift its position. New technology has been dominant in this group historically, which shifts the curve downward. Changes in input cost as a result of labor, capital, or energy transitions, could shift the curve either up or down. These possibilities have historically proven difficult or impossible to predict reliably.

5.9   WATER RESOURCES

Water is a renewable resource, but a limited one nonetheless. It has been said that "water is the oil of the 21st century," and there is every prospect that the availability of water will increasingly limit industrial processes that utilize it. Unlike wood or other renewable resources, the total average quantity of water available to humans is fixed.

TABLE 5.5 Global Uses of Water

Activity	Water consumption (10^3 km^3/yr)
Agriculture	2880
Industry	975
Domestic	275
Other	300
Total	4430

Data source: S.L. Postel, G.C. Daily, and P.R. Ehrlich, Human appropriation of renewable fresh water, *Science, 271,* 785–788, 1996.

What is not fixed, however, is any constancy in rate of supply, since droughts, floods, and average water flows are all relatively common experiences. Because all industrial sectors use water, some much more than others, industrial location, type, and efficiency are important factors in water budgets.

As seen in Table 5.5, roughly 80% of global water use is for crop irrigation by the agricultural sector. All other industries consume nearly 10% additional. The various uses and their rates of change are highly variable from year to year and region to region, but current consumption amounts to more than half of all water that is geographically available to humans. As populations increase and industrial development proceeds, the competition for water to be used in agriculture, industry, and domestic life will intensify in many parts of the world.

A possible future picture of water limitations is shown in Figure 5.5 for the year 2025. It is anticipated that much of Africa, Europe, and South Asia will be in a condition of water scarcity by that time, and China and Southern North America will be under water stress. Because industrial facilities commonly operate for more than a quarter-century, these projections suggest that new industrial facilities should be located in such areas only if their water requirements are modest, because the amount of water available there may not be sufficient to meet the demands placed upon it.

5.10 SUMMARY

Are we running out of resources? This question has been explored from the standpoint of absolute amounts of materials, the form in which they are available to us, their geographical distribution, the energy required for extraction and processing, and the environmental impacts that limit availability. We conclude that each one of these considerations has the potential to limit supplies of certain materials, though at different temporal and spatial scales. Materials whose cumulative supply curves follow predictable supply–cost patterns, as in Figure 5.4a, are unlikely to suffer from significant scarcity for at least several decades, although cost increases may over time change the spectrum of uses. For materials whose cumulative supply curves contain discontinuities (Figure 5.4b) or costs that rise rapidly with small increases in supply (Figure 5.4c), however, virgin material scarcity is a real possibility. For energy resources, gradual transitions to a higher proportion of renewable sources are indicated. In the case of water, many geographical locations appear likely to be increasingly constrained.

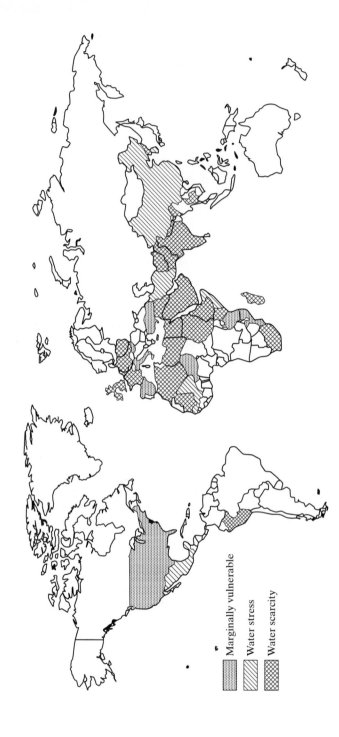

Figure 5.5

A map of the world predicting levels of water vulnerability in 2025 for a scenario incorporating anticipated population growth and industrial development, together with moderate effects of climate change. (Reproduced with permission from S.N. Kulshreshtha, *World Water Resources and Regional Vulnerability: Impact of Future Changes*, Report RR-93-10, Laxenburg, Austria: International Institute for Applied Systems Analysis, 1993.)

Marginally vulnerable

Water stress

Water scarcity

It is very unlikely that any resources will completely disappear, because if they become very scarce the price will become very high and use will decline. However, there appears to be a real potential over the next several decades for low-abundance resources to become scarce enough that their increased price will force a new spectrum of conservation, substitution and more thoughtful use.

FURTHER READING

Campbell, C., and J.H. Laherrère, The end of cheap oil, *Scientific American, 278* (3), 78–83, 1998.

Chapman, P.F., and F. Roberts, *Metal Resources and Energy*, London: Butterworth, 1983.

Gleick, P.H., *The World's Water, 2000–2001: The Biennial Report on Freshwater Resources*, Washington, DC: Island Press, 2000.

Gordon, R.B., T.C. Koopmans, W.D. Nordhaus, and B.J. Skinner, *Toward a New Iron Age?*, Cambridge, MA: Harvard University Press, 1987.

Hodges, C.A., Mineral resources, environmental issues, and land use, *Science, 268*, 1305–1312, 1995.

Kesler, S.E., *Mineral Resources, Economics, and the Environment*, New York: Macmillan, 1994.

Tilton, J.E., *On Borrowed Time? Civilization and the Threat of Mineral Depletion*, publication forthcoming.

EXERCISES

5.1 Lead is a material used in the electronics industry and in batteries. Evaluate lead's availability from the standpoint of abundance, co-occurrence, and geographical occurrence. On the basis of whatever information you can locate on lead, what do you predict for lead as an industrial material in the next few decades?

5.2 The grade of copper ore has decreased over time, as shown in Figure 5.2. If you wished to produce 1 kg of copper in 1900, how much ore was required, assuming a 12% loss of copper to tailings in the mill and a 1% loss of copper to slag in subsequent smelting? How much ore was required in 1980 if loss percentages did not change?

5.3 Using Figure 5.3, estimate the energy required to extract and process copper from ore of grade 3.0%. Repeat for ore of grade 0.3%.

C H A P T E R 6

Society and Culture

6.1 SOCIETY, CULTURE, AND INDUSTRIAL ECOLOGY

What has society and culture to do with industrial ecology? Consider Figure 6.1, showing the concentric levels in a typical industrial ecology system. At the lowest level, that of the component, the environmentally related decisions are those of the design engineer and the interaction with society and culture is negligible. At successively higher levels, however, the degree of interaction increases. A societal preference for privacy and status, for example, is reflected in sprawling communities far from jobs and shopping. A consequence of that preference is an incentive to produce larger, more powerful, and more comfortable vehicles than would be the case in a more condensed urban pattern. Technology and society are thus inherently related, and because environmental impacts result from those relationships, technology, society, and the environment are closely linked as well.

The relationship among social and economic systems, industrial systems, and natural systems is not static. Each co-evolves with the others, changing its internal dynamics and the overall system state as it does so. These interactions occur at three levels.

The first and broadest perspective treats industrial ecology as part of the overall scientific and technological enterprise. At this scale, industrial ecology is, like any human activity, heavily dependent on the current historical, social and cultural context. For example, industrial ecology as an important response to environmental perturbations could not have developed until after the early 1970s, with its explosion of environmental activism. This broad level of industrial ecology is the purview of sociologists and historians, not industrial ecologists themselves.

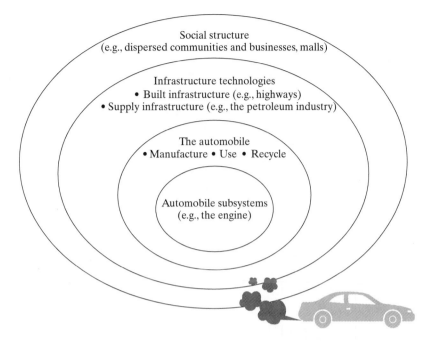

Figure 6.1

The automotive technology system: a schematic diagram.

Similarly, most industrial ecologists will not be directly involved in the second level of societal interaction, where industrial ecology is viewed as a case study from the perspective of the sociology of science.

The third level is that of the *practice* of industrial ecology. Here, cultural and social dimensions are treated as objective elements of an industrial ecology study. For example, any industrial ecology study of the automobile industry would need to take notice of the changes in demand and demographics that led to people buying less fuel efficient vehicles, driving them further, and generating a greater degree of environmental impact.

6.2   CULTURAL CONSTRUCTS AND TEMPORAL SCALES

Stability is generally assumed in analyses of social and cultural frameworks, and most of the time this is a viable and useful assumption. Over long time scales, however, or when ideologies are very powerful, this assumption may not be valid and may result in decisions that cause significant environmental, social, or economic harm. Table 6.1 emphasizes that system structures become much more flexible over time, with additional elements becoming more variable the further out one goes. In the longer term, virtually everything is interdependent and changeable, which is one reason why long-term predictions, frequently based primarily on projections of current trends, are almost always fundamentally flawed.

TABLE 6.1 System Structure over Different Time Scales

Time scale	Endogenous	Exogenous	Principal implementation mechanism	Principal R&D component	Integration of natural and artifactual systems
Short term (ca. 5 years)	Incremental technology evolution within existing major technology systems	Population level, cultural change	Policy	Short term industrial ecology R&D (e.g., design for environment, integrated pest management)	Experimental stage involving small systems (e.g., bioreactors, drug production in genetically engineered sheep)
Medium term (ca. 5–10 years)	Evolution of product and process technology systems, marginal cultural change	Population level, significant cultural change	Changes in legal structures, disciplinary assumptions, based on industrial ecology	Industrial ecology infrastructure (e.g., environmentally preferable materials database)	Partial integration of biological and engineered systems (e.g., commercial energy from biomass; engineered wetlands for flood, pollution control, and waste processing)
Long term (ca. 10–100 years)	Significant evolution of major technology systems; link between quality of life and material consumption; population levels; most aspects of culture	Almost nothing	Metrics, changes in fundamental conservative cultural systems	Industrial ecology systems (e.g., resource and energy maps of communities and regions)	Management of integrated regional and global systems (e.g., water cycles in Yellow River watershed or southwestern United States); Earth systems engineering and management

In the short term, there are many systems components that can legitimately be assumed to be fixed: major industrial systems such as transportation or energy networks, economic structures such as private companies, or many government policies. In the medium term, more elements become flexible: whereas a product redesign might not be feasible in a five-year framework, it may well be possible in a 10-year framework. Fundamental shifts in consumer demand—away from cigarettes or toward larger automobiles—typically occur in such a time frame. In the longer term, major technology systems and their acceptance by the public can change completely—perhaps from reliance on fossil fuel to a nuclear-based hydrogen economy, for example.

The changeable nature of almost everything in the long term illustrates the importance of what sociologists and psychologists call "cultural constructs." These are ideas that are invented for certain purposes within a society, but soon begin to seem absolute and unquestionable within the society that created them, though not necessarily elsewhere. For example, many developing countries are concerned about the effort of environmentalists from developed countries to make environmental standards part of global trading regimes, on the grounds that such standards can become powerful barriers to trade. Similarly, some are concerned that current environmental policies espoused by developed countries represent a powerful but unrecognized drive to favor Earth's present peoples over those who will populate the planet in the future. The global climate change negotiations, for example, have been criticized as seeking to stabilize current climactic conditions, thereby removing an important source of variability that has affected the evolution of life on this planet. The negotiators are, among other things, doing their best to remove an important driver of biological evolution. By the same token, opposition to biotechnology implicitly grants precedence to current genetic structures over what may be evolved in the future. This ideological and ethical statement is unconscious for most participants in these dialogs.

Industrial ecologists have cultural constructs as well, the obvious one being "sustainable development," invented and popularized in the 1987 book, *Our Common Future*. Before the book, no sustainable development; after the book, sustainable development. Moreover, as successful cultural constructs tend to do, "sustainable development" and the looser term of "sustainability" over the past 15 years have become the goal of environmentalism. What began as a cultural construct now defines the desired endpoint for all human activity for environmentalists: the teleology of sustainability. In the process, the contingency of the term—although quite explicit in its history—has vanished for many people. The dangers this presents, while common to many cultural constructs, are apparent when one places sustainable development in a framework of basic political values; it then becomes apparent that the term represents a fairly specific culture: basically, northern European social democratic traditions. The danger of elevating the contingent to the absolute also becomes apparent: Conflicting values that may be of concern to others (individual equality of opportunity for libertarians, for example) are implicitly stifled.

As William Cronon has pointed out in his book *Uncommon Ground*, the prevalence of cultural constructs is not limited to sustainable development, but extends to many other ideas that are seen as absolutes: wilderness, nature, and, indeed, environment itself. That a number of the terms used in industrial ecology are cultural constructs is not necessarily good or bad, but it suggests that the tenets of responsible environmentalism are neither culture-independent nor absolute.

6.3   THE PRIVATE FIRM IN A SOCIAL CONTEXT

Understanding the cultural roles and dimensions of private firms is obviously important to industrial ecology, because firms are where social organization, economic goals, design decisions, and environmental impacts all come together. The two most important points to grasp are that firms in their totality are cultural creations, and that they evolve over time.

Clearly, private firms are pivotal economic agents in any modern economy. They also reflect and create the cultures and economies within which they function, and are creatures of law, created, defined, and modified by law. Individual firms may be components of self-organizing and loosely linked industrial districts, such as Silicon Valley in the United States or the collection of textile firms near Florence, Italy. These companies compete intensely, while at the same time learning from one another about changing markets and technologies through informal communication and collaborative practices. The functional boundaries within firms are porous in a network system, as are the boundaries among firms themselves and between firms and local institutions such as trade associations and universities.

The private corporate enterprise is such an intrinsic part of the modern capitalist economy that few realize its relative youth. It is possible to trace the antecedents of the corporation back to the medieval merchant guild systems, or, more recently, to trading companies enjoying monopolies granted under royal charter, such as the British East India Company. However, the advent of the truly modern firm awaited the development in the early 19th century of laws under which any entity meeting statutorily defined criteria was able to incorporate. The pattern subsequently established in western economies—a complex network of independent firms, frequently competing on the basis of technological and scientific creativity and with successful innovation rewarded in the marketplace—became the basis for modern, materially successful economies. Thus, the modern corporation appeared at a certain stage in the development of the Industrial Revolution, because such a construct was necessary for the continued evolution of the industrial economies characterizing the modern state. Indeed, it can also be argued that such entities as "virtual firms" represent a continuing evolution of the firm into increasingly complex and flexible entities.

6.4   ENVIRONMENTALISM, TECHNOLOGY, AND SOCIETY

Just as industry has cultural dimensions, so does environmentalism. Figure 6.2 indicates some of the many ways of valuing nature. Most are anthropocentric approaches that value nature from the perspective of human utility. Nature-centered approaches, in contrast, treat natural systems as of priceless value. These different perspectives create different ethics by which individuals conceptualize the environment. The industrial ecologist is likely to be seen by a committed environmentalist as representing corporate managerialism, focusing too much on rationality and not enough on emotion and values. If not addressed, this can create conflict and tension that make dialog among different groups difficult, if not impossible.

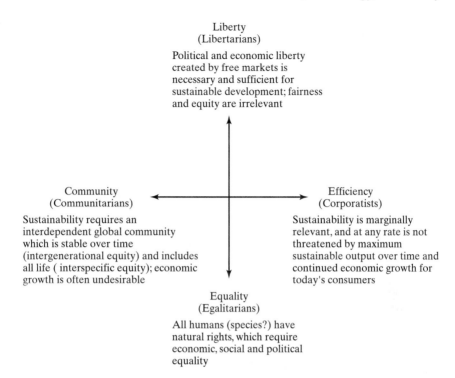

Liberty
(Libertarians)

Political and economic liberty
created by free markets is
necessary and sufficient for
sustainable development; fairness
and equity are irrelevant

Community
(Communitarians)

Sustainability requires an
interdependent global community
which is stable over time
(intergenerational equity) and includes
all life (interspecific equity); economic
growth is often undesirable

Efficiency
(Corporatists)

Sustainability is marginally
relevant, and at any rate is not
threatened by maximum
sustainable output over time and
continued economic growth for
today's consumers

Equality
(Egalitarians)

All humans (species?) have
natural rights, which require
economic, social and political
equality

Figure 6.2

Belief systems regarding the economy, individual liberty, the environment, and sustainability.

Differences in cultural and personal value systems result in the creation of different frameworks from which to view the interactions among environmentalism, technology and society. Table 6.2 illustrates the four most common. Following the discussion in Chapter 1 on the relationship between global technology state and population, Figure 6.3 links these conceptual approaches with the trajectories of population that they most likely imply.

The first option in Table 6.2 is *radical ecology*. It represents essentially a program of preindustrial, low (even anti-) technology, pastoralism. It rejects the use of modern agriculture, electronics, medicine, transportation, and other benefits of technology. Although this option clearly responds to the environmental impacts associated with the Industrial Revolution to date, there are significant costs of implementation, including a greatly enhanced susceptibility to famine, disease, and "acts of God." Wholesale adoption of this option seems likely to result in a world that could not support the current population, much less that of the future. Those groups that support this option have not explained how they would reduce Earth's population to meet their goal, but any rapid removal of technology from society would certainly result in major social and economic upheaval. Moreover, it is difficult to see how this option would be implemented, as there is no indication that it could be reached without the imposition of a completely powerful authoritarian structure that was willing to take the necessary steps.

TABLE 6.2   Options for Technology–Society Interactions

Approach	Effect on technology	Implication
Radical ecology	Return to low technology	Unmanaged population crash; economic, technological, and cultural disruption
Committed environmentalism	Appropriate technology, "low-tech" where possible	Lower population, substantial adjustments to economic, technological and cultural status quo
Industrial ecology	Reliance on technological evolution within environmental constraints; no bias for low-tech unless environmentally preferable	Moderately higher population, substantial adjustments to economic, technological, and cultural status quo
Continuation of current trends	Ad hoc adoption of specific mandates (e.g., CFC ban); little effect on overall trends	Unmanaged population crash; economic, technological, and cultural disruption

The fourth option, *continuation of current trends*, is also a fundamentally flawed, high-cost choice. The most obvious problem with this path, which is essentially a continuation of exponential growth until terminated by extreme environmental and social pressure, is that it can be followed only in the short term (a few decades at most), and only by imposing substantial costs on future generations and perturbing global environmental systems significantly. Continually escalating material flows and rapid growth in capital stock, energy use, and resource consumption cannot be maintained indefinitely; the results of such an effort, ironically, could be similar to that resulting from radical ecology: a dramatic and uncontrolled reduction in human population, with a significant risk of political, economic and social disruption. Fortunately, evi-

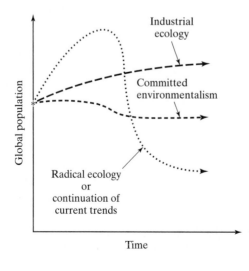

Figure 6.3

Alternative human population scenarios suggested by different options for technology–society interactions. The asterisk marks the present global human population of approximately six billion.

dence such as the successful implementation of the Montreal Protocol and the evolution of private firms toward a broader definition of their role in society indicates that this path is already falling out of favor. Moreover, the major assumption that supported this path—that continuing growth in personal income and consumption was necessary for a superior quality of life—is increasingly difficult to support. While it is difficult to measure quality of life directly, much evidence indicates that increasing personal income—a rough indicator of consumption—does not lead to increases in quality of life. The emergence of a decoupling of material consumption from perceptions of high quality of life is an encouraging indication that society can adjust to environmental challenges in a politically acceptable, humane manner.

The remaining options, *committed environmentalism* and *industrial ecology*, share a recognition that environmental considerations and constraints must be internalized into human cultures and economic activity at all levels. They differ principally in their views of the role of technology in the transition to a more sustainable world. Committed environmentalism tends to view technology with suspicion, and actively discourages the introduction of new technologies. This is accomplished through the implementation of policies such as the precautionary principle, which requires proof of negligible risk before any new technology is introduced, regardless of the risk presented by the extant technology. Accordingly, the implications of committed environmentalism are a lower level of support for population and a reduction in Earth's carrying capacity. As with radical ecology, this is unlikely to be attainable without considerable political stress.

Industrial ecology recognizes the need for continued technological evolution and views the development of environmentally appropriate technologies as a critical component of the transition to a sustainable world. Given that the global technology state and the level of human population are inextricably linked, if the goal is to maintain current population levels (or even to allow for population growth), then evolving an enhanced technological state is critical. Industrial ecology is not merely naïve technological optimism, however. The history of human perturbation of natural systems does not inspire confidence that the human species can migrate itself to a relatively stable and desirable carrying capacity without significant economic, social and cultural dislocations, or even precipitous population fluctuations. Nonetheless, industrial ecology adopts as an operative assumption the *possibility* of a reasonably smooth transition to a stable carrying capacity through a combination of concomitant technological and social progress.

FURTHER READING

Allenby, B.R., *Industrial Ecology: Policy Framework and Implementation*, Upper Saddle River, NJ: Prentice Hall, 1999.

Board on Sustainable Development, *Our Common Journey*, Washington, DC: National Academy Press, 1999.

Cronon, W., Ed., *Uncommon Ground: Toward Reinventing Nature*, New York: W.W. Norton, 1995.

Harvey, D., *Justice, Nature and the Geography of Difference*, Cambridge, MA: Blackwell Publishers, 1996.

McNeill, J.R., *Something New Under the Sun*, New York: W. W. Norton & Co., 2000.

Turner, B.L., W.C. Clark, R.W. Kates, J.F. Richards, J.T. Matthews, and W.B. Meyer, eds., *The Earth as Transformed by Human Action*, Cambridge, U.K.: Cambridge University Press, 1990.

World Commission on Environment and Development, *Our Common Future*, Oxford, UK: Oxford University Press, 1987.

EXERCISES

6.1 As an industrial ecologist, you have been assigned by your company, a lumber and forest products producer, to begin a dialog with environmentalists who are concerned about your forest management practices. You believe that the available scientific evidence supports your practices, and are concerned that the environmentalists are more interested in headlines than forestry. What kind of presentation would you prepare for your first meeting? What role do you think the governments of the areas within which you operate should play?

6.2 Increasing urbanization is a powerful trend in today's world: Within the next several decades, United Nations projections show that over half of the world's population will be located in urban centers. This will require the construction of the equivalent of 8 cities of 10 million inhabitants each every year for the foreseeable future. Discuss what this trend implies for flows of products and residues: food, water, sewage, energy, and so on. Overall, is urbanization environmentally advantageous or disadvantageous? Will it continue in any event, regardless of environmental impacts?

6.3 Using your country as an example, what are the political implications of each of the four scenarios displayed in Figure 6.3 and described in Table 6.2? Are there any options which would be difficult, if not impossible, to implement politically in your country? If so, and you were committed to such an option, what could you do to force its implementation, and what would the associated costs be?

CHAPTER 7

Governments, Laws, and Economics

Governments, especially as they act through the development and implementation of legal and policy systems, can play important roles in the implementation of environmentally responsible technology. The discipline of economics and its metrics are perhaps the most powerful generic underpinnings of these systems. Accordingly, it is important for industrial ecologists to have a working knowledge of applicable aspects of governments, laws, and economics. We present a brief survey of these areas in this chapter; more detailed treatments can be found in the Further Reading.

7.1   NATIONAL GOVERNMENTAL STRUCTURES AND ACTIONS

Governments at all levels dramatically influence the behavior of firms, nongovernmental organizations (NGOs), and communities, and thus their effect on the environment. The most obvious examples in the environmental area include environmental laws and regulations. However, these mechanisms, while necessary, often imply a command-and-control, end-of-pipe approach to environmental issues. That is, they frequently endorse specific environmental control technologies and impose emission constraints that are met not by changing manufacturing processes, technologies, or product design, but by putting air scrubbers and water treatment plants in place. Such traditional environmental approaches are of only minor interest to industrial ecologists. Increasingly, however, governments are developing more sophisticated means of encouraging environmentally appropriate behavior on the part of firms and consumers, centered largely on market instruments and sustainability goals (Figure 7.1). These strategies reflect increased experience with, and understanding of, the interaction of market systems with legal and regulatory structures on the one hand, and impacted environmental systems on the other (Figure 7.2).

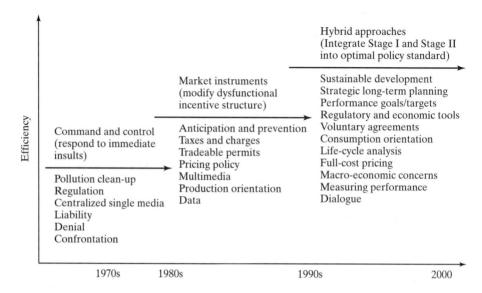

Figure 7.1

The evolution of environmental policy in OECD countries. (Courtesy of B. Long, Organization for Economic Cooperation and Development, paper presented at ECO '97 International Congress, Paris, Feb. 24–26, 1997.)

An obvious way in which governments influence corporations positively is that they are large purchasers of products for their own use. They may also be thought of as "indirect customers," expressing the demand of consumers as formulated in legislation, regulation, and less formal policies and practices. More than is commonly realized, governments as well as customers shape markets, and they have it within their power to create new markets. Similarly, governments are beginning to appreciate the

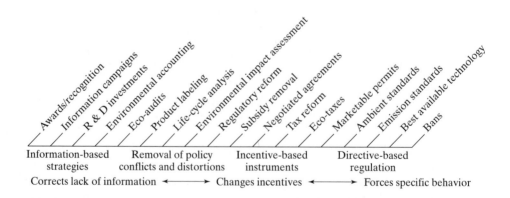

Figure 7.2

Governmental strategies to achieve environmentally responsive behavior in a market economy. (Courtesy of B. Long, Organization for Economic Cooperation and Development, Paper presented at ECO '97 International Congress, Paris, Feb. 24–26, 1997.)

strong coupling between technology and environmental policy systems. Policies can encourage environmentally preferable technologies and industrial ecology principles: product takeback legislation, if properly implemented, is one example. Alternatively, policies can impede progress: Laws that restrict the ability of government entities to purchase products containing refurbished subassemblies or components are examples of the latter. In general, government technology policies that assist the diffusion of new technologies will also indirectly increase the use of environmentally preferable technologies, because newer technologies tend to be more efficient in the use of resources and energy.

Governments also establish incentives through their tax and subsidy systems. Reductions in subsidies that encourage inefficient use of resources and energy, or increases in the taxation of environmentally problematic activities are well studied examples. Most countries subsidize energy consumption, which, all things being equal, leads to people using more energy than they otherwise would. "Green taxes" based on carbon emissions from energy consumption are an attempt to reduce such consumption, but they have generally not been widely adopted at levels which actually affect behavior. Broad proposals for tax reform that would tax environmental problems such as pollution and relieve taxes on social benefits such as employment have been widely proposed, but have not yet been politically successful.

There are a number of dimensions that can significantly affect the ability of nation-states to respond to environmental challenges. Among these are:

(a) *Form of government.* In general, democracies such as those in western Europe and the United States will be more responsive than totalitarian governments such as those that used to exist in eastern Europe.

(b) *Wealth.* Wealthier countries have more resources to respond to environmental challenges than do poorer countries, and the former may be able to place more relative value on environmental benefits.

(c) *Size.* Even very progressive small countries such as Denmark or the Netherlands cannot overlook the fact that much of their industrial production is exported, and thus subject to standards and requirements beyond their direct reach.

(d) *Focus.* Countries emphasize different aspects of environmental protection. For example, Japan is a leader in energy efficiency, and Germany and the Netherlands in developing consumer product takeback approaches.

(e) *Culture.* There is a distinct contrast between Japan, with its parsimonious approach to resources and energy born of its island status, and the former Soviet Union, which possessed a great natural resource base, a focus on industrialization at any cost, and a cavalier attitude towards conservation.

(f) *Attitude toward law.* Some countries have a culture in which equality before the law dominates interactions, especially economic ones. Other countries have more informal systems, where written law is only one of a number of considerations that affect commercial relations. As the latter are less transparent and less responsive to public opinion, they may not foster high levels of environmental quality.

(g) *Interaction of environment with other policy structures.* All countries have existing policy structures—technology policy, financial policy, national security policy, trade policy, consumer protection policy—which have, by and large, developed without considering environmental issues. One of the results of environmental issues becoming strategic for society is that they must be integrated into these pre-existing policy structures. As these will be different from country to country, and region to region (the North American Free Trade Area and the European Union, for example), one would anticipate different results as environmental issues are considered in different countries.

(h) *Transparency and openness of legal processes.* Because legal systems are not just objective, but cultural, creations of various countries, they differ in terms of their transparency (openness) and their consensual nature. In general, nations with legal systems that are transparent and encourage public participation will have stronger environmental laws.

7.2   INTERNATIONAL GOVERNANCE CONSIDERATIONS

While it is true that the international governance system is changing rapidly, it remains the case that the nation-state is the jurisdiction with which most people are familiar. Moreover, nation-states have authority over any activities occurring within their borders, and they are often the primary jurisdiction within which remediation and compliance activities are regulated. However, many environmental issues, including virtually all of the difficult ones, are beyond the borders of any single nation-state: acid precipitation, watershed pollution, ozone depletion, global climate change, loss of biodiversity and habitat. There is thus a mismatch in scale between the political bodies with the most authority and legitimacy and the environmental perturbations with which they must deal. The global scope of many of the anthropogenically perturbed natural support systems, in combination with an unwieldy international law system, raises questions about the appropriate venue for addressing environmental issues and adds to increasing devolution of nation-state obligations to international organizations, transnational corporations, and political subunits.

Existing international environmental treaties and agreements are negotiated, approved and enforced at the nation-state level. Reflecting the shift in environmental focus from compliance and enforcement to industrial ecology, these international agreements are becoming increasingly numerous, as shown in Figure 7.3. Strictly speaking, they apply only to signatory countries that specifically agree to be bound by the requirements, but their influence tends to be broad. Because of their global scope and their application to industrial and economic activities beyond the traditional environmental domain, these treaties and other covenants are of particular interest to the practitioner of industrial ecology. The most obvious examples include the Montreal Protocol, under which production and consumption of CFCs and other ozone-depleting chemicals are being phased out; the Basel Convention, under which transnational shipment of hazardous residues is controlled; and the Kyoto Protocol on Climate Change, designed to deal with the emissions of greenhouse gases.

The Montreal and Kyoto protocols and the Basel Convention are designed solely as environmental initiatives. They are thus fairly straightforward compared to the prob-

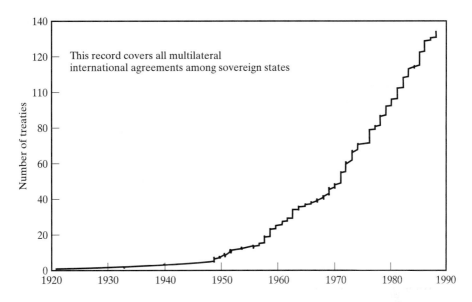

Figure 7.3

The number of international agreements dealing with environmental matters. Such agreements form the basis for current global governance of environmental issues, but they notably exclude firms, NGOs, and communities from the formal governance process. (Reproduced with permission from N. Choucri, The technology frontier: Responses to environmental challenges, in *Global Accord: Environmental Challenges and International Responses*, N. Choucri, ed. Copyright 1993 by the MIT Press.)

lems that arise when environmental considerations are incorporated into international policy structures which heretofore had no environmental dimensions. For example, environmental considerations are beginning to be integrated into national security policy structures at the national and multinational levels (e.g., in the North Atlantic Treaty Organization). A difficult area has been the attempt to integrate environment and trade policy. Potential conflicts arise because environmental laws generally seek to control the means by which goods are made and to forbid or discriminate against environmentally inappropriate processes, products, or technologies, while trade laws in general seek to liberalize the flow of goods among nations. This dynamic has made the interaction of trade and environment extremely complex and contentious, involving difficult questions of interpretation. To complicate matters further, developing countries are very concerned that efforts to inject environmental considerations into trade negotiations are not about environmental protection at all, but simply attempts by developed countries to discriminate against developing country products to protect domestic markets.

7.3   INDUSTRIAL ECOLOGY AND THE LEGAL SYSTEM

Environmental laws are not a recent phenomenon. As early as 1306, London adopted an ordinance limiting the burning of coal because of the degradation of local air quality. Such laws became more common as industrialization created substantial point source emissions. For example, the LeBlanc system for producing soda (sodium carbonate), patented in 1791, resulted in substantial emissions of gaseous HCl to the air in the environs of the facilities. As a result, the English Parliament eventually passed the Alkali Act of 1863, requiring manufacturers to absorb the acid in special towers designed by one William Gossage.

As in these historical examples, much environmental law has reflected the perception of environmental problems as localized in time, space, and media (i.e., air, water, soil). For example, it was not uncommon in the recent past for groundwater contamination by organic solvents to be eliminated by "air stripping," or simply releasing the solvent to the air, where it contributed in many cases to the formation of tropospheric ozone. Similarly, many hazardous waste sites in the United States have been "cleaned" by simply shipping the contaminated dirt somewhere else, which not only does not solve the problem but creates the danger of incidents during the removal and transportation process. Environmental regulation has thus traditionally focused on specific phenomena and adopted the so-called "command-and-control" approach.

7.3.1   Fundamental Legal Issues

The invention of the concept of "sustainable development," together with the recognition that environmental issues are now strategic for firms and society, has enormously complicated the legal structure within which environmental issues are managed. As the example of trade and environment discussed above suggests, this raises a number of fundamental legal issues, as discussed below.

Intragenerational Equity.   The distribution of wealth and power both among nation-states, and between the elites and the marginalized populations within individual nation-states, is one of the principal themes of political science. It is a highly ideological and contentious arena, with the poles in the debate being equality of outcome, called *egalitarianism,* and equality of opportunity, called *libertarianism.* Sustainable development is an egalitarian concept; it contemplates reasonably equal qualities of life for all people as a prerequisite for societal stability and environmental consciousness, and thus implies a substantial shift in resources between rich and poor nations, as well as within nations. Whether and how this should be accomplished are the subjects of extensive debate.

Intergenerational Equity.   This concept, which would impose egalitarian ethics across generations, is also part of the sustainable development concept. In addition to being contentious for its egalitarian implications, it is problematic from a practical perspective. It would require, for example, some representation of the interests of future generations in current disputes—such as storage of radioactive residues from electric power production, deforestation and reforestation, and the extent of climate change mitigation efforts—which might affect those interests. While many legal and philosophical traditions recognize some general requirement of fairness between generations, future generations are not viewed as having enforceable legal rights, in part because it is virtually impossible to identify either future individuals or their interests with sufficient specificity to involve them in adjudication.

Flexibility of Legal Tools.   Because legal systems in many societies tend to be important components of social structure, they are usually conservative and relatively inflexible. One does not want a situation, for example, where inheritance laws are changed every six months: the passage of property between generations, the assurance of free title, and preservation of family harmony and expectations all argue for a reasonably stable inheritance system. The price for this inflexibility is paid in terms of inability to adjust to changing situations. Where change is limited and foreseeable—as it may be with inheritance issues—this inflexibility is relatively unimportant.

Where change is rapid and fundamental, however, as it currently is with environmental issues, such inflexibility can lead to substantial inefficiency. This is especially true as environmental regulations are extended from command-and-control end-of-pipe requirements—which may be expensive, but don't really affect choices of materials, manufacturing technologies, product design, or customer choice—to pollution prevention and product regulation. For example, an overly conservative requirement for scrubbers might cause manufacturing plants to spend marginally more than they should on such technologies. However, an overly conservative process or product standard—or, worse yet, one that subsequent data demonstrates was environmentally inappropriate—skews manufacturing, product design, and consumption patterns for a long time as it becomes embedded in industrial systems, a process known as "lock-in."

The initial inflexibility built into such systems is augmented by the well known tendency of regulation to create and nurture those who benefit from continued regulation and that consequently form a significant barrier to subsequent regulatory rationalization. For example, reform of hazardous waste laws tends to be impeded by a number of interest groups, including associated legal and engineering firms, the firms that produce government-approved technologies and waste management services, and, more subtly, the environmental groups that use the fear of hazardous waste as a membership recruitment and fund-raising device.

Regulatory Management Structure.   Traditional command-and-control regulatory mechanisms have been applied with significant success to easily visible environmental problems. Unthinking extension of such simple regulatory tools to far more complex situations, however, can frequently be both environmentally and economically costly. Rather than continuing to rely primarily on centralized command-and-control, a more sophisticated environmental management system that recognizes the complex nature of the systems at issue should be used.

In general, targeted intervention is appropriate where the environmental impact is significant in spatial terms and potential damage, and difficult or impossible to reverse (use of lead compounds in gasoline, for example). On the other hand, if the environmental impact is the result of diffuse activities throughout the economy, with no easy technological solution, flexible policy approaches that attempt to establish appropriate boundary conditions are likely to be preferable.

Decentralized Mechanisms versus Centralized Micromanagement.   Another common policy choice arises when choosing between reliance on decentralized mechanisms, principally the information and operational efficiency of markets, or centralized micromanagement, principally traditional command-and-control environmental regulation. As a general principle, both environmental and economic efficiency benefit if market mechanisms are used to the maximum extent possible. In part, this is because once the obvious environmental insults to air, water, and human health are addressed, the remaining ones tend to manifest themselves heterogeneously over time and space. Temporal and spatial heterogeneity are also characteristic of relevant human and natural systems, such as resource and sink availability, structure and resiliency of biological communities, infrastructure availability, cultural considerations, and technological capabilities. Such scalar variability tends to argue for caution in relying on centralized regulatory mechanisms in many cases.

Determining Appropriate Jurisdictional Level.   Political jurisdictions are creations of human culture and history, and there is no *a priori* reason why their boundaries should reflect underlying natural systems. It is thus no surprise that many problematic environmental perturbations are not coextensive with political boundaries: Emission of acid rain precursors in the United States or China causes acid rain in Canada or Japan, for example, and watershed degradation involving different nation-states generates enormous legal and political conflict. It is sometimes possible, however, to reduce distortions produced by jurisdictional boundaries that are not coextensive with the scope of the issues under consideration. This can be achieved both horizontally (by harmonization of policy across jurisdictions of equal legal status) and vertically (by integration of policy structures up jurisdictional hierarchies). Harmonization of regulatory management structures across jurisdictions of equal status is desirable for several reasons. First, commerce is increasingly regional and global in scope, so the imposition of substantively different constraints on economic activity at small geographic scales can cause significant economic dislocation or impose protectionist trade barriers. Also, it is desirable to avoid exportation of risk. Where a particular substance or activity is banned or stringently regulated in one area, yet demand for the regulated item remains strong, it will frequently simply be displaced to a jurisdiction where it is less regulated.

7.3.2   Legal Case Studies Relevant to Industrial Ecology

A brief discussion of specific examples where existing legal structures come into conflict with the environment is illustrative. These should be seen as case studies whose stages can be relatively easily characterized. First is the recognition that a particular policy structure has environmental implications that heretofore have been ignored. Second, there is a period of conflict when neither the environmentalists nor the target policy community are willing to jeopardize their goals for the others. Finally, there is a period of dialog that builds trust and gives each side a better idea of the issues. This lays the basis for resolution.

One example is consumer protection law. The general purpose of environmentally relevant consumer protection law is to encourage full disclosure by the producer, and thus avoid fraud. For example, used products, considered to be inferior to new ones, are not to be passed off as new. Thus, in part, consumer protection laws require that any product that is used or contains used parts be prominently labeled. Such a label, not surprisingly, significantly reduces the price that can be charged, and can hurt the trademark of the producer or vendor. The result is the creation of strong incentives against recycling used components, subassemblies, or products. Reuse of products or parts, however, can provide clear environmental benefits and should be encouraged by public policy. Moreover, in some instances—memory chips, for example—there is little difference between a new and a used part, and in many other cases a used or refurbished part is more than adequate for the use for which it is intended. Although this conflict between legitimate policy goals has yet to be resolved, there are several obvious possibilities. As a stopgap measure in the short term, the principle could be established that so long as a product, component, or part meets all relevant specifications it is immaterial whether it is used or not. In the longer term, the issue is perhaps less one of law than of consumer education: Customers have been acculturated to avoid used products, or to value them less, and will need to be educated about their benefits.

Another example concerns government procurement policies that seek to avoid consumer fraud and to reduce the possibility of collusion between the purchasing entity and the vendor to buy inferior goods and split the savings. Nonetheless, there does not appear to be any reason why the possible solutions applicable to the consumer protection situation cannot also apply here: Purchase according to relevant specifications, or purchase functionality or service rather than product. Government procurement practices are also important for their potential positive impacts on industrial behavior. End-use, individual consumers are a broad, unorganized group, and most consumers do not make purchases for environmental reasons. Governments, on the other hand, have substantial buying power centralized in one organization and thus can exercise significant control over a market. To the extent that government procurement practices can be made environmentally preferable, they can exercise significant beneficial impacts on the performance of producers and vendors.

As with trade and environment, there are several fundamental issues involved in the relationship between antitrust and environmental policies. Antitrust seeks to maintain the competitiveness of markets by limiting the market power of firms, which generally means limiting their scope and scale. Many environmental initiatives, such as postconsumer product takeback, however, seek to do the exact opposite: to expand the scope and scale of the firm so that it is responsible for the environmental impact of its product from material selection, through consumer use, to takeback and recycling or refurbishment. Thus, while antitrust seeks a market with no central control, environmental policy seeks to extend the control of firms in the interest of internalizing to them the costs of negative environmental externalities.

The inherent conflict between antitrust and environmental policies is exacerbated by the question of technological evolution. By and large, technological evolution is most rapid in competitive markets with low barriers to the introduction of new technologies. Such market structures may well be fostered by traditional antitrust policies. On the other hand, if firms are to implement environmentally preferable practices across the life cycles of their products, they will generally have to develop means of linking technologies used at various points in the product life cycle. Thus, the technologies used to disassemble the product after the consumer is through with it need to be considered in the initial design of the product (a process called by designers, reasonably enough, design for disassembly). Linking technologies in such a way creates a more complex technological system, and reduces the ability to evolve any single part of that system rapidly. Moreover, it usually requires the development of standards, if not of active managerial control, across the entire process—again conflicting with the basic goals of antitrust.

7.4   ECONOMICS AND INDUSTRIAL ECOLOGY

Economics is perhaps the most powerful discipline in terms of its capability to shape policy. Its analyses strongly influence most national and international policy formulation, and failure to maintain a strong economy can be catastrophic to the quality of life of a country's citizens. Moreover, economic performance is inherent in the concept of sustainable development, and development generally, and the institutions that exist to foster development rely heavily on economic analysis and criteria. It is likely that in the future both economics and industrial ecology will be significantly changed by insights derived from the other: indeed, economic insights are critical to successful implementation of industrial ecology. Concomitantly, the study of economics without an understanding of industrial ecology will grow increasingly sterile.

7.4.1   Valuation

Traditionally, many economists have viewed economics as an objective, not normative, discipline. This has been strongly challenged by some, particularly in the environmental movement, who claim that the foundation of economic analysis—an assumption that all things can be quantitatively valued in terms of money—is a fundamental statement about morality, and an incorrect one. The assumption that everything has a monetary value, that nothing is sacred, is obviously contentious, especially to those who would regard nonhuman species or particular ecosystems from a normative perspective. This dichotomy of views has led to an increasingly well recognized phenomenon wherein "objective" economic studies of environmental issues are simply rejected by large elements of the public based on a strong belief that subjective dimensions are meaningful in such situations.

Even where it can be agreed that quantitative evaluation is acceptable from an ethical perspective, obtaining the proper measures—valuation—can be difficult. A number of tools, or valuation methods, have been developed to quantify such difficult phenomena as the health effects of pollution. Examples include:

1. The *human capital* method, which measures earnings foregone due to illness or premature death as a result of pollution exposure.
2. The *cost of illness* method, which measures lost workdays plus out-of-pocket medical and associated costs resulting from pollution exposure.
3. The *preventive/mitigative expenditure* method, which measures expenditures on activities to mitigate or reduce the effects of pollution, such as putting in new water delivery systems to avoid exposure to contaminated groundwater.
4. The *wage differential* method, which uses wage differentials between areas differing in pollution exposure as a surrogate for the implicit value of less pollution for people.
5. The *contingent valuation* method, which uses surveys to determine what value people say they put on pollution avoidance.

6. The *surrogate actions* method, which infers an economic value to environmental goods by costing the actions people have taken that are surrogates for the good itself. For example, the *travel cost* method examines how much people are willing to travel for a higher quality environmental amenity, such as an uncrowded park.

Although all of these methods offer a means by which dollar values can be assigned to environmental insults, they all have some drawbacks and confounding factors, and must therefore be used intelligently and the results treated with caution.

Quantification methods illustrate a common pitfall in economic analyses. Although such an approach simplifies analysis, and can generate more rigorous and understandable results, it also means that factors that cannot be quantified are, in practice, simply not included in the analysis. Even when qualitative impacts are considered, they tend to be assigned a lesser weight. Thus, both valuation techniques and economic analyses in general should be used with caution where the moral or ethical content of the issue is significant, and where the systems under consideration have important qualitative dimensions.

7.4.2   Discount Rates

Standard economic analysis asserts that money today is worth more than the same amount of money tomorrow, because of inflation and the returns over time that can be anticipated if the money is invested. This is expressed by applying a discount rate to future returns as compared to current returns. Technically, this is represented by an equation which gives the present value, A, of an amount V which will be available t years from now, where i is the discount rate:

$$A = V(1 + i)^{-t} \qquad (7.1)$$

The use of discount rates to value resources and plan investments in business and government is ubiquitous and, in many cases, appropriate. Without such an approach, it would be difficult to compare investments that required expenditures and generated streams of returns in differing time periods. Obviously, however, this approach also provides a strong incentive to use resources as soon as possible rather than save them for the future. If the investment's return is itself properly invested so that it provides a stream of benefits over time, the future may indeed benefit more from the economic growth generated than it would have if the resources were conserved. Thus, the issue does not appear to be one of rejecting the concept outright, but of understanding under what conditions such analyses are useful. In particular, discount rates should be applied cautiously in cases where significant social value issues and externalities are present.

7.4.3   Benefit-Cost Analyses

In its broadest sense, a benefit-cost analysis is any methodology, formal or informal, that attempts to comprehensively evaluate the benefits and costs of a particular action in order to help determine whether the action should be taken. When applied to environmental issues, a usual complication is that difficulties in identifying and quantifying all important factors are endemic. Accordingly, one should consider the appropriateness of the assumptions required for the assessment, the interests of the parties making (or funding) the assessment, the distributional impacts of the proposed action (who benefits and who loses), and the significant uncertainties involved.

There are several subtleties about benefit-cost analyses. It is important to know, for example, whether an assessment includes social costs and benefits or just environmental ones. If the former, it is important to know what values of stakeholders have driven the quantification process, because social costs in particular are difficult to quantify and highly value laden. It is also important to know whether the benefit-cost assessment has been linked to any scientific assessments of risks or probable outcomes, and, if so, whether the uncertainties and probabilities inherent in the underlying studies have been accurately reflected. Understanding these questions of scope aid the industrial ecologist in interpreting the results of the benefit-cost assessment, and in determining what nonquantitative elements should be included as part of the overall decision process.

7.4.4 Green Accounting

Information concerning the economic performance of the firm is generally captured in management accounting systems. Traditionally, such systems have treated environmental costs—even real, quantifiable environmental ones, such as residue disposal costs—as overhead, and have therefore not broken them out by activity, product, process, material, or technology. The result has been that managers, not having access to the environmental cost information concerning their choices, have had neither the incentive nor the data needed to reduce those costs.

The solution, called *green accounting*, is conceptually simple: Develop managerial accounting systems that break out such costs, assign them to the causative activity, and thus permit their rational management. In practice, however, this may be a difficult task. For example, in many complex manufacturing operations, developing sensors and systems to provide the physical data on the contributions of different processes and products to a liquid residue stream is a nontrivial task, and one that involves engineering design and capital investment. Moreover, managers tend to resist additional elements of the business process for which they will be made responsible. Also, the assignment of "potential costs," such as estimates of future regulatory liability for present residue disposal practices, may be resisted for fear of creating unnecessary legal liability (it might be argued that a company which foresaw potential future liabilities was thereby admitting its planned behavior was inappropriate or illegal). Nonetheless, it is clear that development of appropriate managerial accounting systems, and their supporting information subsystems, is critical to completing a necessary feedback loop for environmentally appropriate behavior by corporations, and is currently underway in many of them.

Green accounting differs fundamentally from the idea of *social costing*. The latter implies that firms should internalize into their financial accounting system their share of related environmental costs not currently being captured in prices. Here quantification is difficult and contentious, and resistance by firm managers is likely to be high. In essence, green accounting asks managers to do their existing jobs more rationally, while social costing seeks to redefine the very nature of the firm.

Green accounting procedures are also appropriate at the national level. Traditionally, national economic accounts are dominated by a focus on Gross National Product (GNP). GNP is typically defined as the aggregate money demand for all products, including consumer goods, investments, government expenditures, and export spending. It and similar national account systems are frequently taken as a measure of individual economic welfare, or, more controversially, as a measure of quality of life. As environmental issues have become more important, such metrics have been increasingly criticised for their failure to depreciate so-called "natural capital" as it is used to produce monetarized assets. Technically, systems based on the United Nations System of National Accounts (SNA), the international standard, recognize land, mineral, and timber resources as assets in a nation's capital stock, but do not recognize them in the income and product accounts. Accordingly, if a natural resource is used, the national income and product accounts show no equivalent depreciation. It is increasingly argued, however, that if a forest or a mineral deposit has been depleted, national income accounts should reflect this reduction in value of a natural asset even as they may reflect money income derived from that depletion. Failure to do so in essence values the existing forest as zero until it is destroyed.

7.4.5 Substitutability versus Complementarity

Another assumption common to standard economic analysis is substitutability among resources and economic inputs based on monetary value, which in turn reflects relative scarcity. In general, this is a valid principle that has been demonstrated many times: As one input becomes scarce, and thus more expensive, another is substituted for it. The difficulty with this principle arises not because it is wrong, but because it is right so much of the time that it has become an axiom, rather than an assumption. As applied in practice, in other words, it assumes that there are always substitutes. It has become an embodiment of unquestioning technological optimism. It is in this latter guise that it is questionable. For example, the element gadolinium, which has the property of heating up when placed in a magnetic field, has been used to develop super-efficient refrigerator technology (a conventional refrigerator has a maximum efficiency of about 40 percent; a gadolinium refrigerator has a theoretical efficiency of about 60 percent). Given the burgeoning demand for refrigeration, especially in rapidly developing tropical countries, and the lack of other materials with this property, gadolinium could well be a complement to efficient cooling, with no (as yet identified) substitutes.

7.4.6 Externalities

An externality is simply a cost, either positive or negative, that is not captured within the economic system through prices. For example, when a factory is allowed to dump toxic materials in a river without charge, it generates a negative externality to the extent that it imposes costs on society that it doesn't have to pay. When fish die downstream, the local fishermen are indeed poorer than they otherwise might have been, but the price paid by the factory to dump its waste does not reflect this cost.

Externalities are important because much of the purpose of environmental management and regulation is to reduce externalities, often in less efficient ways than by simply changing prices. For example, postconsumer product takeback regulations may be seen as an effort to internalize to the manufacturing firm the environmental costs associated with management of the environmental impact of used products. Clean air laws that prescribe certain control technologies may be seen as efforts to internalize to the firm the costs associated with the emissions being controlled. Imposing environmental requirements in government procurement contracts is another policy whose basic mechanism is internalizing externalities to the market system.

While internalizing externalities is desirable, there are a number of barriers in practice. Changing any price structure to internalize a cost that previously was not incorporated will almost always adversely affect some stakeholder, who can thus be expected to oppose the action. Thus, internalizing externalities by removing environmentally inappropriate subsidies and increasing taxes on environmentally disfavorable actions so that their market price more accurately reflects their full costs will almost always be politically difficult. For example, any effort to increase gasoline taxes in the United States to more accurately reflect social cost appears politically impossible.

7.5   FINANCE, CAPITAL, AND INVESTMENT

Most environmentalists tend to be suspicious of capital and financial markets, if they think of them at all. Students of industrial ecology cannot afford to be, for it is the availability of capital, and its productive use to support the evolution of environmentally preferable technologies, that offers promise for reducing the environmental impacts of humanity as it moves towards unprecedented levels of wealth and population.

The importance of investment for industrial ecology policy is apparent: If progress toward a more sustainable global economy requires rapid technological evolution, then the new technology must be financed, and thus must attract investment. The capital required for the diffusion of new technologies can be daunting. For example, one estimate by the World Business Council is that bringing the paper products industry up to best industry standards would take a worldwide capital infusion of at least $20 billion—and that is without considering technological evolution off current baselines. Therefore, investment and technological evolution are not just synergistic, but inextricably intertwined. For example, one reason given for the success of Silicon Valley, one of the most technologically innovative regions in the world, is the unparalleled access to venture capital in that area. Conversely, in many European and Japanese financial markets, capital is much harder for new ventures to obtain, so there are correspondingly fewer small firms with their concomitant innovation.

Establishing a robust financial market and being able to attract adequate capital is a complex problem for developing countries. In earlier years, much of this needed capital was provided by foreign investment from developed country governments, or from multilateral lending organizations such as the World Bank. More recently, the bulk of such investment has come from large transnational firms in the form of foreign direct investment (FDI). The amounts involved can be significant: In 1994, some U.S.$90 billion flowed into developing countries; FDI in China alone for the past several years has amounted to some U.S.$30 to 40 billion annually. FDI is important because it not only helps create the capital base for indigenous industry, but because it supports the transfer of modern technology and modern business practices—both of which are usually more environmentally preferable than the alternatives.

The Netherlands Approach to Environmental Policy

People tend to regard Germany, the United States, and perhaps Sweden as governments that are the most advanced in environmental regulation. The most sophisticated, comprehensive approach to the integration of environment and technology, however, has without question been that of the Netherlands, as set forth in the National Environmental Policy Plan (1989), the National Environmental Policy Plan Plus (1990), and the National Environmental Policy Plan 2 (1994), and additional implementation documents released in support of the plans.[1]

The plans are all explicitly based on the goal of attaining sustainable development in the Netherlands within one generation, with sustainable development being defined as in the Brundtland Report: "Development that meets the needs of the present without compromising the ability of future generations to meet their own needs." The approach is comprehensive: Although the lead is taken by the Ministry of Housing, Physical Planning and Environment, the need to include all other sectors, especially transportation and housing, is explicitly recognized by including the relevant ministries and councils in the planning process. Target activities for the plans include agriculture, traffic and transport, industry and refineries, gas and electricity supply, building trade, consumers and retail trade, environmental trade, research and education, and "societal organizations" (environmental groups, unions, etc.).

At all stages of analysis, the economic impacts of proposed changes are explictly considered. Moreover, the emphasis is clearly on collaboration with industry and other stakeholders in developing and implementing specific proposals, the goal being scientifically and technically correct decisions. This collaborative approach is borne out by the use of enforceable agreements (covenants) between industry and the government in order to achieve environmentally desirable ends, rather than the more adversarial legislative process, whenever possible.

The Netherlands, which receives most of its air and water from other countries as a result of regional air and watershed patterns, is not under the illusion that a complex and difficult-to-define concept such as sustainable development can be achieved by a small country within a few short years. But the selection of sustainable development as the policy goal has enabled a far more sophisticated multidisciplinary approach than would otherwise be possible. Thus, it has encouraged a critical focus on the identification of appropriate metrics to determine progress towards sustainability. It has also resulted in a more sophisticated, comprehensive approach to risk than that usually entailed in traditional environmental risk assessments. As the initial plan noted, "Making sustainable development measurable ... is not an easy task, but the results of research on this topic are necessary to enable feedback at the source."

The approach taken by the Netherlands is unique, pathbreaking, and exemplary of industrial ecology. The implicit acceptance of the need for fundamental change in existing economic structures is both commendable and still all too rare at the national government level.

[1]The plans, and other supporting documentation, are available from the Ministry of Housing, Physical Planning and Environment, Department for Information and International Relations, P. O. Box 20951, 2500 EZ, The Hague, The Netherlands. Most of the important material is available in English.

These new sources of capital for developing countries matter for several reasons. First, they imply that one benefit of a globalizing economy is that technological improvements, once adopted by world class transnationals, may be anticipated to diffuse to developing as well as developed countries as part of normal global investment patterns. Second, they focus again on the critical role of industrial firms as agents not only of development but—at least potentially—of environmental efficiency and even sustainability. Third, they affirm a strong potential link between financial institutions and environmental performance: FDI and associated multilateral lending streams are important mechanisms by which environmentally preferable practices can be diffused. In sum, technological evolution, whether in developed or developing countries, must be financed, and it will only be financed if it is profitable in a competitive capital market.

FURTHER READING

Allenby, B.R., *Industrial Ecology: Policy Framework and Implementation*, Upper Saddle River, NJ: Prentice Hall, 1999.

Bhagwati, J., The case for free trade, and H.E. Daly, The perils of free trade, in Debate: Does free trade harm the environment?, *Scientific American, 269*(5), 41–57, 1993.

Costanza, R., ed., *Ecological Economics: The Science and Management of Sustainability*, New York: Columbia University Press, 1991.

Crocker, D.A., and T. Linden, eds., *Ethics of Consumption: The Good Life, Justice, and Global Stewardship*, Lanham, MD: Rowan and Littlefield, 1998.

Environmental Science and Technology, Special Issue on Valuing the Environment, *34*, 1381–1461, 2000.

Gentry, B.S., *Private Capital Flows and the Environment: Lessons from Latin America*, Cheltenham, U.K.: Edward Elgar, 1998.

Weiss, E.B., *In Fairness to Future Generations: International Law, Common Patrimony, and Intergenerational Equity*, Tokyo, Japan: The United Nations University, 1989.

EXERCISES

7.1 You are the environmental officer of a chemical firm producing commodity polymers. The government of the country in which most of your production facilities are located has just proposed a broad energy tax, substantially higher than in any other developed country.

   (a) List and evaluate the positions your firm can take in response to this public policy initiative.

   (b) Which would you choose, and why?

   (c) What data on your firm's operations would be useful in helping you develop your positions?

   (d) What organizational elements of the firm should be involved in helping you develop and implement your positions?

7.2 Your firm owns mineral rights currently estimated to be worth $100,000 in a (politically) relatively unstable country.

   (a) Assuming a discount rate of 7%, how much will your rights be worth in 5 years? In 10?

(b) What factors would you consider in making a decision about when to exploit the mineral rights, from the perspective of (i) the firm that owns them, (ii) the government of the country, and (iii) the community near the ore deposit.

7.3 The Organization of American States (OAS) exists to provide mutual defense against aggressors in North and South America. Unlike the North Atlantic Treaty Organization (NATO), the OAS has not incorporated environmental considerations into its activities in any extensive way. Investigate the two organizations and propose a group of environmentally oriented activities for the OAS.

PART III Design for Environment

C H A P T E R   8

Industrial Product Design and Development

8.1 THE PRODUCT DESIGN CHALLENGE

One of the most daunting challenges in modern technology is that faced by the designer of a new product. The products themselves can be enormously complex, some having many more than a million individual parts (see Figure 8.1). Designing a product that does what it is supposed to do is not the only job for designers, however. Their tasks also include the following:

- *Surveying customers to receive ideas for product characteristics.* A business aphorism is that "products should not merely satisfy the customer, they should *delight* the customer." This is a high standard to live up to, and requires that potential customers and their desires be understood very well.
- *Addressing competitive products.* New designs must meet or exceed those of the competition, or they will be unsuccessful.
- *Complying with regulations.* Product safety, labeling requirements, recyclability, and a host of other legally binding constraints must be considered.
- *Protecting the environment.* Design considerations that have environmental implications are increasingly a topic of interest to customers, regulators, and industrial managers.
- *Producing designs that are attractive, easy to manufacture, delivered on time, and competitively priced.* In today's business world, immediate customer acceptance, efficient manufacturing, and timeliness are crucial.

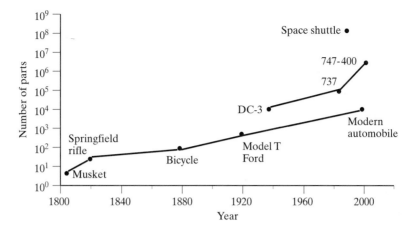

Figure 8.1

The increase in complexity of industrial products over time. (Adapted from a sketch provided by Paul Sheng, University of California, Berkeley.)

   Although there are substantial commonalities among design approaches, there are great diversities as well. An important distinction among the industrial sectors is the lifetimes of their products, as indicated in Table 8.1. Some are made with the anticipation that they will function for a decade or more. Others have lives measured in months or weeks. Still others are used only once. A designer must obviously adopt different approaches to these different types of products so far as features such as durability, materials choice, remanufacturing, and recyclability are concerned.

TABLE 8.1   Manufacturing Sectors and Their Products

Manufacturing	Product examples	Product lifetime
Electronics	Computers, cordless telephones, video cameras, television sets, portable sound systems	Long
Vehicles	Automobiles, aircraft, earth movers, snow blowers	Long
Consumer durable goods	Refrigerators, washing machines, furniture, furnaces, water heaters, air conditioners, carpets	Long
Industrial durable goods	Machine tools, motors, fans, conveyor belts, packaging equipment	Long
Durable medical products	Hospital beds, MRI equipment, wheelchairs, washable garments	Long
Consumer nondurable goods	Pencils, batteries, costume jewelry, plastic storage containers, toys	Moderate
Clothing	Shoes, belts, blouses, pants	Moderate
Disposable medical products	Thermometers, blood donor products, medicines, nonwashable garments	Single use
Disposable consumer products	Antifreeze, paper products, lubricants, plastic bags	Single use
Food products	Frozen dinners, canned fruit, dry cereal, soft drinks	Single use

To accomplish their assigned tasks, designers have available to them a large tool-box, developed over many years. Many of these tools were developed before environmental considerations became as important as they are now. Accordingly, in this chapter and several following, we present and discuss several popular design tools that are environmentally focused or that can incorporate environmental considerations within them. We concentrate not on tools primarily used in product realization, but on those whose focus is on concepts (such as product definition, product decisions, and product development).

8.2   CONCEPTUAL TOOLS FOR PRODUCT DESIGNERS

8.2.1   The Pugh Selection Matrix

The Pugh Selection Matrix is designed to display the characteristics of a potential product design and to assess how well those characteristics are met by current products of the corporation, by competing products, and by alternative new designs. An example of the rather intricate diagram used in this process is given in Figure 8.2, a room air conditioner being used as the example product.

	Product characteristics	Customer importance	Design targets	A	B	C	D	Customer Survey
Performance	Affordable	9	< $253.00	+	+	S	+	
	Easy to handle	3	< 40 lbs	−	+	S	+	
	Quiet	9	Nc30 max	+	+	S	−	
	Not rumbly	9	RC30 max	S	S	S	+	
	Easy to service	1	20 in. lb. torque	−	−	−	+	
	Air flow	3	200 to 3000 CFM	S	S	S	S	
	Sound	9	RC30 max	−	−	+	S	
	Damper leakage	3	100 cfm at 1" SP	+	+	−	S	
	Electric heat	1	1 kw per 200 CFM	−	+	S	−	
DfE	Energy efficient							
	Recycled materials							
	CFC-free							
	Easy to upgrade							
Cost & Schedule	Fabrication cost	9	< $180.00	+	+	+	+	
	Lead time	9	2 weeks	+	+	−	+	
	Production date	3	10-12-94	−	−	+	+	
Controls	UL 1096	3		S	S	S	S	

Customer Survey symbols: ▽ Our company, □ Competitor A, □ Competitor B, ● Target (scale 1–5).

Figure 8.2

The Pugh Selection Matrix for a room air conditioner.

The left side of the diagram lists existing or desired characteristics of room air conditioners. This list can be as extensive as the design team chooses to make it; we provide only enough detail here to illustrate how the tool works. The diagram in Figure 8.2 includes a group of characteristics that pertain to DfE aspects of the design; this is a feature not traditionally included within the Pugh matrix.

Once the list of characteristics is completed, the next step is to develop design targets for each of the characteristics that the design team wishes to address, and to generate several alternative designs. A customer focus group is asked to rate the characteristics for relative importance (on a high, medium, low, or 9, 3, 1 scale), and to evaluate the current product design and those of the principal competitors on a 1–5 scale for its performance relative to the characteristics selected.

When these steps have been taken, the design team is in a position to evaluate each of the alternatives. At each point in the matrix, the design in question is rated + (exceeds the target), S (about the same as the target), or − (does not meet the target). With this information, the team can proceed to designate one alternative as the design target, and proceed to the detailed design process.

8.2.2 The House of Quality

An analytical tool that is an alternative or supplement to the Pugh Selection Matrix is the House of Quality. This tool, shown in Figure 8.3, shares the customer focus of the Pugh matrix, but goes deeper into the design process. To begin, one enters the "customer room," and lists the product characteristics desired by the customer together with their relative importance. Next, the "environmental foundation" is added, in which desirable environmental attributes are listed. Potential engineering characteristics are added in the "design room."

With the foundation and room characteristics established, the importance of each of the characteristics is rated on a high (H), medium (M), low (L) scale. The potential features of the product are then related to the alternative design approach being evaluated in the "relationships room." Here the degree to which engineering approaches respond to the desirable characteristics identified by customers are evaluated on a strong (S), moderate (M), weak (W) scale in order to guide design decisions toward effective responses. Matrix elements with no entries indicate engineering characteristics that do not influence customer or environmental attributes.

The House of Quality is completed by constructing the "roof matrix." The entries in this matrix indicate where a particular design goal reinforces others (X), or where design goals are in conflict (O). If the design room has been constructed in sufficient detail, the roof matrix can often provide insights into design approaches that will satisfy multiple customer preferences and the topics called out by the environmental foundation.

8.3 DESIGN FOR X

Product definition—what the product will be used for, how it will function, what its properties will be, the range of probable cost, and (if appropriate) its aesthetic attributes—provides designers with a substantial range of things to attempt to optimize si-

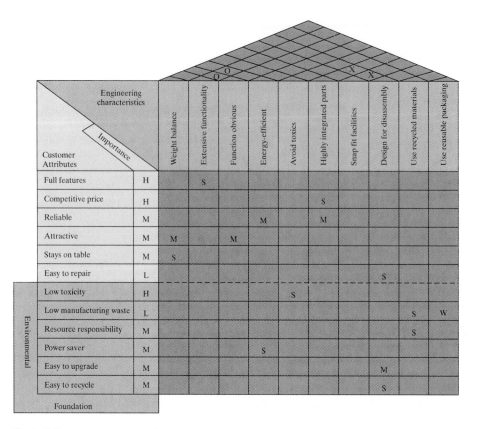

Figure 8.3

The House of Quality for a desktop telephone.

multaneously. Modern designers have a list much longer still, however, since they need to consider related product attributes that may, in the end, determine the product's success or failure. The paradigm for these latter considerations is termed "Design for X" (DfX), where X may be any of a number of design attributes, such as:

Assembly (A): Consideration of assemblability, including ease of assembly, error-free assembly, common-part assembly, etc.

Compliance (C): Consideration of the regulatory compliance required for manufacturing and field use, including such topics as electromagnetic compatibility

Disassembly (D): Consideration of end of life efficiency of product disassembly for refurbishment, recycling, or disposal

Environment (E): This component of DfX, and the philosophy on which it is based, is a principal subject of this book

Manufacturability (M): Consideration of how well a design can be integrated into factory processes such as fabrication and assembly

Material logistics and component applicability (MC): The topic focuses on factory and field material movement and management considerations, and the reliable supply of components and materials

Reliability (R): Consideration of such topics as electrostatic discharge, corrosion resistance, and operation under variable ambient conditions

Safety and liability prevention (SL): Adherence to safety standards and design to forestall misuse

Servicability (S): Design to facilitate initial installation, as well as repair and modification of products in the field or at service centers

Testability (T): Design to facilitate factory and field testing at all levels of system complexity: devices, circuit boards, and systems

The major characteristic that distinguishes DfE from traditional environmental regulatory compliance is that its scope extends far beyond the factory walls. This perspective is captured by considering the entire life cycle of products and processes, as shown in Figure 8.4. Stage 1, premanufacture, is performed by suppliers, drawing on (generally) virgin resources and producing materials and components. Stage 2 is the manufacturing operation directly under the control of the corporation making the product under consideration. Stage 3, product delivery, will generally be under the control of the manufacturer, although complex products containing many components and subassemblies may involve a global web of shippers, dealers, and installers. Stage 4, the customer use stage, is influenced by how products are designed and by the degree of continuing manufacturer interaction. In Stage 5, a product no longer satisfactory because of obsolescence, component degradation, or changed business or personal decisions is refurbished, recycled, or discarded.

DfX practices that address all aspects of the life cycle are already being implemented by leading manufacturing firms. Accordingly, the least difficult way to ensure that environmental principles are internalized into manufacturing activities in the short term is to develop and deploy Design for Environment (DfE) as a module of existing DfX systems. Moreover, the fact that DfE is intended to be part of an existing design process acts as a salutary constraint, requiring that DfE methods and resulting recommendations be implementable in the real world.

An increasingly important aspect of the design process is the degree to which it is linked to the computer. Most modern industrial design teams utilize Computer-Aided Design/Computer-Aided Manufacturing (CAD/CAM) software, which can incorporate standard component modules into a design, check a design for spatial clearances, produce lists of materials, and so forth. To the degree that DfE can be integrated into these design tools, it will become automatically a part of the physical design process. DfE incorporation into CAD/CAM is far from universal, and diligent effort will be needed to bring it to a high level of development.

8.4   PRODUCT DESIGN TEAMS

The focus on the customer, on cost, and on all the topics raised by the DfX process requires that product and process design no longer be performed by an isolated group of engineers. Rather, modern industrial practice is to form product design teams made up

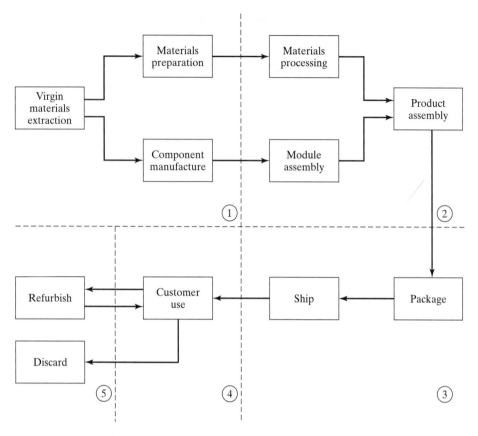

Figure 8.4

Activities in the five life-cycle stages (circled numbers) of a product manufactured for customer use. In an environmentally responsible product, the environmental impacts at each stage are minimized, not just those in stage 2.

of individuals from a wide range of specialties. An example of team structure is given in Figure 8.5. In addition to the appropriate mix of engineering specialists, such teams often include environmental experts, packaging engineers (i.e., those who can determine how safely and inexpensively products can be delivered to customers), manufacturing engineers (i.e., those that understand the intricacies of product fabrication), marketing specialists, business planners, and perhaps financial and purchasing experts. In a reflection of today's increasingly complex economy, many design teams include members from strategic partners such as critical suppliers or customers.

The practice of using diversified teams for product design considerably complicates the process in its initial stages. The benefits, however, arise from the early consideration of the variety of attributes that will ultimately determine the success or failure of the product. Most features of a product design are effectively frozen very early in the design process. The presence of an industrial ecology specialist on every

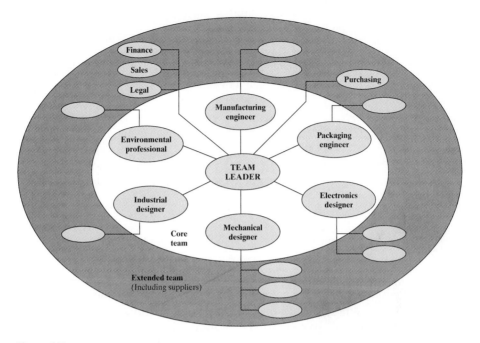

Figure 8.5

The structure of a typical design team for an electromechanical product. *(*Adapted from K.T. Ulrich and S.D. Eppinger, *Product Design and Development,* New York: McGraw-Hill, 1995.)

design team is probably the single most effective way to improve the environmental responsibility of the products of our modern technological society.

8.5   THE PRODUCT REALIZATION PROCESS

Modern industrial managers wish to stimulate their design and development staffs to generate numerous ideas for new products, in the hope that a few really successful products will result. Carrying every product idea through from concept to manufacture is too expensive to be feasible, however, and so a structured process, the product realization process (PRP), has been developed to guide business decisions at each step along the way.

There are a number of versions of PRP, some labeled IDS (integrated development system), some IPD (integrated product development). They vary in level of detail and in the number of sequence steps, and many corporations have developed handbooks to guide their design teams. All share the general approach, if not each specific step, shown in Figure 8.6. Eight steps in the product realization process, from idea to obsolescence, are indicated in the figure. The transition from one step to the next passes through "gates" (circled numbers in the diagram): opportunities for the managers to decide whether to permit the product development to proceed. In the formal structure of the product realization process, a review is held when a product idea under development reaches each gate in the sequence. The review team typically includes

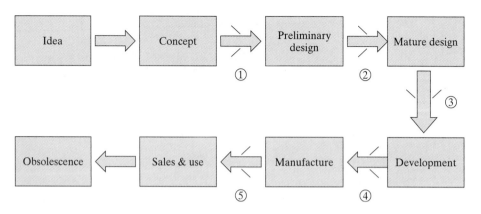

Figure 8.6

The product realization process. The successive points at which "go–no go" decisions are reached ("gates") are circled numbers.

representatives from design, manufacturing, purchasing, marketing, and other appropriate organizations within the corporation. The outcome of the review determines whether or not gate passage will be allowed.

The items considered at each gate review include marketability (do we still think our customers want this product?), manufacturability (can we make the product as envisioned?), economics (can we make a profit on this item?), strategy (are we ahead of our competitors?), and a variety of other factors. Cost is a major influence on decision making, since the financial investment required to move to the next step of the product development increases as one moves from early gates to later gates. By gate 4 or 5, if the product is then judged unpromising, a substantial unrecoverable investment will have been made. The goal of the oversight process is to let promising products move quickly to manufacture, but to close gates early on projects that will consume investment dollars without the probability of substantial financial return.

Information of all kinds becomes more detailed as a product progresses from early to later stages of development: concepts are transformed to designs, materials are specified, sizes and features are determined, costs can be more and more accurately calculated, and customer response can be better estimated.

Gate 1: From Concept to Preliminary Design The first gate controls the transition from concept to preliminary design. The business questions at this gate are very basic: Does this concept appear to meet a customer need? Is it consistent with the corporate product line? Does it have the potential to compete effectively?

Gate 2: From Preliminary Design to Mature Design The initial or concept stage of product development typically involves only a handful of people, and the only expense is their time. At the next stage, preliminary product design, the size of the group expands but its activities are still limited to computers, communications, and pencil sketches, so the embedded development expense is still modest. At Gate 2 the major design decisions have been made, but few details are available.

The typical business questions at Gate 2 are formulated from the perspective of the preliminary design: Do the estimated performance specifications meet the product goals? Is the design visually attractive? Is the product likely to be profitable? This is often the most critical decision point in the product development process, because the corporate investment in a product that passes the second gate begins to increase rapidly.

Gate 3: From Mature Design to Development   At Gate 3, the design team presents a fully worked product design together with moderately detailed information on manufacturing processes. The product itself can then receive a reasonably detailed environmental review. In the case of processes, if the product is relatively similar in type and materials to other products of the corporation, there may be little need for a new review of the environmental implications of manufacturing. For a new process or set of processes, however, the manufacturing review will be extensive.

The Gate 3 product review is in all cases quite detailed. From a business standpoint, the questions become more focused than at earlier stages: Are there technical impediments to development? Are the manufacturing processes satisfactory? Are the electrical and mechanical goals for the product fully realized? Will the product have customer appeal?

Gate 4: From Development to Manufacture   By the time the Gate 4 review committee meets, the design is finalized, the manufacturing process set, all materials and components chosen, all suppliers at least tentatively identified, and all costs established. The decision at this gate is whether to proceed with manufacture, the most costly of all the stages.

The business decisions at Gate 4 are obvious and important: Have the cost estimates been met? Is product manufacturability satisfactory? Has a reliable set of suppliers been identified? Does the product as it will emerge from the manufacturing process retain the desirable characteristics identified at Gate 3?

Gate 5: From Manufacture to Sales and Use   The Gate 5 review is often ceremonial, especially if decisions at previous gates have been sufficiently thoughtful and comprehensive. Provided that no unexpected and unwelcome information has arisen, the product is released for sale and use. The business questions involve a review of the degree to which manufacturing meets expectations and the ways in which the marketing campaign should move forward.

FURTHER READING

Billetos, S.B., and N.A. Basaly, *Green Technology and Design for the Environment*, Washington, DC: Taylor & Francis, 1997.

Hauser, J., and D. Clausing, The house of quality, *Harvard Business Review*, 63–73, May–June, 1988.

Hendrickson, C.T., A. Horvath, L.B. Lave, and F.C. McMichael, Industrial ecology and green design, in *A Handbook of Industrial Ecology*, R.U. Ayres and L.W. Ayres, eds., Cheltenham, U.K.: Edward Elgar Publishers, 457–466, 2002.

National Academy of Engineering, *Design in the New Millenium*, Washington, DC: 1999 *http://books.nap.edu/catalog/9876.html*

Pugh, S., *Total Design*, Reading, MA: Addison-Wesley, 1990.

Ulrich, K.T., and S.D. Eppinger, *Product Design and Development*, New York: McGraw-Hill, 1995.

EXERCISES

8.1 Select a wrench or other simple hand tool and, to the degree possible, evaluate it qualitatively for each of its "Design for X" features.

8.2 Repeat Exercise 8.1 for a kitchen electrical appliance of your choice.

8.3 Form a design team with four other students. Choose one of the following products: a 10-cup coffee maker, overhead projector, bicycle, power lawn mower. Each member of your team will play one of the following roles: mechanical designer, manufacturing engineer, environmental specialist, marketing specialist, corporate lawyer. Develop the House of Quality for the product of your choice, and describe the diagram and the resulting product design concept in a four- to five-page report.

CHAPTER 9

Industrial Process Design and Operation

9.1   THE PROCESS DESIGN CHALLENGE

An industrial process is a sequence of operations designed to achieve a specific technological result, such as the manufacture of a telephone from metallic and polymeric starting materials. As with the design of products, the design of industrial processes can be enormously complex and challenging. Typical goals for an industrial process designer have traditionally included the following:

- Accomplish the desired technological result.
- Achieve high precision by manufacturing products that consistently fall within desired tolerance limits.
- Achieve high efficiency by manufacturing products in a minimum amount of time.
- Design a process for high reliability over a long period of time.
- Make the process safe for the workers who will use it.
- Design the process to be modular and upgradable.
- Design for minimum first cost (equipment purchase and installation).
- Design for minimum operating cost.

Industrial ecology imposes on the process designer several additional goals:

- Prevent pollution.
- Reduce risk to the environment.
- Perform process design from a life-cycle perspective.

It is seldom that each of these goals can be optimized independently. Rather, the aim is to achieve the optimum balance among the goals. The industrial ecology goals are more recently conceived than most of the others, and generally not as well understood. Accordingly, we attempt in this chapter to provide environmentally related guidance to the process designer.

9.2   POLLUTION PREVENTION

One of the central approaches to industrial process design and operation is termed *pollution prevention* (often referred to as P^2) or *cleaner production*. The objective of this activity is to reduce impacts or risk of impacts to employees, local communities, and the environment at large by preventing pollution where it is traditionally first generated. The sequence is to identify a problem or potential problem, to locate its source within the manufacturing process, and to change the source so as to reduce or eliminate the problem.

Process evaluation in P^2 deals with sequence flow, the consumption of materials, energy, water, and other resources, the manufacture of desired products, and the identification and quantification of residues. An example flow chart for this *Process Characterization* is shown in Figure 9.1. No attempt is made at this stage to quantify any of the flows. The more complete this diagram, the easier the subsequent assessment will be.

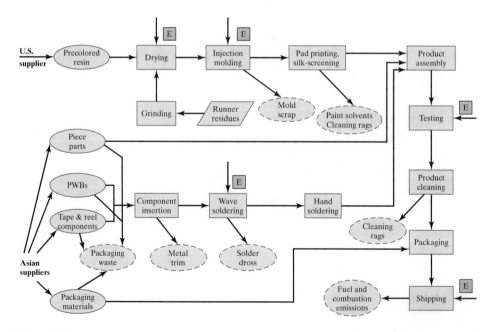

Figure 9.1

A process characterization flow chart for the manufacture of a desktop telephone. (Courtesy of Lucent Technologies.)

P^2 techniques for dealing with issues identified by process characterization include:

1. *Process modification:* Changing a process to minimize or eliminate waste generation
2. *Technology modification:* Changing manufacturing technology to minimize or eliminate waste generation
3. *Good housekeeping:* Changing routine maintenance or operation routines to minimize or eliminate waste generation
4. *Input substitution:* Changing process materials to minimize quantity or potential risk of generated waste
5. *On-site reuse:* Recycling residues within the facility
6. *Off-site reuse:* Recycling residues away from the original facility

The health and environmental risks from different flow streams and processes are, of course, far from equivalent. For example, a detailed study of the petrochemical intermediate industry showed that only a small number of compounds were responsible for most of the potential toxicity. The results of that study, shown in Figure 9.2,

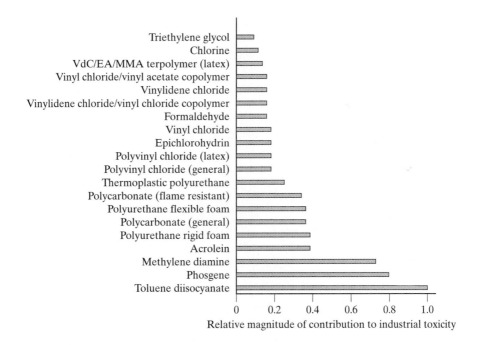

Figure 9.2

Compounds that are major contributors to the overall potential toxicity of the chemical intermediates industry. (Adapted from S. Fathi-Afshar and J.C. Yang, Design the optimal structure of the petrochemical industry for minimum cost and least gross toxicity of chemical production, *Chemical Engineering Science,* 40, 781–797, 1985.)

demonstrate that toluene diisocyanate, phosgene, and methylene diphenylene are obvious targets for P^2 efforts.

David Allen, now of the University of Texas, has pointed out that P^2 can be addressed at different spatial scales. At the microscale, or molecular level, chemical synthesis pathways and other material fabrication procedures can be redesigned to reduce waste and lower process toxicity. This approach is sufficiently important that it has its own name, *green chemistry*. At the process line level, or mesoscale, design considerations include adjustments in temperature, pressure, processing time, and the like, with energy and water use, by-product generation, and inherent process losses as foci. Activities at the mesoscale level are those most commonly termed P^2. Finally, the macroscale, at the sector or intersectoral level, can be addressed through industrial ecology systems perspectives in which by-products find uses outside the facility in which they are generated. Microscale P^2 activities are generally possible only where chemicals are being synthesized; mesoscale and macroscale P^2 can be undertaken anywhere chemicals or other materials are being used.

Once a chemical or process of interest has been identified, *materials accounting* can be employed to determine the best avenues for action. For a single process, the result would be a diagram of the form of Figure 9.3. Here the quantities of the target species are listed and compared.

With the target identified and its flows determined, reduction or elimination are approached in a multistep manner:

- Redesign the process to substitute low toxicity materials for those that are highly toxic, or to generate high toxicity materials on site as needed.
- Minimize process residues.

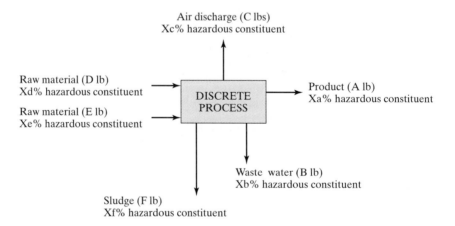

Figure 9.3

A material-centered process analysis directed toward the use, retention, and residue production of a hazardous constituent. (Reprinted with permission from D.G. Willis, Pollution prevention plans—A practical approach, *Pollution Prevention Review*, 347–351, Autumn 1991.)

- Reuse process residues.
- Redesign the process so that unwanted residue streams become streams of useful byproducts.

In the case of a process already existing or newly begun, a *waste audit* is generally useful. The approach is to study *all* waste flows from the facility to determine which can be decreased, and how. Industrial solvents, cleaning solutions, and etchants are often good places to begin. An approach that has proven beneficial in a number of cases is the regeneration of chemical solutions, which can often be accomplished by filtration, changes to relax purity requirements, the addition of stabilizers, redesign of process equipment, and so on. An example of the improvement that can be achieved is shown in Figure 9.4. A peroxide bath, initially used once and discarded, was gradually redesigned over a period of years until it was eventually replaced only every week. The reduction in cost and decrease in liquid residues that resulted were very large.

A further lifetime-extension technique for solutions is that of recuperative rinsing. This is most often used where parts are sprayed or immersed to clean them after electroplating or surface finishing. Subsequently, the rinsewater containing process chemicals can be returned to the process tank to replace fluid lost during evaporation rather than being discarded. Recuperative rinsing can be successful where a high degree of solution monitoring and control is used to maintain quality.

9.3   THE CHALLENGE OF WATER AVAILABILITY

Many industrial processes are heavy users of water. The extraction and processing of metal and the synthesis of chemicals are particularly water intensive on an overall consumption basis, but on a normalized basis sectors such as electronics are surprisingly large users: where metal ore mining requires around 5 grams of water per gram of ore processed, computer chip making requires around 1000 grams of water per gram of final product manufactured.

In a number of the more developed countries, industrial use of water has been declining, especially on a normalized (amount of water per unit of product) basis. Figure 9.5 (a) demonstrates that in the United States, for example, industrial use was

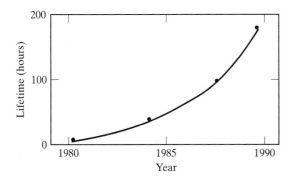

Figure 9.4

The extension in lifetime of a peroxide solution used in the manufacturing of electronic components. (Courtesy of E. Eckroth, AT&T Microelectronics.)

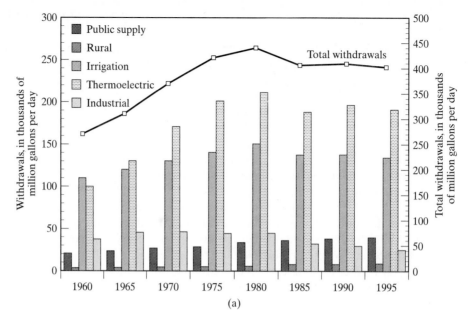

(a)

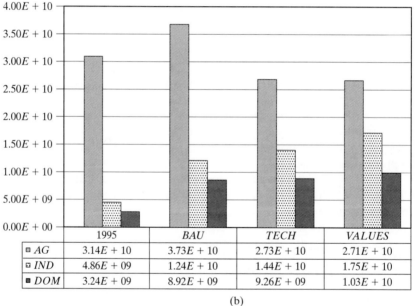

	1995	BAU	TECH	VALUES
AG	3.14E + 10	3.73E + 10	2.73E + 10	2.71E + 10
IND	4.86E + 09	1.24E + 10	1.44E + 10	1.75E + 10
DOM	3.24E + 09	8.92E + 09	9.26E + 09	1.03E + 10

(b)

Figure 9.5

(a) Trends in water use in the United States, 1960–1995. Reprinted from W.B. Solley, R.R. Pierce, and H.A. Perlman, *Estimated Use of Water in the United States in 1995,* Circular 1200, Denver, CO: U.S. Geological Survey, 1998. (b) Current water use in the Yellow River Basin, China in 1995, and water use predicted for three scenarios for the year 2025: Business as usual (BAU), Technology, Economics, and the Private Sector (TECH), and Values and Lifestyles (VALUES). (Adapted from K. Strzepek and A. Holt, Local scale implications of the world water scenarios: A case study of water management in the Yellow River Basin in China, *China Water Vision,* The Hague, The Netherlands: Second World Water Forum, pp. 77–88, 2000.)

roughly constant in the 1970s and decreased by a third or more in 1980–1995. In China, however, industrial water use has been on the increase. Industrial water use in the Yellow River Basin, for example, about 12% of the total in 1995, is anticipated to increase to between 21% and 32% of the total by 2025 [Figure 9.5 (b)]. Developing countries thus face a special challenge: Water is crucial to serve the individual needs of their growing populations and to irrigate the agricultural crops that feed them, leaving increasingly less for industrial consumption. Worldwide, but especially in the developing countries, the design of industrial processes in ways that use water very sparingly will become an increasingly high priority.

9.4   THE PROCESS LIFE CYCLE

As with products, industrial processes have life cycles, though the components are different. Process life stages comprise three epochs (Figure 9.6): Resource provisioning and process implementation occur simultaneously; primary process operation and complementary process operation occur simultaneously as well; and refurbishment, recycling, and disposal is the end-of-life stage. The characteristics of these stages are described below.

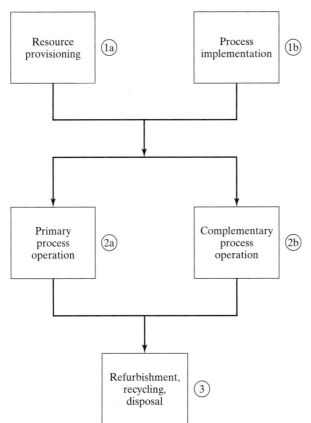

Figure 9.6

The life-cycle stages of a manufacturing process.

9.4.1 Resource Provisioning

The first stage in the life cycle of any process is the provisioning of the materials used to produce the consumable resources that are used throughout the life of the process being assessed. One consideration is the source of the materials, which in many cases will be extracted from their natural reservoirs. Alternatively, recycled materials may be used. Doing so may be preferable to using virgin materials because recycled materials avoid the environmental disruption that virgin material extraction involves, and often require less energy to recover and recycle than would be expended in virgin material extraction. In addition, the recycling of materials often produces less solid, liquid, or gaseous residues than do virgin materials extractions. Tradeoffs, especially in energy use, are always present, however, and the choice must be considered on a case-by-case basis. Another consideration is the methods used to prepare the materials for use in the process. Regardless of the source of a metal sheet to be formed into a component, for example, the forming and cleaning of the sheet and the packaging of the component should be done in an environmentally responsible manner. Supplier operations are thus a topic for evaluation as the process is being developed and, later, as it is being used.

9.4.2 Process Implementation

Coincident with resource provisioning is process implementation, which looks at the environmental impacts that result from the activities necessary to make the process happen. These principally involve the manufacture and installation of the process equipment, and installing other resources that are required, such as piping, conveyer belts, exhaust ducts, and the like.

9.4.3 Primary Process Operation

A process should be designed to be environmentally responsible in operation. Such a process would ideally limit the use of hazardous materials, minimize the consumption of energy, avoid or minimize the generation of solid, liquid, or gaseous residues, and ensure that any residues that are produced can be used elsewhere in the economy. Efforts should be directed to designing processes whose secondary products are salable to others or usable in other processes within the same facility. In particular, the generation of residues whose hazardous properties render their recycling or disposal difficult should be avoided. Because successful processes can become widespread throughout a manufacturing sector, they should be designed to perform well under a variety of conditions.

An unrealizable goal—but a useful target—is that every molecule that enters a manufacturing process should leave that process as part of a salable product. One's intuitive perception of this goal as unrealistic is not necessarily accurate: Certain of today's manufacturing processes, such as molecular beam epitaxy, come close, and more will do so in the future.

9.4.4 Complementary Process Operation

It is often the case that several manufacturing processes form a symbiotic relationship, each assuming and depending on the existence of others. Thus, a comprehensive process evaluation needs to consider not only the environmental attributes of the primary process itself, but also those of the complementary processes that precede and follow. For example, a welding process generally requires a preceding metal cleaning step, which traditionally required the use of ozone-depleting chlorofluorocarbons. Similarly, a soldering process generally requires a post-cleaning to remove the corrosive solder flux. This step traditionally required the use of chlorofluorocarbons. Changes in any element of this system—flux, solder, or solvent—usually require changes to the others as well if the entire system is to continue to perform satisfactorily. The responsible primary process designer will consider to what extent his process imposes environmentally difficult requirements for complementary processes, both in their implementation and their operation.

9.4.5   Refurbishment, Recycling, Disposal

All process equipment will eventually become obsolete. It must therefore be designed so as to optimize disassembly and reuse, either of modules (the preferable option) or materials. In this sense, process equipment is subject to the same considerations and recommended activities that apply to any product: use of quick disconnect hardware, identification marking of plastics, and so on. Many of these design decisions are made by the corporation actually manufacturing the process equipment, which may well not be the user, but the process designer can control or frustrate many environmentally responsible equipment recycling actions by his or her choice of features in the original process design.

A classic example of the consequences of failing to design for recycling is the Brent Spar oil platform. This North Sea installation was designed and built in the 1960s, well before development of a design for recycling consciousness, as a temporary storage reservoir for crude oil awaiting transfer to refineries. In 1995, Royal Dutch Shell, after a careful review of alternatives, decided to dispose of the then-obsolete platform by scuttling it in the Atlantic Ocean off the west coast of Scotland. This action was seen by Greenpeace, other environmental organizations, and most of western Europe's citizens as personifying the wasteful society and the disregard of industry for the environment. For several weeks, as options were debated and Greenpeace boats circled the platform, Shell gasoline sales dropped sharply. Finally giving in to the pressure, Shell agreed to bring the platform to shore and pay large sums to have it cut up, the petroleum residues properly treated, and the metal recycled. It made little difference that the overall environmentally preferable choice (if energy, exposure to hazardous material, and use of recycled material were considered) may well have been the scuttling originally proposed. The fact was that Shell had paid an enormous price, both monetarily and from a public-relations point of view, for its failure to consider the final life stage in the design of the platform.

9.5   THE APPROACH TO PROCESS ANALYSIS

Process analysis begins with data gathering, and the focus is in three areas: (1) the primary process itself, (2) the equipment used in the process, and (3) information concerning any complementary processes involved. The information should be as quantitative as possible, provided quantification can be done quickly, but much can be done with qualitative data and the analyst should not get bogged down in attempting to precisely quantify everything.

After the data are assembled, checklists can be used in making assessments. As before, the goal is to effect improvements, not to get mired in the quicksand of attempting to do a perfect job of evaluation. A list of recommendations should then be generated and those recommendations prioritized.

9.5.1   The Process Itself

The analysis of a process begins with the study of the actual operation that the process performs. Almost certainly, the process requires energy. What is the energy source, and can less energy or a more benign source of energy be used? Are energy recovery or cogeneration used? Often the process requires chemicals. If so, what are the chemicals used, and are they hazardous to humans, other biological species, or ecosystems? Are the chemicals from virgin sources or from recycling streams? If they are from the former, what recycling streams or outputs might be available for use?

Output streams also require analysis. What are the chemical by-products of the process, if any? (Note that by-product production can occur without an input chemical stream because the by-product may be derived from the incoming component, as in turnings from lathe operation.) If energy is consumed, it is very likely that heat is given off. Is the process well insulated so that little of that heat is lost? Alternatively, is the utility of the heat captured in any way and reused, say, to heat nearby offices? (It is worth pointing out that most households, offices, and light commercial operations avoid industrial areas. In fact, zoning laws may dictate separation. Thus, the best potential customers may be excluded from consideration by geography and public policy.)

The mechanical and relational arrangements of the processes can also be a useful item to review. Is the process batch or continuous? If it is batch, are energy and materials requirements for startup minimized? Is the process located in proximity to other processes or to flows of incoming or outgoing components so as to minimize transport requirements? If a substantial by-product stream is generated, is there a nearby process that can receive and use it (either within the corporation or in another corporation nearby)?

A major area of concern for many manufacturing firms is that of surface coating—painting, plating, anticorrosion treatments, etc. Volatile organic carbon emissions to the air and metalworking oils to the water are specific areas for attention. No generic advice can be given, because individual processes may or may not include emission controls, frequent equipment cleaning requirements, or oil mist reduction approaches. Nonetheless, any facility engaged in surface coating would do well to begin its environmentally related efforts by examining opportunities to improve its coating processes.

9.5.2   The Process Equipment

Process equipment should be analyzed as though one were analyzing a product. (It is, of course, the product of the process equipment manufacturer, and the purchaser of the process equipment assumes the equipment's environmental attributes, good and bad.) The analysis should treat the materials used to manufacture the equipment, the methods by which the equipment is assembled, its modularity, the ease with which the equipment can be disassembled, and the degree to which the materials are identified. Most process equipment is made from steel. Depending on the process, stainless steel or organic or ceramic surface coatings may be required. Process equipment is also generally painted on exterior surfaces. Some process equipment, especially outside the heavy machinery industries, includes or is made entirely of plastic. Computer control, and the associated electronics, are common.

If the process is being installed or soon will be, examine the techniques and materials used for packaging, shipping, and installation.

Energy use is an important attribute of a product made to be used in a manufacturing process. Any component that draws current should be designed to be capable of shutdown when not in active use. Motors should generally be the variable-speed type, the speed being load controlled.

9.5.3   Complementary Processes

Examine whether the primary process being assessed requires preceding or subsequent processes, and whether those processes are of a particular type or use a particular chemical. If the complementary processes are thus defined, is the primary process itself environmentally responsible? If not, can it be modified to improve its characteristics? If any of several complementary processes can be used, have the designers chosen ones that are environmentally responsible? If not, can alternatives be suggested?

9.6   GUIDELINES FOR PROCESS DESIGN AND OPERATION

It is useful to list guidelines for process design and operation activities. At the microscale, green chemistry is enhanced by implementing several principles:

- To the extent possible, addition reactions are preferred over displacement and elimination reactions, because the latter automatically generate by-products.
- Synthetic methodologies should be designed to preserve efficacy of function while using and generating substances that possess little or no toxicity to human health and the environment.
- Feedstocks derived from renewable materials are generally preferable to those derived from nonrenewables.
- The use of auxiliary substances (solvents, separation agents, etc.) should be eliminated where possible and designed to be innocuous if used.

The focus at the mesoscale is at the overall process level, typically involving a sequence of reaction, treatment, assembly, and packaging stages. Here the advice is more generic, and improvements may sometimes be gained by tradeoffs among individual process elements, following these guidelines:

- Minimize consumption of materials, water, and energy.
- Minimize the number of cleaning processes.
- Minimize secondary processes such as surface coating.

- Match waste streams with feed streams, and transform waste streams into usable feed streams.
- Plan for and perform regular preventive maintenance.
- Eliminate redundant processes.
- Capture residues for reuse where possible.
- Capture wastes for treatment or proper disposal.

Guidelines for macroscale P^2 (termed industrial symbiosis) are given in Chapter 22.

9.7   IMPLICATIONS FOR CORPORATIONS

An obvious and appreciated benefit of environmentally responsible process design and operation is financial—3M's Pollution Prevention Pays program and those of many other corporations result in savings of large amounts of money each year from lower waste handling and treatment costs, reduced purchases of process materials, reduced liability costs, and reduced compliance costs. A second benefit is managerial in nature—it is easier to practice green product design if those around you are practicing green process design and green process operation. Integrating the efforts toward corporate environmental superiority is a powerful way to link approaches and stimulate achievement.

Establishing performance targets is an effective way of monitoring and evaluating performance. In the process operation area, DuPont has set its target at the limit by advertising "The Goal is Zero," meaning that DuPont would like to completely eliminate emissions to the environment. This goal, like that of zero defects, is not possible, of course, but it sends a clear signal that an increasingly close approach to zero is a goal worth striving for.

Industrial processes have substantial inertia, much more so than manufactured products. The capital and personnel costs of installing processes are high, and they often remain in place for decades. The costs of modifying or retrofitting them is high as well, and processes installed without much attempt to take long-term environmental considerations into account may eventually require substantial investments in pollution control add-ons and treatment facilities. It is thus very important, from both operational and environmental standpoints, to do it right the first time.

FURTHER READING

Allen, D.T., Pollution prevention: Engineering design at macro-, meso-, and microscales, *Advances in Chemical Engineering, 19*, 21–323, 1994.

Allen, D.T., and D.R. Shonnard, *Green Engineering: Environmentally Conscious Design of Chemical Processes,* Upper Saddle River, NJ: Prentice Hall PTR, 2002.

Anastas, P.T., and J.C. Warner, *Green Chemistry: Theory and Practice*, Oxford, U.K.: Oxford University Press, 1998.

Cano-Ruiz, J.A., and G.J. McRae, Environmentally conscious chemical process design, *Annual Review of Energy and the Environment, 23*, 499–536, 1998.

Diwekar, U., and M.J. Small, Process analysis approach to industrial ecology, in *A Handbook of Industrial Ecology*, R.U. Ayres and L.W. Ayres, eds., Cheltenham, U.K.: Edward Elgar Publishers, 114–137, 2002.

Graedel, T. E., *Streamlined Life-Cycle Assessment*, Chapter 8: Process assessment by SLCA matrix approaches, Upper Saddle River, NY: Prentice Hall, 1998.

Nguyen, N., ed., *Green Engineering: Environmentally Conscious Design of Chemical Processes*, Upper Saddle River, NJ: Prentice Hall, 2001.

U.S. Environmental Protection Agency, *Sector Notebooks Project*, A series of more than 20 volumes that provide sector introductions, process descriptions, and pollution prevention opportunities. Available on the web by searching the EPA Online Library System for "sector notebook" at *http://www.epa.gov/natlibra/ols.htm*.

van Berkel, R., E. Willems, and M. Lafleur, Development of an industrial ecology toolbox for the introduction of industrial ecology in enterprises-I, *Journal of Cleaner Production, 5*, 11–25, 1997.

EXERCISES

9.1 Choose a manufacturing process, historic or modern, for which you can locate considerable detail concerning its implementation at a specific industrial facility. To the extent applicable and possible, evaluate the process, pointing out its strengths and weaknesses from an industrial ecology standpoint.

9.2 You are the industrial ecologist for a manufacturing company whose leading product is cables for personal computers. The principal components of the cable are copper wire, flexible plastic wire coating, and rigid plastic connectors. What by-product or residue streams do you anticipate? About which should you be most concerned?

9.3 Assume that the peroxide bath whose lifetime is shown in Figure 9.4 is used to make 50,000 silicon wafers per year. The cost of the chemicals for the bath is U.S.$12/liter and the bath is five liters in volume. Ten silicon wafers per hour can be processed. How much depleted peroxide bath was generated in 1980, 1983, 1988, and 1990? At an on-site processing cost of U.S.40¢/liter, how much was the cost in each of those years? Was the expenditure of U.S.$3,500 for a filtration system in 1988 and U.S.$9,400 for a replenishment system in 1990 justified?

CHAPTER 10

Choosing Materials

10.1   MATERIALS SELECTION CONSIDERATIONS

Materials influence the functioning of a product, its ruggedness, its appearance, and numerous other characteristics. In many cases, any of a number of different materials could be chosen for a particular application. The initial considerations of the product designer so far as materials choice is concerned are obvious and important:

- Does the material have the desired physical properties (strength, conductivity, index of refraction, etc.)?
- Does the material have the desired chemical properties (solubility, photosensitivity, reactivity, etc.)?
- Is the cost reasonable?

In the modern, increasingly complex world, the designer must also pay attention to a number of additional considerations:

- Is the material an environmental hazard? (This broad area includes concerns about human toxicity.)
- Is the material a safety hazard? (For example, is it flammable?)
- Is the embodied energy in the material high?
- Is the material under potential supply constraints?
- Is a recycled supply of the material available?

- Is the material readily substitutable? (That is, are realistic alternative materials available should the primary choice become less desirable due to changes in costs, environmental concerns, or some other factor?)

In this chapter, we explore two topics related to this latter set of considerations— environmental hazards and resource availability—and suggest materials that appear to present the most benign environmental characteristics.

10.2   MATERIALS AND ENVIRONMENTAL HAZARDS

Materials choices are often limited by toxicity concerns. Other things being comparable for a specific application, a designer's objective should be to select materials that have the least significant toxic properties. Government environmental agencies generally define those materials that merit concern from a toxicity standpoint, and those lists are a good starting point for the physical designer. In Table 10.1, we reproduce the 17 chemicals or chemical groups targeted for reduction in the U.S. EPA's Industrial Toxics Project, a voluntary effort by industrial corporations to reduce emissions of targeted chemicals. Most of these materials have also been restricted by the European Union and other governmental bodies. The list includes both product chemicals and process chemicals. Cadmium, chromium, lead, mercury, nickel and their compounds are sometimes used industrially such that the metals end up as part of the products that are produced, often as platings or coatings. Most of the remaining materials are process chemicals which may be used as either solvents or cleaners. Chlorinated solvents and monoaromatic species comprise most of the listed items. Cyanide solutions generally employed in metal plating also appear. The physical design team thus needs to consider two facets of materials choices involving toxic materials: the potential for materials substitution in products and the potential for process changes.

More detailed lists of suspect chemicals could be given. The U.S. Clean Air Act and its amendments have now defined more than 600 chemical species as "hazardous air pollutants." The same chemical groups appear as on the list above: heavy metals and their compounds, cyanides, and halogenated solvents. A number of pesticides are included, as are several organic nonhalogenated compounds that have been shown to have toxic properties. Also present are solids such as asbestos (magnesium silicate

TABLE 10.1   Chemicals Identified in EPA's Industrial Toxics Project

Benzene	Cadmium and compounds
Carbon tetrachloride	Chloroform
Chromium and compounds	Cyanides
Dichloromethane	Lead and compounds
Mercury and compounds	Methyl ethyl ketone
Methyl isobutyl ketone	Nickel and compounds
Tetrachloroethylene	Toluene
Trichloroethane	Trichloroethylene
Xylenes	

minerals) whose toxicity is related to their physical properties, not their chemical ones. Good toxicology is obviously an important prerequisite to an adequate appreciation of the implications of these species lists.

Radionuclides, whether natural or artificial, have the potential to cause substantial health problems and are closely controlled as a consequence. Typical annual radiation doses demonstrate that natural and medical sources comprise virtually the entire concern for anyone not occupationally related to radionuclides. Nonetheless, small amounts of radionuclides are used industrially in specific nonmedical applications such as radioluminous products (watches, clocks), smoke detectors, electronic devices, antistatic devices, and scientific instruments. The governmental restrictions on the extraction, use, and disposal of radioisotopes are probably sufficient to ensure that the materials are seldom used unless their properties are essential to the particular product involved.

Because the restrictions on manufacture, use, and disposal of products containing significant amounts of radionuclides are so onerous, the use of radionuclides should be avoided by designers unless specific characteristics make such use necessary. If used, the amounts should be stringently minimized; a good example is the improved targeting of pharmaceutical uses of radionuclides so that the required medical benefits can be achieved with ever smaller quantities of radioactive material.

Even if a material on one or more hazardous lists is important to a product or process and can be used with safety in a manufacturing facility, it is necessary to consider the implications of disposal of the inevitable residues. Disposal regulations vary widely around the world and undergo constant change. Many of the materials on the toxic or hazardous lists have disposal constraints, but the lists include in addition classes of compounds having particularly undesirable physical characteristics such as instability, high flammability, high acidity, and high alkalinity. Finally, one should avoid the use of any ozone-depleting substances subject to phase-out under international protocols.

A final caution for design engineers selecting materials is to attempt to anticipate future restrictions on materials whose use is not now constrained. As an example, concerns have been raised regarding the use of chlorine and chlorinated organic compounds. Among the issues are the tendency of chlorinated organics to be health hazards, manifesting this property through cancer production and modifications to endocrine function in both humans and animals. Individual compounds show a wide range of behavior on specific toxicity tests, however, and producers and users of chlorine-containing materials defend the undeniable utility of the materials and suggest a compound-by-compound approach. It seems unlikely that a comprehensive ban on chlorine-containing compounds will ever result, but targeted policies are possible. Thus, the astute practitioner of industrial ecology may wish to consider the following:

- Investigate alternatives to chemicals that contain chlorine in the final product.
- Attempt to develop alternative synthesis routes for processes in which chlorinated compounds are currently used as intermediates.
- Minimize and control, but continue to use, chlorinated compounds in processes where their utility is especially suitable.

Arsine Generation

The use of a toxic material or a material chemically closely related to a toxic material as a feedstock or intermediate in an industrial process generally requires that it be shipped to the manufacturing facility and stored there until consumed. Storage of a toxic material or its precursors was the major cause of the worst chemical accident in history: the deaths of some 2000 people due to the dispersal from a storage tank of methyl isocyanate in Bhopal, India on December 3, 1984. An alternative design approach is the generation of a required chemical from nontoxic precursors as it is needed. For example, on-demand generation of arsine, a toxic gas widely used in the manufacture of electronic materials, has been demonstrated by Jorge Valdes and coworkers at AT&T Bell Laboratories. The technique is based on an arsenic metal cathode in an electrolytic cell and is pictured schematically in Figure 10.1.

10.3 MATERIALS SOURCES AND PRINCIPAL USES

10.3.1 Absolute Abundances

Ultimately, the prerogative of the designer to choose materials with physical and chemical properties suitable to the purpose at hand is limited by the supplies of those materials and their associated costs. Many of the issues surrounding this question were

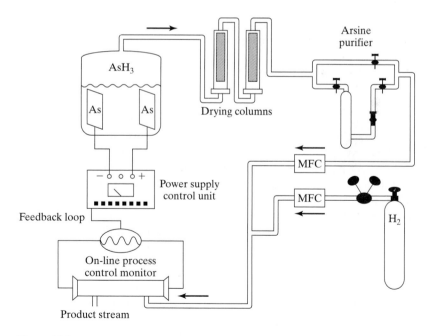

Figure 10.1

A schematic diagram of the on-demand electrochemical arsine generator. MFC = material flow controls. (Courtesy of J.L. Valdes, AT&T Bell Laboratories.)

discussed in Chapter 5, and the information presented there made it clear that a number of widely used resources may become less available in the future because of a variety of factors limiting their extraction and processing. Questions of global resource availability have seldom been concerns in the past (the Arab oil embargoes of the 1970s being obvious exceptions), but will become more and more significant at some time in the future as populations increase dramatically, as the global standard of living rises, and as resource use associated with those developments places increasing stress on materials resources.

What is the proper interpretation of this perspective on abundance and supply? It is not that materials in potentially short supply should be avoided, especially on economic grounds. In fact, there turns out to be no robust relationship between physical abundance and cost, cost being so heavily influenced by stockpiling, cartelization, global economic activity, and the like. Nonetheless, it seems reasonable that materials in potentially short supply should be used only in those cases for which their properties are uniquely suited. This is particularly true for low-value material use.

Materials from renewable sources, such as biomass, are not subject to the absolute supply limitations of nonrenewables, and should be used where feasible. If not, to the extent that efficient and acceptable means of recovery can be devised, such efforts will ameliorate materials supply limits on the growth of the technological society. Further, where a particular material is required, the minimum amount of that material should be used. In addition, there is at least the suggestion that waste management practices that today comingle materials resources in landfills be modified to permit eventual access to minimally mixed materials, should favorable economic and technological conditions eventually make it profitable to use them. In this way, landfills would become materials storage sites for the future. In summary, therefore, we are not suggesting prohibitions so far as using materials is concerned, but rather the use of materials with care and perspective.

10.3.2 Impacts of Materials Extraction and Processing

The extraction of raw materials from Earth's crust generally involves the movement and processing of large amounts of rock and soil. To recover one ton of copper, for example, requires the removal of some 350 tons of overburden and 100 tons of ore. As a result, extraction of materials is extremely energy-intensive and tends to be destructive of local ecological habitats. Some feel for the enormous magnitude of material that is involved is given by Table 10.2, and the transition into products in Figure 10.2.

Among the environmentally related actions that often need to be taken at mines, wells, and other extraction sites are the following:

- Retain topsoil removed from the site so that it can later be replaced.
- Control surface runoff in sedimentation ponds.
- Line any working pits with impermeable material to reduce groundwater contamination.
- Monitor trace metal concentrations, pH, and total suspended solids in any water discharges.

TABLE 10.2   Global Materials Flows Associated with Major
Minerals, 1991

Mineral	Ore (Tg)	Avg. grade (%)	Residues (Tg)
Copper	910	0.91	900
Iron	820	40.0	490
Lead	120	2.5	115
Aluminum	100	23.0	77
Nickel	35	2.5	34
Others	925		850
Total	2910		2460

Abstracted from J.E. Young, *Mining the Earth*, Worldwatch Tech. Paper 109,
Worldwatch Institute, Washington, DC, 1992.

- Control acid drainage from mines (produced when sulfur in exposed tailings reacts with air and water).
- Collect water from underground operations in sumps, and pump it to the surface for treatment.
- Restore the site to its former appearance and productivity at the conclusion of extraction operations.

Following extraction, the material that results must be processed and purified in order to yield the metal or chemical or starting mixture that is needed. This sequence of steps can be highly energy-intensive, especially if high pressures or high temperatures are required. In the case of metals, processing and purification are carried out in the molten state, and the energy consumption is therefore related to the melting point. The energy requirements for acquisition in usable form from virgin stocks of a number of common materials (the "embodied energy") are shown in Figure 10.3. Together with environmental impact considerations related to extraction and processing of virgin stocks, these provide a compelling case for the consideration of resource recycling and reuse.

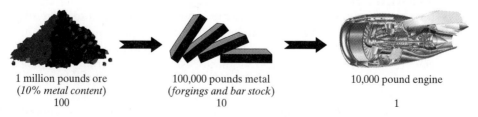

1 million pounds ore	100,000 pounds metal	10,000 pound engine
(10% metal content)	*(forgings and bar stock)*	
100	10	1

Figure 10.2

The weight of the mining and metal residues that result from the manufacture of a single jet engine are one hundred times the weight of the engine itself. (Courtesy of R. Tierney, Pratt & Whitney, Inc.)

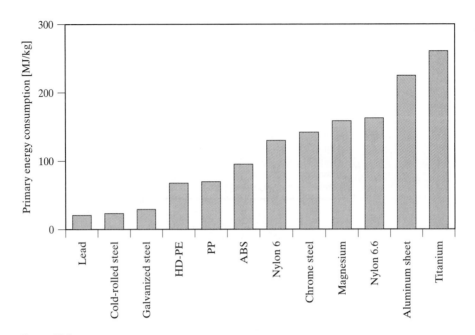

Figure 10.3

The primary energy consumption required to produce one kilogram of various materials. (Adapted from M. Schuckert, H. Beddies, H. Florin, J. Gediga, and P. Eyerer, Quality requirements for LCA of total automobiles and its effects on inventory analysis, *in Proceedings of the Third International Conference on Ecomaterials,* Tokyo: Society of Non-Traditional Technology, 325–329, 1997.)

10.3.3 Availability and Suitability of Post-Consumer Recycled Materials

In contrast to extraction and processing, an efficient recycling operation may be able to provide adequate quantities of a needed material at much lower expenditures of cost and environmental impact. Some materials have not, as yet, been efficiently recovered and recycled, but there are many that have been. Table 10.3 gives current statistics on the percentages of certain materials recycled in the global economy. Many metals are recycled with reasonable efficiency and can generally be re-refined to the desired purity. Paper recycling is prevalent as well, but is complicated by the fact that each stage of recycling shortens the paper fibers and thus often restricts the material to lower quality uses, a sequence known as *cascade recycling.* In the case of plastics, the difficulties of separating and reprocessing have made progress slower, but vigorous efforts to improve that picture are now underway. It is important to further develop efficient resource recycling, because virgin stocks of many materials cannot be guaranteed in perpetuity, and every time a resource becomes unavailable to the technological community, designers lose a potentially important degree of freedom.

A sometime impediment to the recovery and reuse of recycled materials is the specification by a designer or the designer's customer of virgin material, generally in an attempt to avoid receiving unsuitable product. Such concerns should be addressed not by specifying the source of the material, but by specifying its properties. Alternatively, one

Sony's Lead-Free Solder: Solving One Problem, Creating Another

Microchips, capacitors, resistors, and other electronic components are held in place by solder, a low-melting alloy of approximately 60 weight % tin, 40 lead. Because of lead's toxicity, work has been underway for a decade to develop a lead-free solder, if possible one that could be directly substituted for the present alloy in existing equipment. In 1999, Sony announced such a solder, a composite of 93.4% tin, 2% silver, 4% bismuth, 0.5% copper, and 0.1% germanium. The material performs the soldering function well, and does so in contemporary soldering machines. As it turns out, however, the new formula has potential materials supply limitations.

The world's annual production of virgin tin is about 200,000 metric tons, about 21% of which is used for 60Sn/40Pb solder. If Sony's new solder were to completely replace this alloy, the tin use would increase by a factor of 93.4/60, or from about 21% to about 33%. The tin industry could probably respond to this requirement; in fact, it would probably welcome doing so. No significant change in copper use rate would occur, and silver use would increase by only 11%. However, bismuth use would increase by 89% and germanium use by 103%; it would severely strain the extraction and processing systems to respond to these demands, especially if the transition to the new solder took place rapidly.

A more serious concern is that depletion times of all these materials are rather short at present, and would become more so under these new rates of consumption. As the table shows, silver, bismuth, and germanium all would have nominal depletion times (i.e., depletion times computed without accounting for changes in prices or reserves) less than 20 years, not a very promising outlook for a replacement technology.

Finally, consider the sources of the materials. Tin, silver, and copper are drawn from their own ores. Bismuth, however, is a hitchhiker, primarily with lead. Not only does the use of bismuth require mining lead (the toxic material the new solder was devised to avoid), but lead itself has a depletion time of 20 years. The situation is similar for germanium, a hitchhiker primarily with zinc, whose depletion time is 19 years.

Thus, from a resource availability standpoint, Sony's technically excellent solder fails miserably. Its widescale use would require extremely rapid supply ramp-up, the depletion times of its constituents are low, its supplies are potentially hitchhiker-limited, and it could not be produced without mining the material one was seeking to avoid in the first place. It is unlikely that the new solder would have been pursued had the development engineers looked at the composition from a resource availability perspective.

Ingredient	Current t_D (yr)	t_D with full use of new solder (yr)
Tin	27	20
Silver	16	14
Bismuth	30	16
Copper	35	35
Germanium	35	17

TABLE 10.3   Global Percentages
                of Recycled Materials

Material	Recycle percentage
Aluminum	28
Cobalt	2
Copper	38
Lead	53
Molybdenum	11
Nickel	34
Steel	64
Tin	13
Tungsten	10
Zinc	28

Data sources: J.L.W. Jolly, J.F. Papp, and P.A.
Plunkert, *Recycling-Nonferrous Metals*, Washington, DC: U.S. Bureau of Mines, 1991; M.E. Henstock, *The Recycling of Non-Ferrous Metals*,
Ottawa, Canada: International Council on Metals
and the Environment, 1996; Steel Recycling Institute, Pittsburgh, PA, *www.recycle-steel.org*, accessed 05/01/01.

can in many cases require suppliers to provide a fixed percentage of purchased materials from post-consumer scrap sources, as Eastman Kodak Corporation has done with its plastic containers and steel drums. If these steps are taken, a more informed design will often emerge, and the number and type of suppliers of recycled materials will be increased.

10.4   MATERIALS SUBSTITUTION

Even when a material has proven satisfactory for a particular application, materials substitution should always be a consideration. Substitution may provide a reduction in cost or an opportunity to improve a design. Substitute materials must, of course, fully satisfy all economic, physical, and chemical requirements of the application.

Materials substitution is often related to the development of new approaches to extraction or processing. Because of embedded capital costs and a general reluctance to change, process transitions have historically been slow to occur: the U.S. interval for steel manufacturing process transformation is about 50 years, for example (Figure 10.4). This pattern influences the quantity of material that is available, and may prove especially significant where unusual composite materials are involved.

A common situation is that a product will contain many materials and have the potential for substitution of some of them. This is the case with automobiles, where over the recent past the use of carbon steel, iron, and zinc die castings has dropped significantly and high strength steel, aluminum, copper (mostly electrical), and plastics use has risen substantially (Table 10.4). Where it makes sense, natural fibers are now being used in fiber-reinforced plastic components, reducing the weight over glass-reinforced components, and increasing recyclability.

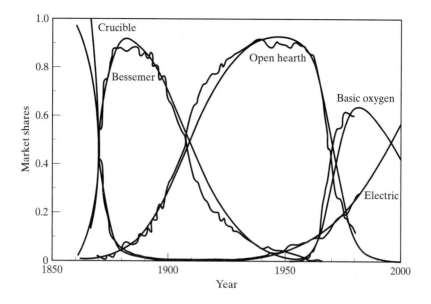

Figure 10.4

Changes in processes used to manufacture steel in the United States during 1850–2000. The jagged lines are historical data and the smooth curves are estimates from coupled equation models. (Reprinted with permission from A. Grübler, *Technology and Global Change,* Cambridge, U.K.: Cambridge University Press, 1998.)

TABLE 10.4 Material in a Typical U.S. Automobile (kg)

Material	1978	1988	% change
Carbon steel	870	654	−25
High-strength steel	60	105	74
Stainless steel	12	14	19
Other steels	25	20	−19
Iron	232	207	−11
Plastics	82	101	23
Fluids	90	81	−10
Rubber	67	61	−8
Aluminum	51	68	32
Glass	39	38	−2
Copper	7	22	320
Zinc castings	14	9	−33
Other	62	57	−9
Total	1621	1437	−11

Data apply to sedans, vans, and station wagons and are from H.A. Stark, ed., *Ward's Automotive Handbook*, Ward's Communications, Detroit, MI, 1988.

Bismuth Shotgun Shells

In an earlier chapter we presented the industrial lead budget for the United States, pointing out that some 5% of annual lead use in 1988 was for ammunition. Soon thereafter, lead-loaded shotgun shells were banned for waterfowl hunting because substantial numbers of birds that were not shot were dying from lead poisoning after ingesting pellets dispersed into the environment. The initial replacement for these shells was shells loaded with steel pellets. These shells were not a success, primarily because the density of steel is so much less than that of lead. This difference had several consequences: different forward allowances were required while aiming at moving birds (old marksmanship techniques are hard to relearn), the steel shot dispersed more rapidly so that effective shooting ranges decreased, and modest initial dispersal of the pellets within the gun barrel caused damage to some older weapons.

A design for environment solution to this situation was clearly called for, and one emerged from John Brown of Ontario, Canada: the substitution of bismuth for lead. Bismuth is nearly as dense as lead, so the shells perform similarly to those they replace, but bismuth is nontoxic, so the principal negative characteristic of lead is avoided. The shells are now being manufactured by the Bismuth Cartridge Company of Dallas, TX.

It is worth pointing out that the substitution of bismuth for lead is not a long-term solution, since bismuth and lead occur in the same ore bodies and bismuth cannot be recovered from its natural deposits without recovering lead as well. Thus, the bismuth shotgun shell substitution works only because ammunition is a minor use of lead. Were most uses of lead to be phased out, the supply of bismuth would be constrained. Further, ammunition is obviously an application in which the pellets are lost when used. No recovery of the pellet material is possible, so that even nontoxic pellet supplies can only be sustained if other bismuth uses permit recovery and recycling.

10.5   MULTIPARAMETER MATERIALS SELECTION

Given the considerations related to material use as discussed above, how can guidance on materials choice be offered to designers? The proper approach is not to concentrate on one material characteristic or another, but to treat selection from a multiparameter perspective. Accordingly, we recommend the use of information on the absolute abundance of a material, whether it originates as a primary product or as a byproduct, whether it is widely available from a geographical perspective, the degree to which its extraction and processing is energy intensive, and its inherent toxicity.

In Table 10.5, we display ordinal ratings for each of these five characteristics of the natural elements. The sums of these rankings permit us to examine the resource consumption and environmental impact potentials of the use of any of the elements, and to compare one element with another from those perspectives. Given physical and

TABLE 10.5   Composite Availability Status of Resources

#	Element (symbol)	Abundance[1]	Hitchhiker[2]	Geography[3]	Energy[4]	Toxicity[5]	Overall score	Use rating[6]
1	Hydrogen (H)	3	2	2	2	2	11	N
2	Helium (He)	2	2	2	1	2	9	N
3	Lithium (Li)	3	2	2	2	1	10	N
4	Beryllium (Be)	1	2	0	1	0	4	H
5	Boron (B)	3	2	2	2	1	10	N
6	Carbon (C)	3	2	2	2	2	11	N
7	Nitrogen (N)	3	2	2	1	2	10	N
8	Oxygen (O)	3	2	2	2	2	11	N
9	Fluorine (F)	3	2	2	2	2	11	N
10	Neon (Ne)	3	2	2	1	2	10	N
11	Sodium (Na)	3	2	2	2	2	11	N
12	Magnesium (Mg)	3	2	2	2	2	11	N
13	Aluminum (Al)	3	2	2	0	2	9	N
14	Silicon (Si)	3	2	2	1	2	10	N
15	Phosphorous (P)	2	2	2	2	2	10	N
16	Sulfur (S)	0	2	2	1	2	7	M
17	Chlorine (Cl)	3	2	2	2	2	11	N
18	Argon (Ar)	3	2	2	1	2	10	N
19	Potassium (K)	3	2	2	2	2	11	N
20	Calcium (Ca)	3	2	2	2	2	11	N
21	Scandium (Sc)	3	0	2	2	1	8	M
22	Titanium (Ti)	3	2	2	0	2	9	N
23	Vanadium (V)	3	2	2	1	2	10	N
24	Chromium (Cr)	3	2	2	0	1	8	M
25	Manganese (Mn)	1	2	1	1	2	7	M
26	Iron (Fe)	3	2	2	1	2	10	N
27	Cobalt (Co)	3	2	2	1	1	9	N
28	Nickel (Ni)	2	2	2	0	1	7	M
29	Copper (Cu)	1	2	2	1	1	7	M
30	Zinc (Zn)	0	2	2	1	2	7	M
31	Gallium (Ga)	2	0	2	0	2	6	M
32	Germanium (Ge)	1	0	1	1	1	4	H
33	Arsenic (As)	0	1	1	1	0	3	E
34	Selenium (Se)	2	1	2	1	0	6	M
35	Bromine (Br)	3	2	2	2	2	11	N
36	Krypton (Kr)	3	2	2	1	2	10	N
37	Rubidium (Rb)	2	1	2	1	1	7	M
38	Strontium (Sr)	1	2	2	1	1	7	M
39	Yttrium (Y)	3	2	2	1	1	9	N
40	Zirconium (Zr)	2	1	2	1	1	7	M
41	Niobium (Nb)	3	2	1	1	2	9	N
42	Molybdenum (Mo)	2	2	1	1	2	8	M
43	Technetium (Tc)	0	0	0	0	0	0	E
44	Ruthenium (Ru)	3	0	2	1	2	8	M
45	Rhodium (Rh)	2	0	2	1	2	7	M

(*continued*)

TABLE 10.5 *Continued*

#	Element (symbol)	Abundance[1]	Hitchhiker[2]	Geography[3]	Energy[4]	Toxicity[5]	Overall score	Use rating[6]
46	Palladium (Pd)	2	0	2	1	2	7	M
47	Silver (Ag)	0	2	2	2	2	8	M
48	Cadmium (Cd)	1	0	2	1	0	4	H
49	Indium (In)	0	0	2	1	1	4	H
50	Tin (Sn)	1	1	2	1	2	7	M
51	Antimony (Sb)	2	1	2	1	2	8	M
52	Tellurium (Te)	2	0	2	1	1	6	M
53	Iodine (I)	3	2	2	2	2	11	N
54	Xenon (Xe)	3	2	2	1	2	10	N
55	Cesium (Cs)	1	1	2	1	0	5	H
56	Barium (Ba)	2	2	2	1	1	8	M
57	Lanthanum (La)	3	2	2	1	1	9	N
58	Cerium (Ce)	3	2	2	1	1	9	N
59	Praseodymium (Pr)	3	2	2	1	1	9	N
60	Neodymium (Nd)	3	2	2	1	1	9	N
61	Promethium (Pm)	3	2	2	1	1	9	N
62	Samarium (Sm)	3	2	2	1	1	9	N
63	Europium (Eu)	3	2	2	1	1	9	N
64	Gadolinium (Gd)	3	2	2	1	1	9	N
65	Terbium (Tb)	3	2	2	1	1	9	N
66	Dysprosium (Dy)	3	2	2	1	1	9	N
67	Holmium (Ho)	3	2	2	1	1	9	N
68	Erbium (Er)	3	2	2	1	1	9	N
69	Thulium (Tm)	0	2	2	1	1	9	N
70	Ytterbium (Yb)	3	2	2	1	1	9	N
71	Lutetium (Lu)	3	2	2	1	1	9	N
72	Hafnium (Hf)	2	1	2	1	1	7	M
73	Tantalum (Ta)	2	1	2	0	2	7	M
74	Tungsten (W)	2	2	2	1	2	9	N
75	Rhenium (Re)	2	0	2	0	1	5	H
76	Osmium (Os)	3	0	2	1	2	8	M
77	Iridium (Ir)	3	0	2	1	2	8	M
78	Platinum (Pt)	3	2	2	1	2	8	M
79	Gold (Au)	0	2	2	1	2	7	M
80	Mercury (Hg)	0	2	1	1	0	4	H
81	Thallium (Tl)	0	0	1	1	0	2	E
82	Lead (Pb)	0	2	2	1	0	5	H
83	Bismuth (Bi)	1	1	2	1	1	6	M
84	Polonium (Po)	0	0	1	0	0	1	E
85	Astatine (At)	0	0	2	0	0	2	E
86	Radon (Rn)	2	2	2	1	0	7	M
87	Francium (Fr)	0	0	2	0	0	2	E
88	Radium (Ra)	1	1	1	1	0	4	H
89	Actinium (Ac)	0	0	2	1	0	3	E

TABLE 10.5   *Continued*

#	Element (symbol)	Abundance[1]	Hitchhiker[2]	Geography[3]	Energy[4]	Toxicity[5]	Overall score	Use rating[6]
90	Thorium (Th)	1	1	2	1	0	5	H
91	Protactinium (Pa)	0	0	2	0	0	2	E
92	Uranium (U)	1	2	2	1	0	6	M

The key to the scoring on this table is as follows:

[1] t_D 3 = abundant, 2 = plentiful, 1 = constrained, 0 = scarce (Data source: Table 5.1)

[2] *H* 2 = not a hitchhiker, 1 = acquired largely as hitchhiker, 0 = total hitchhiker (Table 5.2)

[3] *G* 2 = widely available (sources on several continents), 1 = modest geographic constraints (sources on only two continents), 0 = substantial geographical constraints (sources on only one continent) (Data source: Table 5.3.)

[4] *EN* 2 = low process energy, 1 = moderate process energy, 0 = high process energy (Data sources: S.E. Kesler, *Mineral Resources, Economics, and the Environment*, New York: Macmillan, 1994; M. Schuckert, H. Beddies, H. Florin, J. Gediga, and P. Eyerer, Quality requirements for LCA of total automobiles and its effects on inventory analysis, in *Proceedings of the Third International Conference on Ecomaterials*, Tokyo: Society of Non-Traditional Technology, 32–329, 1997.)

[5] *EV* 2 = high toxicity, 1 = moderate toxicity, 0 = low toxicity (Data sources: Table 10–1; R.C. Weast, ed., *CRC Handbook of Chemistry and Physics*, 60th ed., Boca Raton, FL: CRC Press, 1974)

[6] Use rating *N* = no supply limitations (total = 9–11), *M* = moderate supply limitations (total = 6–8), *H* = high supply limitations (total = 4–5), *E* = extreme supply limitations (total <4)

chemical properties adequate for the desired application, this table can be used as an aid in materials choice by product and process design engineers.

Most designers do not work with pure elements, however, but with alloys or polymers or composites. Extensive efforts have been made, particularly in Japan, to determine what properties might make some materials more environmentally friendly than others, that is, to define the characteristics of "ecomaterials." It is clear from the start that the use of any material at all cannot be environmentally benign—acquisition and processing inevitably require energy and produce environmental impacts. As an operational definition of an ecomaterial, therefore, we can say:

> An ecomaterial is one whose acquisition and use cause minimal environmental impacts, minimal resource depletion, and minimal regulatory constraints to use.

With this definition as a guide, seven properties of ecomaterials can be enumerated:

- An abundant supply of the material exists.
- A recycled supply of the material can be utilized.
- The material requires low energy consumption in extraction, processing, and manufacturing.
- The material has little or no associated environmental impact.
- The material has no existing or anticipated legal restrictions.
- The material can be used over extended time periods.
- The material can be renewed and/or recycled.

In some cases, the degree to which a property is satisfied by a given material can be evaluated quantitatively. For others, the evaluation must of necessity be more qualitative. It is therefore useful to apply an ordinal approach, as described below, to the evaluation of candidate ecomaterials.

We will assess prospective ecomaterials by using a seven-pointed star diagram, with one point for each of the seven properties listed above (Figure 10.5a). In each case, we assign a rating of A (the poorest), B, C, or D (the best), following the guidelines given in Table 10.6. We then plot the ratings on the star diagram, with the A ratings plotted at the furthest location along the stellar point axes and D the closest. Thus, a highly rated ecomaterial would be one in which most of its ratings were plotted near the stellar core.

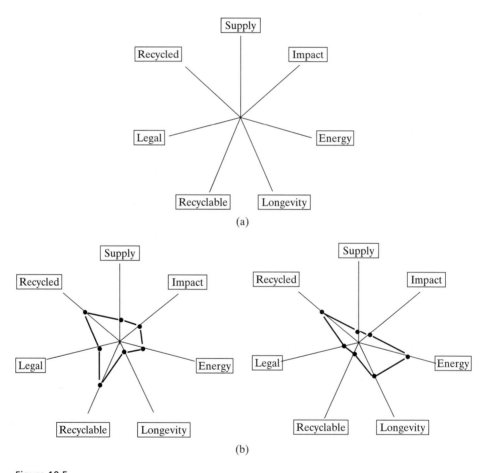

Figure 10.5

(a) The star diagram for ecomaterials evaluation. (b) A star diagram analysis for the use of aluminum (left) or plastic composite (right) for automotive applications in tropical climates.

TABLE 10.6   Ratings Criteria for Ecomaterials Properties

The Supply Property*
      A—Depletion time <25 years
      B—Depletion time 25–50 years
      C—Depletion time 51–100 years
      D—Depletion time >100 years

The Recycled Supply Property
      A—Made of totally recycled material
      B—Made with >50% recycled content
      C—Made with <50% recycled content
      D—Made with totally virgin material

The Energy Property
      A—Embedded energy <50 MJ/kg of material
      B—Embedded energy 50–99 MJ/kg of material
      C—Embedded energy 100–200 MJ/kg of material
      D—Embedded energy >200 MJ/kg of material

The Environmental Impact Property**
      A—Scorecard hazard rating <25
      B—Scorecard hazard rating 25–50
      C—Scorecard hazard rating 51–75
      D—Scorecard hazard rating >75

The Legal Property
      A—Environmentally benign
      B—Little likelihood of legal restrictions
      C—Potential legal restrictions
      D—Currently under legal restrictions

The Longevity Property
      A—Not lifetime limited
      B—Slowly degrades in the environment in which it will be used
      C—Moderately degrades in the environment in which it will be used
      D—Quickly degrades in the environment in which it will be used

The Recyclable Property
      A—Completely recyclable
      B—More than 50% recyclable
      C—Less than 50% recyclable
      D—Completely unrecyclable

*Depletion times for elements and common minerals are given in S.E. Kesler, *Mineral Resources, Economics, and the Environment,* New York: Macmillan, 1994.
**Hazard ratings for most common chemicals are available at *www.scorecard.com.*

To illustrate ecomaterial evaluation, Figure 10.5b compares aluminum and a plastic composite for use in an automobile in the tropics. Aluminum scores well on supply (it is abundant), impact (it is relatively benign), recyclability, and legal status. It does poorly on recycled supply (virgin aluminum is used in automobiles), energy (aluminum processing requires substantial amounts of energy), and longevity (aluminum corrodes in salt air). The plastic composite does well on supply, environmental impact, and energy consumption, but poorly on the use of recycled material and on recyclability. In this example neither material is clearly environmentally superior, but on balance the aluminum appears slightly preferable.

10.6 DEMATERIALIZATION

No matter what materials are chosen for a product, the amount that is used can generally be minimized by careful designing involving stress analysis. Thinner walls and supporting members can often be made suitable by such techniques, especially if common physical design rules are applied. These include:

- Avoid sharp corners to permit thinner walls (Figure 10.6a).
- Use a greater number of smaller supporting ribs rather than a few large ribs (Figure 10.6b).
- Where sheets of metal or plastic are used, achieve strength by providing support with bosses (protruding studs included for reinforcing holes or mounting subassemblies) and ribs, not by using thick sheets (Figure 10.6c).
- Gussets (supporting members that provide added strength to the edge of a part) can aid in designing thin-walled housings (Figure 10.6d).
- Metal inserts should be avoided in nonmetal assemblies. If that cannot be accomplished, install them on break-off bosses (Figure 10.6e).

After the minimization of first-try materials has been studied, consider as well the substitution of nontraditional materials. A simple example of the success of this approach is the ongoing transition from the use of platinum in automotive catalytic converters to the use of palladium as the active agent. Since palladium is significantly more abundant and cheaper than platinum and rhodium, formerly used with palladium in such catalysts, both resource sustainability and market advantage are achieved with the new approach.

10.7 MATERIAL SELECTION GUIDELINES

When considerations of supply and toxicity are merged to give a common perspective, one finds a clear preference for some materials over others. Because their supplies are ample (and/or the potential for recycling is good) and because they have no significant toxicity problems, we recommend that designers investigate using the following materials: Al, C, Fe, Mn, Si, and Ti. Conversely, because they promise to be in short supply and/or have significant toxicity problems, we recommend that designers attempt to limit or avoid the use of the following elements, all of which have use ratings of High or

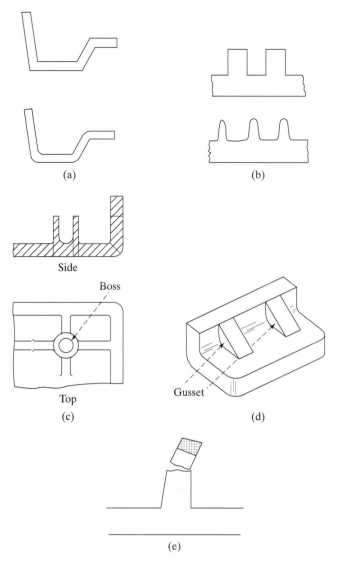

(a)

(b)

Side

Boss

Top

(c)

Gusset

(d)

(e)

Figure 10.6

Design approaches that minimize the amount of materials needed to accomplish a desired function: (a) wall thickness transitions (bottom approach preferred); (b) rib design (bottom approach preferred); (c) use of bosses for reinforcement of a thin wall; (d) use of gussets to support a thin curved section; (e) metal insert on break-off boss. (Adapted from sketches devised by J.R. Kirby and I. Wadhera, IBM Corporation.)

Extreme on Table 10.5: As, Au, Be, Cd, Cs, Ge, Hg, In, Pb, Re, Tc, Tl, and Zn. With the possible exception of uranium in nuclear power applications, radioactive elements—Ac, At, Fr, Pa, Po, Ra, Rn, and Th—should be avoided; most are very scarce anyway. Less specific advice is possible in the case of the many molecular and composite materials that serve so much of modern technology. For those, an ecomaterials analysis as described earlier in this chapter provides the means for choosing among materials whose varied characteristics present a spectrum of environmental challenges.

The materials selection process can be summarized in four short goals for physical designers:

- Try to get most of the needed materials through recycling streams rather than through raw materials extraction.
- Choose abundant, nontoxic, nonregulated materials if possible. If toxic materials are required for a manufacturing process, try to generate them on site rather than by having them made elsewhere and shipped.
- Design for minimum use of materials in products, in processes, and in service.
- Design for longevity, refurbishment, and recycling, to increase the utility of materials in any particular use, and enable recovery of materials when that use ceases.

FURTHER READING

Ashby, M.F., *Materials Selection in Mechanical Design*, London: Butterworth Heinemann, 1992.

Furuyama, T., Ecomaterials selection guide and green procurement system at Toshiba, *Proceedings of the Fifth International Conference on Ecomaterials*, Tokyo: Society for Non-Traditional Technology, A2–3, 2001.

Grubler, A., *Technology and Global Change*, Cambridge, U.K.: Cambridge University Press, 1998.

Kesler, S.E., *Mineral Resources, Economics, and the Environment*, New York: Macmillan, 1994.

Kosbar, L.L., J.D. Gelorme, R.M. Japp, and W.T. Fotorny, Introducing biobased materials into the electronics industry: Developing a lignin-based resin for printed wiring boards, *Journal of Industrial Ecology, 4* (3), 93–105, 2001.

Wernick, I.K., R. Herman, S. Govind, and J.H. Ausubel, Materialization and dematerialization: Measures and trends, in *Technological Trajectories and the Human Environment*, J.H. Ausubel and H.D. Langford, eds., 135–156, Washington, DC: National Academy Press, 1997.

EXERCISES

10.1 In 1991, nearly 2.5 Pg of residues was produced worldwide as a consequence of ore processing. If the typical density of this ore is 4.5 g cm^{-3}, how long would be a line of trucks carrying this ore if each truck holds 2.5 m^3 of ore and is 6 m long?

10.2 Ignoring cost as a factor, which one of each of the following pairs of materials should be preferred by product and process designers: (a) titanium or tin; (b) toluene or heptane; (c) tin or bismuth; (d) beryllium or tin; (e) titanium or vanadium? Why?

10.3 Form a design team with two or three of your classmates. Choose a moderately complex product such as a coffee machine, an overhead projector, or a lawn mower. Inspect the product (disassembling it if necessary) and propose design alternatives that would retain the function but dematerialize the product to some degree.

CHAPTER 11

Designing for Energy Efficiency

11.1 ENERGY AND INDUSTRY

Industry uses substantial amounts of energy and, as a consequence, contributes significantly to energy-related environmental problems. In the United States, for example, manufacturing activities account for some 30% of all energy consumed, and much of that energy is very inefficiently employed. Figure 11.1 shows that the use of electricity (mostly generated from fossil fuels) is concentrated in a few industry sectors, such that six industry groups consume more than 85% of total industrial energy or energy equivalents. This information says nothing about the efficiency with which that energy is used, however. A useful index of industrial energy use, though not a wholly satisfactory one, is energy intensity, the energy consumption per dollar of gross domestic product. Individual corporations within the same industry vary widely in energy intensity. In general, industries dealing largely with raw materials rather than finished or semifinished products have higher energy intensities.

Although the focus of this chapter is to examine how energy is used and how to reduce energy use while maintaining industrial operations, it is also of interest to mention the consequences of energy generation, since plant and process engineers sometimes have optional energy sources upon which to draw. In this connection, we list in Table 11.1 the types of air pollutants released by a variety of energy generation processes. The table illustrates the well known fact that energy supplied by fossil fuel combustion has the potential to be more environmentally harmful than energy produced by nuclear power or by renewable energy sources, at least so far as the atmos-

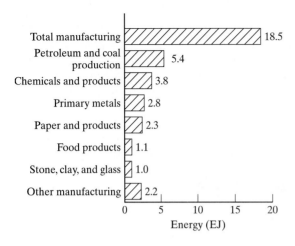

Figure 11.1

Consumption of energy in selected manufacturing industries. (*Source:* U.S. Department of Energy, Energy Information Administration, *Manufacturing Energy Consumption Survey: Changes in Energy Efficiency 1980–1985*, DOA/EIA-0516(85), Washington, DC, 1990.)

phere is concerned. Incineration is intermediate in impact, and its ability to serve as a waste disposal alternative as well as an energy source makes it suitable in some industrial applications. To the extent possible, energy obtained for industrial use should be from environmentally preferable sources.

Energy efficiency must often be balanced with toxicity concerns. For example, more efficient lighting relies on mercury vapor in lamps, and many catalysts which reduce energy requirements in industrial and consumer applications are either toxic or are scarce, nonrenewable resources. Superconducting materials may offer the possibility of significant energy savings, but are likely to contain toxic materials. Identifying and resolving these "impact-balancing" situations is a difficult but necessary task as products and processes evolve toward sustainability.

TABLE 11.1 Atmospheric Pollutant Species Emitted by Energy Generation Processes

Process	Species						
	CO_2	CH_4	NO_x	SO_2	H_2S	HCl	Particles
Fossil fuel energy sources							
Coal	•	•	•	•			•
Petroleum	•	•	•	•			
Natural gas	•	•	•				
Other anthropogenic energy sources							
Nuclear power							
Refuse incineration	•					•	•
Biomass incineration	•	•	•				•
Natural energy sources							
Hydrothermal steam					•		
Solar power							
Hydropower							
Wind power							

11.2   PRIMARY PROCESSING INDUSTRIES

Although the materials extraction and processing sectors have the highest energy intensity, these industries are suppliers to the intermediate processing industries, so one cannot plan to decrease industrial energy use by eliminating the extraction industries. Rather, one needs to investigate opportunities within the extraction industries for reductions in energy intensity. Of the industry groups shown in Figure 11.1, the one with the largest energy use is petroleum and coal production. Most of this energy use is attributable to the refining of petroleum. The trend toward desulfurization of crude oil and the production of high octane gasoline without the use of metal-containing additives place ever-increasing energy demands on the refining operation. Refinery operations are generally subject to careful supervision and continuing engineering effort to improve efficiencies, but increased attention to cogeneration, heat exchange, and leak prevention are likely always to offer opportunities for further improvements.

Chemicals and chemical products rank second among the industries in Figure 11.1, although about a third of the amount shown represents petroleum and natural gas used as feedstocks for products rather than fuel that is consumed to produce energy. Of the remaining two-thirds, a substantial amount is used in the generation or removal of process heat as a result of temperature differences between the process streams and the heating and cooling streams. The production of compressed gases is another energy-intensive area: About 70% of the cost of the gases represents electricity costs, and improving the energy efficiency thus has the potential to pay rich benefits for that industry. Physical designers are developing processes that minimize these temperature differences, by better in-plant use of process heat, by process redesign involving different feedstock materials or improved catalysts, or by capture of process heat for subsequent use or sale. Their success can be demonstrated by the commendable decrease in the U.S. chemical industry in power consumed per unit of product produced (Figure 11.2). Over the period 1970 to 1999, this measure of power efficiency improved by nearly 50%.

Primary metals is the third industry listed in Figure 11.1. Although the extraction of ore from the ground and its shipment are quite energy intensive, the bulk of the energy use is in crushing rock and recovering the target ore, and in generating the large amounts of process heat needed to extract metal from ore and to produce ingots and other purified products. Historically, major changes in the use of energy in the primary

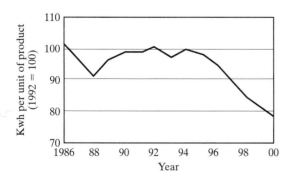

Figure 11.2

The amount of electricity needed by the U.S. chemical industry to produce a unit of product, 1986–2000. (Reproduced with permission from W.J. Storck, Chemical industry is more energy efficient, *Chemical & Engineering News, 79* (6), 19, 2001.)

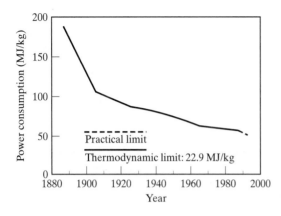

Figure 11.3

The history of electric power consumption in the production of aluminum. (Adapted from P.R. Atkins, D. Willoughby, and H.J. Hittner, in *Energy and the Environment in the 21st Century*, J.W. Tester, D.O. Wood, and N.A. Ferrari, eds., 383–387, Cambridge, MA: MIT Press, 1991.)

metals industry have occurred as a result of the introduction of new processes. In the case of steel, for example, the relatively new electric arc furnaces are much more efficient than the older open hearth or basic oxygen processes. Another example is the large historical decrease in electrical energy consumption needed to produce aluminum, shown in Figure 11.3. It is worth noting that for aluminum and other metals the practical and thermodynamic limits to the energy needed for processing are beginning to be approached, suggesting that major gains from process changes alone may have already occurred among the more advanced manufacturers.

11.3 INTERMEDIATE PROCESSING INDUSTRIES

The intermediate processing industries are too diverse to be discussed individually, but several general techniques for improving their energy efficiency can be described. The most straightforward is the use of computerized systems for the management of energy use. The overall concept is that energy should be used only when needed, and not because inattention or lack of on-site personnel make it impractical to exercise control. Thus, equipment should be started and stopped as dictated by time of day or by sensors that monitor product stream characteristics. Among the types of energy-using equipment that can be controlled in this way are motors, boilers, fans, and lights.

A second technique, previously discussed in connection with the chemical industry, is the utilization by the corporation or by its infrastructure partners of residual heat from process streams, product streams, exhaust streams, and the like. Often these actions will take the form of increased attention to process redesign so that the exchange of heat among material flow streams can be optimized. Alternatively, the heat can serve unrelated processes, as in Nova Corporation's use of residual heat from a natural gas compressor station in Alberta, Canada to provide heat for greenhouses producing flowers, plants, and tree seedlings.

Third, increased use can be made of modern-design motors, especially those with variable-speed drives. The gains that can be expected are quite dependent on the application, but 20–50% decreases in energy use have been realized in several test cases.

11.4   ANALYZING ENERGY USE

It is often the case in industry that the amount of energy required to operate a facility is well monitored, while the energy required for each individual operation or set of operations within a facility is not known. In such cases, an energy audit is advisable to show where the opportunities for gains might lie, as well as to provide data for "green" accounting systems. Figure 11.4 shows such an audit for a facility that uses oil, coal, and electricity to provide energy for three different industrial processes, as well as for lighting and heating. The diagram demonstrates that more than enough energy is available in losses from the Process A energy stream to operate Processes B and C, and to heat and light the entire factory in the bargain. The diagram also suggests that boiler losses would be the highest priority target for improvement, and that steam losses also appear to constitute a substantial opportunity.

For a particular process, one wishes to audit the energy use at each stage of material extraction, processing, and manufacturing. In the production of aluminum cans, for example (Figure 11.5), the major energy use is in the separation and purification of aluminum contained in the ore. Production of sheet and of cans is also significant, but at a much reduced level. The transport of material between stages is a minor contributor to total energy use. With this information as a basis, one might choose to increase the amount of recycled material used to produce metal products rather than extract metals directly from ore. Although aluminum presents the greatest opportunity for energy savings through recycling, the use of many other kinds of scrap material can also result in energy savings of 30% or more.

To examine the energy use implications of virgin material and recycled material, consider the process sequence shown in Figure 11.6. Each processing step has associ-

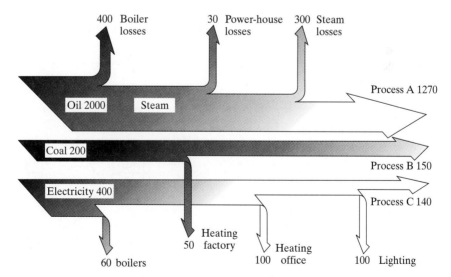

Figure 11.4

A "Sankey diagram" of energy sources, uses, and losses for a typical industrial facility. The units are arbitrary. (Reproduced with permission from *Climate Change and Energy Efficiency in Industry.* Copyright 1992 by International Petroleum Industry Environment Conservation Association.)

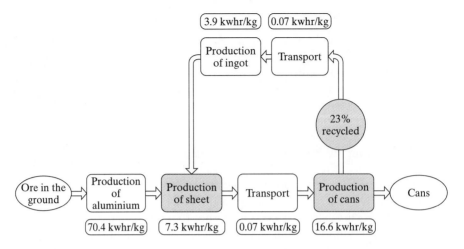

Figure 11.5

A process energy use diagram for the production of aluminum cans. (Reproduced with permission from *Climate Change and Energy Efficiency in Industry*. Copyright 1992 by International Petroleum Industry Environment Conservation Association.)

ated with it an energy per unit of throughput. For simplicity, we choose the amount of output material to be 1 kg. β is the fraction of throughput that is immediately reused as "prompt scrap" rather than output material: rejected material, sprues, runners, lathe turnings, and so forth. The energy consumed per kg of output material is then given by

$$\Phi = E_p + E_f(1 + \beta) + E_m(1 + \beta)$$
$$= E_p + (E_f + E_m)(1 + \beta)$$

$$(11.1)$$

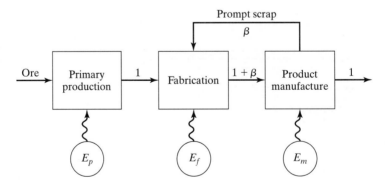

Figure 11.6

Schematic diagram of a metals processing system using only virgin material. (Adapted from P.F. Chapman and F. Roberts, *Metal Resources and Energy*, Boston, MA: Butterworths, 1983.)

It is obvious from this equation that manufacturing operations that produce a smaller fraction of scrap will require less energy per unit of output than those where a large fraction of material must be refabricated.

A more relevant case for industrial ecology is a manufacturing sequence that uses both virgin and consumer recycled material. The latter need undergo only secondary production, which is generally much less energy intensive than primary production. The situation is illustrated in Figure 11.7, where ϕ is the fraction of output material from primary production, Ω is the amount of material entering the process in the ore, and Ψ is the amount of the material entering the process as consumer scrap. The energy consumed by this system per kg of output material is given by

$$
\begin{aligned}
\Phi &= E_p(\phi)(1 + \beta) + E_s(1 - \phi)(1 + \beta) + E_f(1 + \beta) + E_m(1 + \beta) \\
&= (\phi E_p + (1 - \phi) E_s + E_f + E_m)(1 + \beta)
\end{aligned}
\tag{11.2}
$$

Since $E_p \gg E_s$, total energy use is minimized by making ϕ and β as low as possible. In this connection, it should be noted that product designers who specify virgin materials in their products may not be directly paying the high energy cost that results, but the virgin material specification forces the cost to be borne at some point within the industrial system.

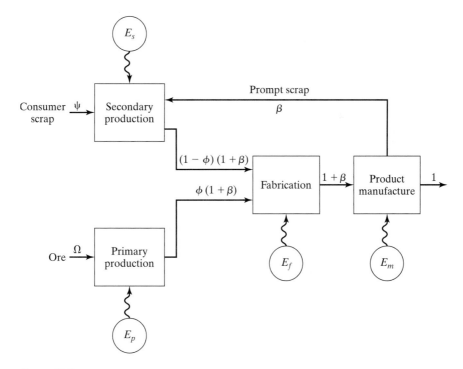

Figure 11.7

Schematic diagram of a metals processing system using both virgin material and consumer scrap. (Adapted from P.F. Chapman and F. Roberts, *Metal Resources and Energy*, Boston, MA: Butterworths, 1983.)

11.5   GENERAL APPROACHES TO MINIMIZING ENERGY USE

Energy conservation in all its facets is good management, responsible action, and progress toward increased corporate profitability, and every little bit helps. In the paragraphs above, we have mentioned those aspects of energy conservation dealing with specific types of industries. In this section, we discuss some more general approaches to industry's use of energy that can be applied across all industrial sectors.

11.5.1   Heating, Ventilating, and Air Conditioning (HVAC)

The "lighter" the industry, the greater the percentage of energy use tending to be attributable to HVAC. This is not only because light industry is inherently less energy-hungry than heavy industry, but also because its manufacturing operations often involve precision control of the in-plant environment. Substantial energy savings may be available by improving the "shell efficiencies" of buildings, i.e., weatherstripping, window treatments, proper planting of trees and shrubs, and the like. Detailed maintenance of HVAC equipment is an often-overlooked action that can be very beneficial. Major gains are possible by replacing aging HVAC equipment with modern, computer-controlled varieties, which can use 30–90% less energy depending on the specific application.

11.5.2   Lighting

The provision of adequate lighting often accounts for 20% or more of industrial energy use. As Figure 11.8 demonstrates, there are many possible approaches to providing adequate light. The traditional use of incandescent lights, or of fluorescent lights without high-reflectance fixtures, electronic ballasts, and high-efficiency bulbs can readily be improved upon, often with payback times of two years or less. In general, the same amount of light can be provided at a 10th or 15th of the energy consumption of incandescent lights.

11.5.3   On-Site Energy Generation

While decreased use of energy is a primary goal of energy assessment, an alternative beneficial activity is the use of energy that is present within an industrial facility but is not used. A common form for such energy is process heat that is not captured. If not utilized in heat exchangers, it can often become available to generate electric power on site. One way to do so is by using both the heat and power from a single thermodynamic cycle, a practice termed cogeneration. There are many variations, one of which is illustrated in Figure 11.9. In these integrated energy systems (IES), the energy stream is a desired output, just as is the product stream. Any excess energy that is generated can be sold back to the electric utility and become a small contribution to the integrated power grid, or perhaps be captured in chemical compounds for subsequent liberation and use.

In a number of industrial facilities, including steel mills, petrochemical complexes, and oil refineries, it is possible to use a residual process stream such as combustible hydrocarbons as a feedstock for power generation. Depending on the type of process and the availability and cost of commercial energy, designing and constructing

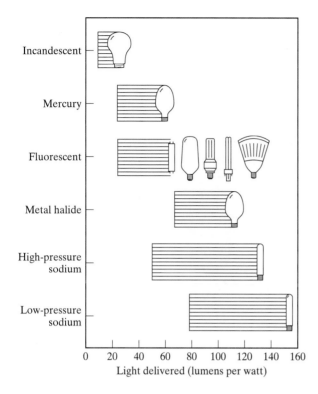

Figure 11.8

Ranges of energy efficiency of various light sources. (U.S. Department of Energy, *Energy-Efficient Lighting*, DOE/CE-0162, Washington, DC, 1986.)

an IES facility may be a sound way to utilize available resources for a variety of purposes.

Environmental impacts in IES facilities can sometimes be minimized by careful choice of energy feedstocks. Biomass fuel may be an option in some places, for example, hydropower in others. An unusual but commendable example of energy feedstock switching is that of Monsanto's Sauget, Illinois facility, which produces energy by burning a mixture of shredded scrap tires and coal. The mixture is cheaper and cleaner burning than coal alone, and disposes of some one million scrap tires per year.

Sometimes energy feedstock switching among more usual supplies is warranted if the total environmental impact can be lowered. Choices can thus be made among fossil fuel options or other potential sources. A successful example of this principle is that of the Lucent Technologies manufacturing facility in Columbus, Ohio, which buys and burns methane from a nearby landfill site. This gas, a product of anaerobic biological degradation of landfill materials which would otherwise be vented to the atmosphere and enhance the potential for global warming, is used in place of fossil natural gas.

11.5.4 Energy Housekeeping

Good energy housekeeping involves taking the industrial situation as it exists and devising ways to modify or change it to make it more energy efficient. Opportunities abound. For example, personal computers now account for more than 5% of all energy used by business, so more efficiently designed computers or, in the interim, better con-

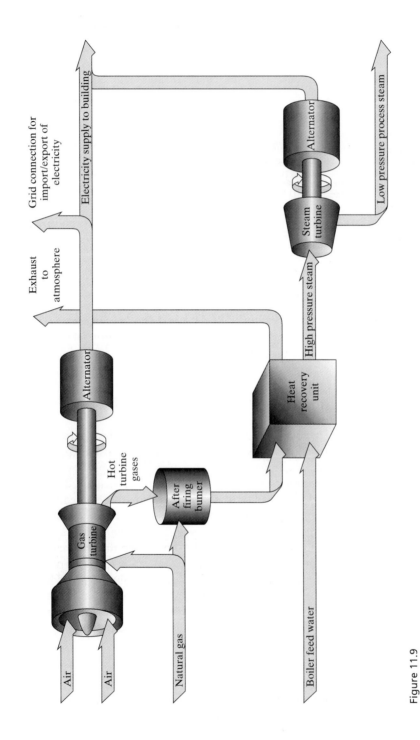

Figure 11.9

A schematic diagram of a cogeneration facility for the simultaneous production and use of heat and power. (Reproduced with permission from *Climate Change and Energy Efficiency in Industry.* Copyright 1992 by International Petroleum Industry Environment Conservation Association.)

TABLE 11.2 Energy Contest Results: Lousiana Division, Dow Chemical

	1982	1984	1986	1988	1990	1992
Winning projects	27	38	60	94	115	109
Average return on investment	173%	208%	106%	182%	122%	305%

Data from K.E. Nelson, Practical techniques for saving energy and reducing waste, *Industrial Ecology and Global Change*, R. Socolow, C. Andrews, F. Berkhout, and V. Thomas, eds., Cambridge U.K.: Cambridge University Press, 1994.

trol of their use of energy, can reduce consumption notably. It is important to realize that since all employees use energy in their jobs, all employees can make useful contributions to energy housekeeping in offices, laboratories, and production facilities.

A particularly successful energy conservation contest for employees was initiated by the Louisiana Division of the Dow Chemical Company in 1982. Many of the improvements embodied techniques useful industry-wide, such as heavy insulation on pipes carrying hot fluids, cleaning heat exchanger surfaces often to improve heat transfer efficiency, and employing point-of-use fluid heaters where storage or long pipelines create the potential for heat loss. The company's energy contest results are summarized in Table 11.2.

There are two central messages in the table. One is that all the good projects were not thought about in the first year. Rather, good ideas kept coming. The second point is that the return on investment is substantial and easily demonstratable. Over the 1982–1992 period, the savings to the corporation, computed after subtracting the relatively small costs involved in implementing some of the ideas, were some $170 million!

11.6 SUMMARY

Energy provides an example of a situation in which the process designer plays an equivalent or greater role than does the product designer. As with most situations, collaborative efforts between the two are likely to produce the greatest energy savings while still promoting efficient and effective manufacturing. Perhaps more than with most of the other topics discussed in this book, energy minimization is a particularly appropriate arena for incremental change as well as for complete process change. As we write this, some countries are imposing a tax on carbon emissions, and others are imposing energy use requirements on industrial products. How these requirements will evolve remains to be seen, but there is little doubt that energy will become increasingly expensive as resources dwindle under the demands of a rapidly growing global population and as environmental concerns inspire more and more legislative activity. Reductions and pattern changes in energy use are thus extremely likely to be good investments in future corporate profitability as well as environmental responsibility.

FURTHER READING

Brown, M.A., M.D. Levine, J.P. Romm, A.H. Rosenfeld, and J.G. Koomey, Engineering–economic studies of energy technologies to reduce greenhouse gas emissions: Opportunities and challenges, *Annual Review of Energy and the Environment, 23*, 287–385, 1998.

Hoffman, J.S., Pollution prevention as a market-enhancing strategy: A storehouse of economical and environmental opportunities, *Proceedings of the National Academy of Sciences, 89*, 832–834, 1992.

Ross, M., Improving the efficiency of electricity use in manufacturing, *Science, 244*, 311–317, 1989.

Special Issue, Energy for Planet Earth, *Scientific American, 263* (3), September, 1990.

Tester, J.W., D.O. Wood, and N.A. Ferrari, Eds., *Energy and the Environment in the 21st Century*, Cambridge, MA: MIT Press, 1991.

EXERCISES

11.1 Assume a materials processing system as shown in Fig. 11.6, with $E_p = 31$ GJ/t, $E_f = 5$ GJ/t, $E_m = 5$ GJ/t, and $\beta = 0.1$. Compute Φ.

11.2 To the system of the previous problem, add a secondary production component to reprocess consumer scrap with $E_p = 9$ GJ/t and $\phi = 0.7$. Find Ψ, Ω, and Φ.

11.3 In the system of problem 11.2, a fraction λ of the material entering the primary production process is irretrievably lost to slag. Reformulate Eq. (11.2) to take this loss into account. If $\lambda = 0.2$, compute Ψ, Ω, and Φ.

11.4 An office building in your community has 50 offices, each with an average of four desks. Each desk has a desk lamp that can use either a 60 watt incandescent bulb or a 13 watt fluorescent unit. The average use of a lamp is seven hours per day. How much power is required for the building per year for each of the two options? Given your local energy cost, what is the annual cost of each of the two options? If the price of an incandescent bulb is $0.88 and that of a fluorescent unit is $12, how long will it take to justify the purchase of fluorescent units, assuming everything is newly purchased?

CHAPTER 12

Product Delivery

12.1 INTRODUCTION

Where detailed assessments have been made, some 30% of all municipal solid waste has been found to be packaging material. Indeed, it has been estimated that about one-third of all plastics production is for short-term disposable use in packaging. For many products—convenience food items, for example—packaging is the primary residue of consumer use. The use of toxic materials such as heavy metals in packaging inks may be a first-order environmental impact for some products. Proper packaging of products of all sorts, from large-volume chemicals to small consumer personal care items, thus plays an important role in maintaining environmental sustainability. It is worth noting that many environmental standards programs, such as Germany's "Blue Angel" program, require the use of packaging that is fully recyclable, has maximum recycled content, does not contain any toxics, and, if paper, is not bleached. While packaging systems have had a substantial and beneficial impact on consumer health and safety, and have greatly reduced waste, especially in food products, it is in many cases possible to retain these benefits while reducing the potential environmental impacts.

An important perspective is that more than a third of all goods and services are not purchased by individual consumers, but by other business or governmental agencies. Packaging in these cases is on a business-to-business basis, and the potential for negotiating packaging reductions and improvements is substantial.

Although packaging is an important topic in industrial ecology, it is often thought not to be within the province of the physical designers who specify the processes or products that generate the packaging need. Rather, in our age of specialization, product delivery is often under the control of an engineer specializing in packaging or transportation technology. The product and process designers must thus work closely with these specialists to see that environmentally responsible products are shipped in environmentally responsible ways; the best way to do this is to involve a packaging engineer in design team activities.

12.2   GENERAL PACKAGING CONSIDERATIONS

There are several possible levels of packaging. For some products, no packaging at all may be required. In other cases, only *primary packaging*, i.e., packaging that is in physical contact with the product, is needed. Less certain is the need in many cases for *secondary packaging* (a supplementary shipping container), or *tertiary packaging* (the outer shipping container and associated material). Product packaging should always aim to use the minimal number of stages. However, different applications impose different packaging requirements. Some food packaging, for example, is quite complex: A potato chip bag may be a "sandwich" of seven or eight different components and many layers, each with a separate function (Figure 12.1). To the degree that any packaging stage can be eliminated or simplified, the residue stream will be reduced and shipping and storage expenses for the producer and consumer will be minimized.

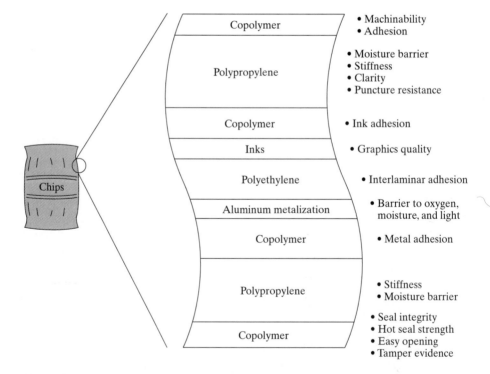

Figure 12.1

The packaging involved in a snack chip bag. (Office of Technology Assessment, *Green Products by Design: Choices for a Cleaner Environment*, Washington, DC: U.S. Congress, 1992.)

The use of packaging does not necessarily involve the generation of residues that must be discarded or reused, since some packaging can serve as an integral part of the product it protects. For example, glass jam jars that serve as beverage containers after the jam is consumed have existed for many years, and part of the packaging of some computer keyboards can double as a dust cover during the product's lifetime. Packaging usually must be recycled or reused, however. That is most easily done if it is made from a single material, such as a cardboard carton assembled without staples. Next best are packaging designs that use more than one material but make them easily separable, such as the bottle cap of a different material than the bottle itself or the styrofoam insert in a cardboard box. Less desirable are commingled materials that are difficult to separate cleanly, such as the polyethylene overwrap fastened by adhesive atop a screwdriver mounted on a cardboard backing. Worst of all are dissimilar packaging materials bonded together so that separation is essentially impossible, such as the aluminized bags often used for electronic circuit boards.

A suggested order of precedence for approaches to packaging, in decreasing order of preference, is:

- No packaging
- Minimal packaging
- Consumable, returnable, or refillable/reusable packaging
- Recyclable packaging

This list is only a guide, since innovative packaging solutions for specific products may outweigh the precedence order, but is a good starting point for the packaging engineer.

Manufacturers should expect to work with their customers in deciding how to package their products. From the perspective of customers, packaging entering their facility or home is material that they would rather not have. To the extent that packaging engineers can reduce packaging, or develop packaging that is easier to recycle or reuse, the entire product becomes more attractive. Alternatively, many customers and some countries are now encouraging suppliers to take back the packaging on items sold by them. Such take-back arrangements can take several forms. One is the negotiated agreement, in which containers for consumed supplies (chemical drums, for example) are returned to the supplier when a new shipment is delivered. The second is a legislated requirement such as exists in Germany, where all sellers are required to accept from customers the packaging in which their products were delivered. This exchange can occur at purchase, as can happen with external packing, or at a later time, as when the tube holding toothpaste can be returned when exhausted. In either the negotiated or legislated case, when a corporation gets its own packaging back it has a very strong incentive to minimize the amount of that packaging and to make it easy to recycle or reuse.

12.3 SOLID RESIDUE CONSIDERATIONS

The first goal in designing solid packaging for products is to minimize it as much as possible consistent with other packaging requirements. Among minimization options that have been implemented are the substitution of soft pouches and paper cartons for rigid plastic bottles and cans in consumer products such as detergent and coffee. For fabric softeners and other liquids that can be sold in concentrated form, customer dilution can permit significant reductions in packaging size; if marketed properly, many consumers are willing to take this small extra step. Analogous approaches in new packaging and concentration of commodity chemicals can and are being used for industrial and commercial products, such as the growing practice in Europe of shipping bulk chemicals in plastic-lined paper bags rather than in drums. Still more commendable is a DuPont program to ship pellets in bags that are chemically compatible with the pellets themselves. Once the pellets are used up, the bag can be crumpled up and thrown into the compounding machine.

Compared with the volume of packaging that has traditionally been used, shrinking the amount may not be difficult. An example is shown in Figure 12.2, which illustrates old and new packaging for a personal computer keyboard. The 30% volume difference was achieved after a series of tests to verify that the new packaging was sufficiently protective. This change alone justified the effort, as the smaller package was much cheaper to move, ship, and store. In addition to the volume reduction, the materials diversity of the packaging was substantially decreased by using a cardboard interior cushioning in place of styrofoam. A factory-wide project with similar goals by NCR's facility in Oiso, Japan produced reductions in package costs alone of 3–30% in the mid-1990s. Pneumatic packaging, in which plastic inserts are inflated to fill whatever shape box is used, can dematerialize packaging considerably.

A list of causes for overpackaging has been presented by the Institute of Packaging Professionals in Herndon, VA:

- An overly cautious approach to the protection of the packaged contents
- Increasing the package size to deter shoplifting

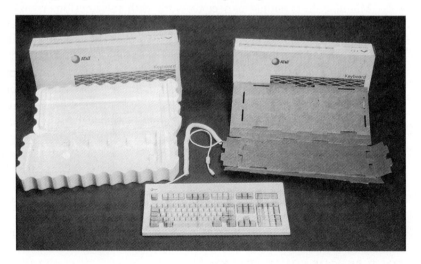

Figure 12.2

Packaging for the AT&T Model 6386 Keyboard. *Left*: 1988 packaging, with cardboard exterior and expanded polystyrene foam insert. *Right*: 1990 packaging, all of cardboard. This packaging is 30% smaller than the packaging it replaces. (Photograph courtesy of AT&T.)

- Overly conservative environmental test specifications
- Requirements of packaging machinery
- Decorative or representational packaging
- Increasing packaging size to provide space for regulatory information, customer information, or bar coding

Many of these causes are capable of being surmounted by thoughtful packaging design. For example, the use of electronic theft protection systems is able to mitigate the need for much of the overpackaging that has traditionally been employed.

Once the amount of packaging is minimized, the next thing to consider is whether the packaging can be reused. Reusable packaging is not limited to foam "peanuts"; it includes innovations such as Ametek Corporation's "couch pouch," a composite of polypropylene sheet foam and polypropylene fabric that can be repeatedly reused for shipping furniture.

One general guide for designing packaging that is environmentally responsible is to mimic nature. To the extent possible, using natural materials avoids problems with toxicity and lack of degradability when discarded. In the case of many food products, edible packaging is an option that is being considered in some cases.

A high degree of materials diversity in packaging is not uncommon, and occurs because of the different physical and chemical properties of packaging materials, in combination with a lack of concern about packaging recyclability. An example is sketched in Figure 12.3 for a large electronic equipment module. First, the module is placed on a wooden pallet that is constructed with polymer cushioning to minimize shock damage during shipment. Next, a plastic sleeve is placed over the equipment for dust protection. Styrofoam blocks are added to cushion the top of the module and restrict its motion. Over the styrofoam and plastic is placed a corrugated cardboard cover to protect the module during shipment. The cover is held in place by steel strapping. The overall packaging system is very effective at protecting the product during shipment. At installation, however, the customer must deal with wood, cardboard, steel, two types of foam, and plastic sheeting. The chances of most of this material being recycled are small. It is likely that a packaging engineer working with the product design team and striving for both safe delivery and environmental superiority in the packaging operation could have reduced the materials diversity substantially.

Since most packaging materials must eventually be discarded, they should be made of materials that have good recyclability. To the maximum extent possible, the materials used should themselves be made of recycled rather than virgin stock. This is generally easy for paper-based products items such as cartons, instruction cards, customer information booklets, and the like. It is less easy for packages that come into intimate contact with the product, such as plastic bottles holding either food items or manufacturing chemicals. Some of these containers can be reused after cleaning; some can be recycled only in degraded form. In any case, plastic packaging materials should be thermoplastics, not thermosets, and their composition clearly marked.

It has long been an act of faith to assume that forest products have less environmental impact than do plastic products. This assumption, based largely on the belief

PACKAGING AN ELECTRONIC SYSTEM

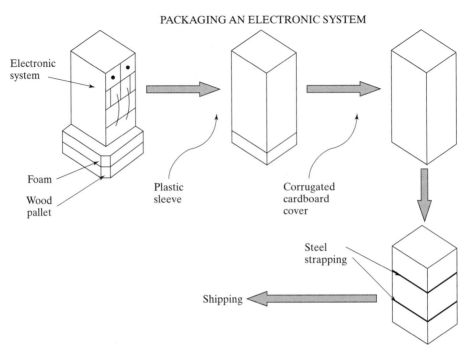

Electronic
system

Foam

Wood
pallet

Plastic
sleeve

Corrugated
cardboard
cover

Steel
strapping

Shipping

Figure 12.3

The sequence of packaging, circa 1995, for a large electronic equipment module.

that forest products degrade more rapidly than plastics, has now been shown to have a questionable foundation, since very little biodegradation occurs in most landfills. Even were biodegradability favorable to forest products, the topic is not central if designers plan for cyclization rather than disposal of residues. Plastics are usually lighter than the forest products materials they replace, consume much less energy in manufacture, and often have better physical properties. However, plastics are not renewable resources on reasonable time scales unless they are produced from biomass precursors. Hence, forest products will continue to play important roles, particularly when long-term resource sustainability is considered.

Does each product require its own package? Not necessarily, say packaging engineers, who can often minimize the total number of packages needed. A common example is the sale of window cleaner and liquid soap refills in a size inappropriate for normal use but suitable for replenishing containers in service. The use of refills also eliminates the need to manufacture, use, and discard spray heads for each bottle of product.

A final rule in packaging is to put package recycling information on (or in) each container, and to help provide an infrastructure for the return of packaging for reuse or recycling.

12.4   LIQUID AND GASEOUS EMISSION CONSIDERATIONS

Although liquid and gaseous emissions are not major problems in connection with product packaging, it is important that those issues not be forgotten in cases where they may be important. A simple example is the steel drum used for shipment of liquid chemicals. Traditionally, the disposal of these drums involved the disposal as well of significant amounts of residual chemicals. New drum designs now allow virtually all of the chemicals within to be drained before disposal. (Even better, where possible, are reusable drums.)

Another example of emissions from packaging is the heavy metals traditionally used to provide color in printing inks. This practice contributes to heavy metal pollution during the manufacture of the packaging and again upon packaging disposal, generally by leaching of the metals into ground or surface water.

12.5   TRANSPORTATION AND INSTALLATION

In many ways it is inherently somewhat contradictory to advise minimum use of materials in packaging and then to advise safe and trouble-free shipping, since products must be protected in some way from such in-transit conditions as shock, vibration, condensation, corrosive gases and liquids, insect infestations, and temperature extremes. One way to approach the problem is to realize that only large products are packaged and shipped individually; most of the rest are shipped in quantity in larger cartons, shipping containers, plastic shrink-wrap sheeting, and the like.

Because product packaging and transportation packaging are often distinct, designers have two opportunities to approach packaging from an industrial ecology viewpoint. Thus, packaging and transportation engineers need to work together to optimize the combination of multiple product packaging for shipment and individual product packaging for the final consumer. For example, it may be the case that good shock protection in the product package can minimize the need for shock protection of the shipping package, or vice versa. The decision on appropriate packaging may depend on the mode of product shipment and on the environmental stresses that the product encounters. A team approach that involves working together with the transportation firm provides another avenue for optimizing a design (a packaging design, in this case) from the industrial ecology perspective. The use of multimodal shipping containers, where feasible, has the potential to eliminate several steps in the common packaging/unpacking/repackaging sequence, with savings in both efficiency and in the generation of packaging waste.

The transporting process itself, whether by company-owned vehicles or by transportation contractors, offers opportunities to minimize environmental impacts. If a product is itself hazardous, contains hazardous constituents, or is potentially subject to spilling or venting, transportation routes should be chosen to minimize possible contact with humans or with sensitive environmental areas. Drivers should be well trained in avoiding problems or dealing with them if they do occur. Finally, no matter what the product, delivery offers the opportunity to collect packing materials and reuse or recycle them.

The impacts of transporting components into a facility and then transporting products from the facility to customers are too frequently regarded as not being directly relevant to a product's environmental impacts. In many cases in today's economy, components and subassemblies will be produced in many countries around the world, then shipped to one or several locations for final assembly into a product which is, in turn, shipped around the world. Japan's otherwise efficient JIT (just-in-time) manufacturing and parts stocking systems have significantly increased the number of delivery vehicles in Tokyo, as well as daily vehicle duty cycles. The result is a substantial and undesirable increase in traffic congestion and direct contributions to air pollution.

In the case of large or complex equipment, installation is often accomplished by the manufacturer. Depending on the product, the installation step may carry with it the potential for environmental degradation. Examples include underground storage tanks for fluids, pipelines for liquids and gases, and the laying of intercontinental communications cables. The most straightforward advice for these situations is to minimize environmental disruptions and to avoid sensitive areas as sites for large projects, especially those that produce significant emissions during use. The ideal industrial ecology solution, however, is to design products or construct societal networks that avoid the necessity for such installations altogether. An example, now in rapid deployment, is cellular mobile telephony. By using digital radio signals, the designers may be moving toward a world in which communications need not move by buried or elevated wire and cable at all.

Transporting Chemicals by Rail

A large fraction of bulk chemicals move from manufacturer to customer by train. Since chemicals can be spilled during loading and unloading operations, and since rail accidents occasionally occur, an important facet of the life-cycle assessment for those chemicals involves the degree to which the impact from those activities can be avoided or minimized. In many parts of the world, manufacturers have entered into partnerships with the railroads to promote environmentally responsible transportation of their products. In the case of the Southern Pacific Railroad in the United States, joint training programs involving chemical manufacturers and their customers help to ensure that cars are loaded and unloaded properly. For the transport phase of the operation, the railroad keeps current logs of disaster response organizations along all its routes, together with maps showing emergency equipment locations and access routes to its right-of-way.

As a result of these activities, accidents attendant to chemicals moving by rail are much less frequent than was formerly the case, and response to incidents that do occur is much quicker and more efficient.

12.6   DISCUSSION AND SUMMARY

Packaging and shipping is the stage at which the manufacturer has the opportunity to indicate to the customer the environmentally responsible characteristics of its products. The performance of many manufacturers in this regard has not been salutary.

Their approach has too often been to trumpet on their product packages unjustified or at least unsupported claims for environmental benefits. In contrast, products that are responsibly designed are seldom advertised or sold with enough information to inform and instruct the consumer. We recommend that the outer package and any more detailed information, such as operating manuals, discuss why the manufacturer considers a design to be environmentally responsible and how these activities compare with what might be done.

In packaging as in other aspects of industrial ecology, designer decisions are aided if a structured method of comparing alternatives is available. Merck and Co. assigns numerical ratings to such factors as use of recycled material, capability of the packaging to be recycled, energy used to produce the package, amount of material required, toxicity of the packaging materials, and potential energy recovery from incinerating the packaging. Competing packaging designs are then ranked for environmental impact, and special justification must be presented if the highest ranking design is not chosen. A more highly structured but narrower approach is that of the Swiss firm Migros-Genossenschafts-Bund, whose computer analysis program allows comparisons of energy use and air and water impacts of alternate packaging concepts. The result of either of these programs is a straightforward, understandable way to encourage environmentally responsible packaging.

FURTHER READING

Russell, P., and others, *Handbook for Environmentally Responsible Packaging in the Electronics Industry*, Herndon, VA: Institute of Packaging Professionals, 1992.

Stilwell, E.J., R.C. Canty, P.W. Kopf, and A.M. Montrone, *Packaging for the Environment: A Partnership for Progress*, New York: American Management Association, 1991.

EXERCISES

12.1 Critique the packaging of six food products in a grocery store, selecting them for maximum diversity. Discuss the good and bad points of each approach to packaging to the extent applicable and possible. If competing products adopt different approaches, point out the good and bad aspects of each.

12.2 Discuss options for packaging of the following consumer products: motor oil, grapefruit juice, toothpaste, magazines, shirts, decongestant tablets.

12.3 Visit one or more hardware or "home center" stores and compare different approaches to packaging several common items such as nails, light bulbs, and screwdrivers. Consider the need for brand identification, theft protection, and other nonenvironmental factors. What are the good and bad points of each product's packaging? Can you think of packaging techniques better than any of those you saw?

Environmental Interactions During Product Use

13.1 INTRODUCTION

Design with the environment in mind involves much more than designing so that products may be manufactured with minimal environmental impact. Another important consideration, sometimes an overriding consideration, is the amount of environmental impact produced by products when they are used. Unlike the extraction or manufacturing environments, which are under the direct control of corporations, the use and maintenance of a product after it passes to the customer is largely constrained only by the product design. This circumstance places special responsibilities on the designer to envision aspects of design that minimize impacts during the entire useful life of the product.

13.2 SOLID RESIDUE GENERATION DURING PRODUCT USE

There are a number of types of solid residue generation during product use. For example, products that in themselves may be long lived and recyclable may nonetheless generate solid residue from consumables. Common examples include computer printers (print cartridges) and cameras (plastic containers from photographic film). To the degree that solid residues from consumables can be eliminated by innovative design, the environmental benefit is obvious. In any case, consumables should have little or no inherently toxic material within them.

Less desirable than eliminating solid residues from consumables, but still meritorious, is the design of consumables to encourage reuse or efficient recycling. There are two requirements for such a recycling program. One is a design that permits ready recycling once the consumable item has been returned. The second is a procedure and supporting infrastructure to encourage the user to return the item for recycling. An example of the latter is the approach used by a number of corporations to recycle the cartridges from laser printers. Users are encouraged by clear instructions to return the cartridges for regeneration and reuse. An additional incentive in some cases is a charitable donation for each cartridge returned. Parcel delivery systems provide much of the necessary infrastructure. Not only is this approach environmentally meritorious, but companies have found that cartridge reuse is much more profitable to them than using a new cartridge every time.

13.3   LIQUID RESIDUE GENERATION DURING PRODUCT USE

Some products generate regular or occasional liquid residues as a result of use. Examples include soil- and detergent-laden water from washing machines, coolants from large industrial motors, and lubricants from internal combustion engines. The ideal designs in this connection are those that require the customer to use no consumable fluids at all. Next best is a design that minimizes the quantity of fluids and uses only fluids with a modest environmental impact. Finally, extensive efforts should be made to recycle fluids. An example of the latter program is that originated by suppliers of solvents to work with independent automotive garages, furnishing solvents in special containers, reclaiming them when dirty, completing the necessary shipping manifests, and cleaning and reusing the solvents as part of an integrated systems approach to solvent cleaning of metal parts. It is worth noting that this system is largely the result of organizational initiative, and that many such programs are quite profitable. This is also an example of substituting a service-intensive offering for a product, an option we discuss in more detail in Chapter 21.

A final consideration is that liquid products are often held in underground or aboveground storage tanks prior to use. Leaks from storage tank plumbing or loss of integrity of the tank itself can result in unplanned dissipative emissions of liquids to water or soil. For such products, containers and/or products should be designed to minimize the potential for inadvertent loss from storage facilities. For example, where possible and safe, tanks and pipes should be above ground to allow for easy visual detection and repair of malfunctions or leaks.

13.4   GASEOUS RESIDUE GENERATION DURING PRODUCT USE

Products whose use involves such processes as the venting of compressed gas or the combustion of fossil fuels require the industrial ecologist to explore design modifications to minimize or eliminate these emissions. The automobile's internal combustion engine is perhaps the most common example of such a product, and one whose cumulative emissions are very substantial, but anything that emits an odor during use is, by definition, generating gaseous residues; examples include vapors from carpet adhesives, polymer stabilizers from plastics, and vaporized fluids from dry cleaners. Replacements for the volatile chemical constituents will often be available if the designer looks for them.

An example of exploring options for reducing the impacts of products in use is provided by Volvo Car Corporation, which has studied emissions from alternative

TABLE 13.1 Equivalent CO_2 Emissions (g/km) from a Volvo 740 GL Using Various Fuels

Fuel	Extraction	Refining	Distribution	Operation	Total
Diesel	17	10	7	205	239
Gasoline	18	15	8	225	266
Methanol	13	51	12	186	263
Methane	10	7	41	187	245

Data source: Volvo Car Corporation, *The Volvo Car Corporation's Fuel Database*, Environmental Report 26, Göteborg, Sweden, 1991.

fuels in its vehicles. The analysis included the complete life cycle of the vehicle, although most emissions occur during operation rather than during extraction, refining, and distribution. The result (Table 13.1) is that, on a CO_2 basis, the engines using diesel fuel and methane are superior to those using methanol or gasoline.

13.5 ENERGY CONSUMPTION DURING PRODUCT USE

A number of products use energy when in operation. The energy may be electrical, as with refrigerators or hair dryers, or furnished by fossil fuel, as with power lawn mowers or chain saws. Recent redesigns have produced lower energy consumption during use for many products, and legislation in an increasing number of countries will enhance such efforts. Energy-efficient designs may sometimes involve new approaches; the result provides not only lower cost operation but also improved product positioning (from a sales standpoint), particularly in areas of the world that are energy poor.

Energy use can also be a function of the way in which a product is designed to use replenishable supplies. To continue our example with Volvo automobiles, Table 13.2 summarizes the energy use of different types of engines. It is interesting to compare this table with Table 13.1, which shows gaseous emissions from the same engines. Diesel engines are seen to be far more energy efficient than other types, as well as being relatively low CO_2 (and CO and VOC) emitters. From an environmental standpoint, there appears to be much to recommend the diesel engine, which in recent years has become much cleaner than used to be the case.

TABLE 13.2 Energy Consumed (MJ/km) by a Volvo 740 GL Using Various Fuels

Fuel	Extraction	Refining	Distribution	Operation	Total
Diesel	0.15	0.14	0.08	2.52	2.89
Gasoline	0.17	0.22	0.10	2.87	3.36
Methanol	0.12	0.85	0.15	2.42	3.54
Methane	0.12	0.12	0.29	2.87	3.40

Data source: Volvo Car Corporation, *The Volvo Car Corporation's Fuel Database*, Environmental Report 26, Göteborg, Sweden, 1991.

13.6   INTENTIONALLY DISSIPATIVE PRODUCTS

Many products are designed to be dissipative in use, that is, eventually to be lost in some form to the environment with little or no hope of recovery. Examples include surface coatings such as paints or chromate treatments, lubricants, pesticides, personal care products, and cleaning compounds. Attempts are being made to minimize both the packaging volume and the product volume in some of these situations, as in the recent introduction of superconcentrated detergents (though this does not change their basic dissipative nature). Alternatively, some liquid products that are dissipated when used can be designed to degrade in environmentally benign ways. Over the past several years this approach has successfully been adopted for a number of pesticides and herbicides. A recent demonstration of design for biodegradability is the development of a biodegradable synthetic engine oil, designed specifically for inefficient two-cycle engines that in operation emit approximately 25% of the gasoline–oil mixture unburned to the environment.

Another common example of a potentially dissipative product is fertilizer for crops, where any excess spread on fields is dissipated to local and regional ground and surface waters. The amount of fertilizer used by farmers has traditionally shown great variation, as seen in Figure 13.1. For the same yield, fertilizer application has ranged over some two orders of magnitude, or, to look at the data in another way, the same

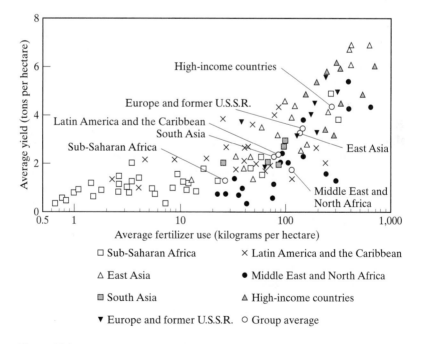

Figure 13.1

Fertilizer input and cereal yields for countries throughout the world, 1989. (Reproduced from *World Development Report 1992*, Copyright 1992 by The International Bank for Reconstruction and Development, The World Bank. Reprinted by permission of Oxford University Press, Inc.)

level of fertilization can produce manyfold variations in yield. Much of this variation is due, of course, to the qualities of the different soils and differences in climate, but it is generally agreed by experts that farmers in the developed countries tend to use more fertilizer than can be justified by the resulting crop yields, and farmers in the developing world tend to use too little (often because of the expense). To the extent that excess fertilizer is used, it has a triple negative impact: The use results in excessive extraction of raw materials, financial penalties accrue to the farmers, and dissipative release of the excess fertilizer may have negative impacts on proximate water supplies.

The concept of intentionally dissipative products is a poor one, but better alternatives are not always readily available. Many gear systems require lubrication, for example. To the degree that they can be sealed so that lubrication need not be replenished often or ever, the dissipative nature of the gear lubricant is ameliorated. Cadmium provides an example of dissipative design related to purity, because zinc oxide is used as a constitutent of automobile tires and because cadmium is an impurity in the refining of zinc. To the degree that the zinc is not well purified during processing, dissipative emissions of zinc oxide from tire tread wear produces dissipative emissions of cadmium as well.

An example of efforts to minimize the environmental impacts of a dissipative product is that of Proctor & Gamble's research on disposable diapers. Such diapers make up about 2% of the United States solid residue stream. Cloth diapers, the alternative, consume substantial amounts of energy and water in repeated washings, however, so that the environmentally desirable alternative for diapers appears to depend on local costs for energy, water, and solid residue disposal (the actual analyses are quite sensitive to the assumptions used). Biodegradable diapers do not constitute a solution, since it has been shown that little biodegradation occurs in a modern, well covered landfill. Proctor & Gamble's approach has been to develop a diaper that is almost entirely compostable.

13.7   UNINTENTIONALLY DISSIPATIVE PRODUCTS

Unintentional dissipative emissions occur when products that are exposed to the environment suffer destructive degradation. Calcium carbonate in disintegrating concrete provides an example. Another is that of cadmium present as an impurity in the zinc used for galvanizing steel. When the zinc sacrificially corrodes, as it was designed to do, both the zinc and its cadmium impurity are dissipated to the environment. Unintentionally dissipative products are often difficult to recognize, and each product engineer needs to ask the question, "What can possibly happen to the materials in my product while it is in use?" Loss or corrosion during shipment or storage also falls in the unadvertent dissipating category, but such losses can be minimized or prevented by good planning and diligent oversight.

13.8   DESIGN FOR MAINTAINABILITY

A final consideration in product use is related to the requirements placed on product maintainability by the designer. Several key principles are involved:

- Components and subassemblies should be repairable and/or replaceable, preferably by the customer.

- Upgrading the system should be possible with exchange of modular parts, and should not involve the purchase of redundant components.
- Residues generated as a result of routine maintenance or repair should not contain hazardous substances, and the amounts should be minimized.
- Manufacturers should ensure that infrastructures exist for the proper handling of maintenance residuals, including used modules.

Maintainability of in-service products may involve the use of cleaners or lubricants. To the degree possible, products should be designed so that maintenance procedures involving dissipative materials are performed infrequently; a good example of this approach is the intervals required between automobile oil changes, which have lengthened markedly in recent years. Where the replacement of worn components or subassemblies may be anticipated, those components should be made so that they are easy to remove and replace. If possible, manufactured systems should be designed to indicate when maintenance is necessary rather than to rely on (necessarily conservative) maintenance schedules to ensure satisfactory performance. If a system requires maintenance involving materials with substantial environmental impact, such as CFCs in many air conditioning units, the design and product support should encourage maintenance only by trained technicians so that loss can be minimized and recovery and recycling optimized.

Design for maintainability often involves a greater degree of commitment to customers than would otherwise be the case. This commitment implies to customers the provision of worry-free product function rather than an abdication of product responsibility by the manufacturer. Modular design and the ability easily to replace defective or worn modules encourage maintenance contracts or other cooperative agreements, as does a commitment to proper treatment of liquid or gaseous residues generated as a part of maintenance procedures. These close relationships can encourage not only proper maintainability, but also proper treatment of packaging and proper recycling of obsolescent products. It is apparent that a systems approach to industrial ecology and customer relationships is the manner in which intelligent corporations should proceed.

Energy Consumption and Solid Residue Generation During Customer Use of Women's Knit Polyester Blouses

Although many products consume no energy and produce no residues when in use, some cause the bulk of their environmental impact after manufacture and before disposal. A common example is clothing, which undergoes many washing cycles during its lifetime, consuming energy in heating water for washing and air for drying as well as requiring resources for the production and disposal of the detergents. An assessment of these environmental interactions for the case of a women's knit polyester blouse has been prepared by Franklin Associates for the American Fiber Manufacturers Association.

Figure 13.2a shows the total energy requirements per million wearings (this turned out to be a convenient comparative unit) for the blouses. The energy used during manufacture (Figure 13.2b) is divided as would be anticipated, resin production and fabric pro-

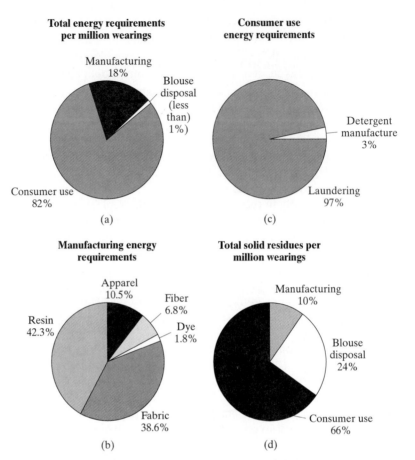

Figure 13.2

Data from an inventory assessment analysis of women's polyester blouses. (a) Allocation of total energy requirements per million wearings. (b) Allocation of manufacturing energy requirements. (c) Allocation of consumer use energy requirements. (d) Allocation of total solid residues per million wearings. (Reproduced with permission from American Fiber Manufacturers Association, *Resource and Environmental Profile Analysis of a Manufactured Apparel Product: Woman's Knit Polyester Blouse*. Copyright 1993 by American Fiber Manufacturers Association.)

duction requiring roughly equivalent amounts and other activities being much less important. Manufacturing energy is less than a fourth of that required for consumer use. The latter is allocated in Figure 13.2c, and is seen to be almost entirely due to laundering operations. In fact, if energy alone were a consideration, blouses should be replaced after every fourth wearing. Solid residues also occur predominantly during the customer use phase; these are the municipal sludge attributable to the washing operation and the ash related to off-site energy generation.

What changes in customer use patterns could improve the environmental responsibility of the blouse design? The study found that use of a cold wash cycle, thus eliminating the need for heating water, reduced overall laundering energy consumption by 60%.

A less extreme alternative is lowering the water temperature but retaining a warm wash; a 10 degree temperature reduction reduced laundering energy use by 14%. Line drying was also a very beneficial activity, reducing overall energy consumption by 31%.

What do these results have to say to the product designer? An obvious answer is that she or he should try to develop blouses that can be cleaned effectively in cold water, perhaps the easiest consumer behavior change that may benefit the environment. A second potential change is to modify the product so that it line-dries quickly or dries mechanically in a shorter period of time, while retaining an attractive appearance. Probably less effective, but still worth doing, is encouraging on the product label the use of lower temperature water if a warm water cycle is used, and of air drying.

FURTHER READING

Saf-T-Clean Corporation, Chicago, IL, pamphlets, available from Ronald H. Mulholland, Corporate Accounts Manager (telephone: 312-697-8460 or 800-323-5740).

Stahel, W.R., The utilization focused service economy—Resource efficiency and product life extension in utilization, in *The Greening of Industrial Ecosystems*, B.R. Allenby and D.J. Richards, eds., Washington, DC: National Academy Press, 1994.

Stigliani, W.M., and S. Anderberg, Industrial metabolism at the regional level, in *Industrial Metabolism—Restructuring for Sustainable Development*, R.U. Ayres and U.E. Simonis, eds., United Nations University Press, 1993.

Volvo Car Corporation, *The Volvo Car Corporation's Fuel Database*, Environmental Report 26, Göteborg, Sweden, 1991.

EXERCISES

13.1 For one week, audit the emissions to air, water, and soil directly relating from your activities. Could changes in the designs of the products that produced those emissions have minimized them? How?

13.2 As a user of many emissive products, you are a valuable data source. Based on the data you and your classmates generated in the above problem, determine which products have emitted the most material over the one-week monitoring period. What changes in the collective consumption or use patterns of the class would reduce these emissions?

13.3 Conduct an audit of the in-use impacts of a photocopy machine convenient to you. Can you suggest design changes to minimize these impacts?

13.4 Conduct an audit of the in-use impacts of a washing machine convenient to you. Can you suggest design changes to minimize these impacts?

13.5 Conduct an audit of the in-use impacts of a passenger or freight train for which you can acquire data. Include any dissipative emissions that are present. Can you suggest design changes to minimize these impacts?

CHAPTER 14

Design for End of Life

14.1 INTRODUCTION

The concept of industrial ecology is one in which products that have reached the end of their useful life reenter the industrial flow stream and become incorporated into new products. As Kumar Patel, formerly at AT&T Bell Laboratories, puts it, "The goal is cradle to reincarnation, since if one is practicing industrial ecology correctly there is no grave." The efficiency with which cyclization occurs is highly dependent on the design of products and processes; it thus follows that designing for recycling (DfR) is one of the most important aspects of industrial ecology.

The consequences of not having considered DfR in earlier periods of industrial design are graphically illustrated by a Carnegie Mellon University research project on personal computer disposal. It was estimated that by year 2005 some 55 million obsolete PCs, none with readily recoverable materials, would be landfilled each year. The required annual landfill volume is nearly three million cubic meters, and the annual cost some U.S. $150 million. These very large numbers apply to only a small fraction of the amounts that may eventually result, however, since research at the Fraunhofer Institute and at Tufts University suggests that three-quarters of all obsolete computers are consigned to storage. Even if a computer is not being used, it appears to be psychologically hard to dispose of something as costly as a PC.

If we consider also washing machines, refrigerators, automotive plastics, and all the other products now in use and not designed for recycling, the embedded stock of unrecoverable materials is enormous. Recovering much of this material currently is more expensive or difficult than necessary, however, because most products have not been designed for refurbishment and recycling. Hence, DfR may not only be advisable, it may be crucial to the ability of societies to continue to use materials in ways to which we have become accustomed.

In Chapter 11 we made the point that the lower down one goes toward the beginning of the chain of materials flows, the more energy must be invested to recover a unit of material. A designer should not merely make it possible to recycle a particular material, but to degrade that material as little as possible during recycling, thus avoid depleting its embodied utility. Having stated this goal, it is nonetheless true that it is often difficult to avoid some level of degradation in a recycled material. In most cases, reuse of a material, even at a degraded level, is far better than discarding it. Hence, the reuse of polystyrene cafeteria trays as foamboard insulation or of polyethylene terephthalate soda bottles as carpet fiber is to be commended even while researchers attempt to develop levels of reuse that retain more of the embodied utility of the original material.

Alternative end of life strategies are shown in the "comet diagram" of Figure 14.1. Users are at the perihelion point in the product's orbital path. During the approach to perihelion (the portion of the orbit at the top of the diagram), materials are formed, converted into components, then to products, and then marketed. During the retreat from perihelion, the products or their constituents are either reused or discarded.

A closed loop is obviously preferable from an environmental viewpoint, and the shorter the loop the better, because short loops retain the materials and energy embodied in products during their manufacture. Most of the loops require processors who provide various services to enable resources to move from the outgoing segment of the orbit to the incoming segment.

The original designer of a product defines the loop options available to the user and potential recycler. Their preferred approach is to practice preventive and therapeutic maintenance for as long as possible, including upgrading to capture efficiency and performance gains resulting from technological innovation. Sooner or later, however, the impracticality of further maintenance or simple product obsolescence will call for major product renovation or replacement. The ideal design permits renovation and enhancement to be accomplished by changing a small number of subassemblies and recycling those that are replaced. Next best is a design that requires replacement of the product but permits many or most of the subassemblies to be recovered and recycled into new products. If subassemblies are unlikely to be reused as such, attempts should be made to design subassembly components for recovery and use through several product cycles. Usually the least desirable of the alternatives is complete disassembly followed by recovery of the separate materials in a product (or perhaps some of the embodied energy, if the product is best incinerated) and the injection of the materials or energy back into the industrial flow stream. Disposal of the product without the possibility of any of these recycling options is generally an unacceptable and unnecessary alternative from the industrial ecology viewpoint.

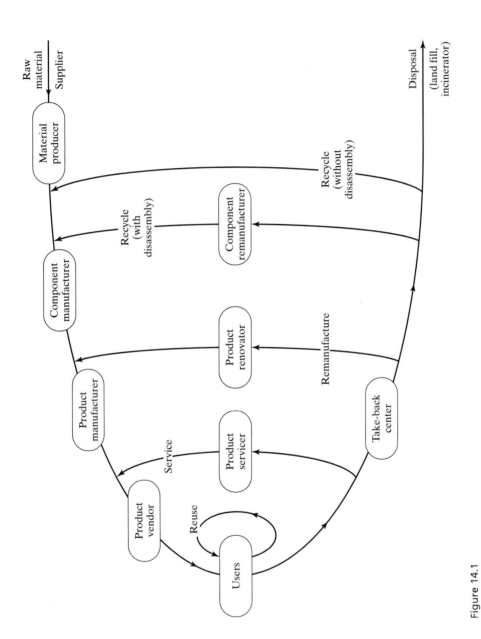

Figure 14.1

The "comet diagram", showing end of life reuse, refurbishment, and recycling strategies at different stages in the product life cycle. (Adapted from C.M. Rose, *Design for Environment: A Method for Formulating Product End-of-Life Strategies*, Ph.D. dissertation, Stanford University, 2000.)

The Automobile Recycling Sequence

Many aspects of Design for Recycling are illustrated by the recycling of automobiles and their components. This recycling occurs in several stages, each stage having its own actors. The sequence is shown in Figure 14.2. It begins with the transfer to a dismantler of a vehicle deemed no longer suitable for service. The dismantler removes components for which a market exists: usable body panels, the lead-acid battery, wheels and tires, radiator, alternator, and so forth. The remaining vehicle (the hulk) is then sold to the shredder operator. The shredding operation is accomplished by large machinery that chops the hulk into small pieces 10 cm or so in size and a kilogram or so in weight. These pieces are then sent through a variety of operations that produce three output streams: the "ferrous fraction" (iron, carbon steel, stainless steels), the nonferrous fraction (aluminum, zinc, copper), and the remainder, termed "automotive shredder residue" (ASR). The ASR is largely metal- and fluid-contaminated polymer materials. Each of the three output streams goes to a further actor in the recycling sequence, the ferrous fraction to a steel mill; the nonferrous fraction to a nonferrous separator, where the several metals are separated for resale; and the landfill operator, who receives the valueless ASR. In some countries and in some cases, the ASR undergoes pyrolysis for the recovery of energy.

The automobile recycling system is surprisingly efficient at recovering vehicles at their end of life and reusing at least some of their parts and materials. In more developed countries, about 95% of all vehicles are eventually involved in recycling, compared with an estimated 63% of aluminum cans, 30% of paper products, 20% of glass, and less than 10% of plastics. How does the automobile recovery system work, and what are its good and bad points?

The automobile recovery participant with the most intricate organizational operation is the dismantler, whose activities are shown in expanded form in Figure 14.3. The intricacy is due to the fact that the components the dismantler recovers from an old car are of value to a number of different industries and businesses, and/or that they present diffi-

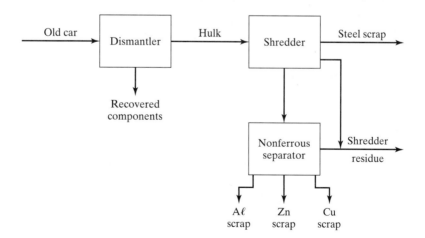

Figure 14.2

The automobile recycling sequence.

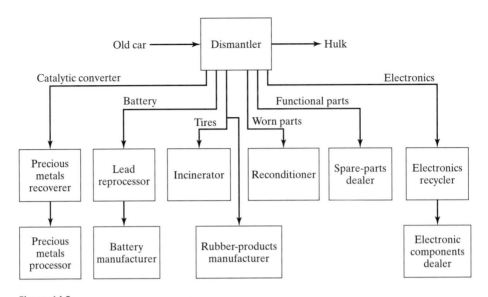

Figure 14.3

Flows of components from automobile dismantlers to component reconditioners, materials reprocessors, and energy recovery operations.

culties at later vehicle recycling stages. One of the first items recovered is the lead-acid battery. Sometimes the battery itself can be returned to a used-parts market, but generally it is of little value as such and is instead sold to a lead reprocessor, who extracts the lead and resells it to a battery manufacturer. A similar process occurs with the catalytic converter and with electronics components: Each go to specialists in the recovery of precious metals or other useful chips or parts, and the recovered materials move on to processors or dealers. Tires that are still road-worthy are reused. If not, they are recycled or incinerated.

Some components of old automobiles can be reused directly, and are immediately sold to spare-parts dealers: wheels, motors for power windows and seats, perhaps the radiator. Others are reusable after reconditioning: alternators, air conditioners, even entire engines. This is particularly important where the engines and components are no longer being manufactured, so recovered and reconditioned parts are the only effective way of keeping older vehicles in service. The wide availability of reconditioned components and the modular design of automobiles are major factors in extending the life of vehicles.

More specialized and more complex parts also see recycling activity. The recent development of automotive electronics has resulted as well in limited recycling of electronic components, sometimes for the components themselves, sometimes for the precious metals contained in them. The platinum group metals in catalytic converters are quite valuable, and constitute some 30–35% of all the use of these materials. As a result, they have been collected and recycled ever since converters became widespread.

Thus, depending on how far one wants to follow separated metals and resold parts, automobile recycling is a linked activity of between a dozen and two dozen independent participants, each with different roles, different technologies, and different mixes of automotive and nonautomotive business. This system has arisen spontaneously as a result of, and is maintained by, economic incentives, not regulatory fiat.

14.2   GENERAL END-OF-LIFE CONSIDERATIONS

A directly practical reason for all industries to practice design for recycling is the trend for governments and other consumers to require or give preference to products incorporating the DfR philosophy. In 1991, for example, U.S. Government Executive Order 12780 was promulgated. It requires all agencies of the government (when combined, the agencies are the country's largest consumer) to buy products made from recycled materials and to encourage suppliers to participate in residue recovery programs. In the same year, the State of New York issued a Request for Proposal for personal computers for its offices in which it stated that recyclability would be a factor in choosing the successful bidder. The German "Blue Angel" environmental seal routinely includes such requirements in its assessments. Such actions provide a graphic and easily communicated rationale for DfR.

Perhaps the most important consideration in DfR is to minimize the number of different materials and the number of individual components used in the design. (This design strategy is independently known as Design for Simplicity.) To get a sense of the importance of this recommendation, picture yourself responsible for recycling hundreds of television sets or photocopy machines or refrigerators every week. If you need to locate, sort, clean, and provide efficient recycling for two or three metals and two or three plastics, you are far more likely to be successful than if you must deal with five metals, four alloys, twelve plastics, and miscellaneous items such as glass or fabric. The functional and aesthetic demands of design sometimes make it difficult to limit materials diversity or complexity too greatly, but minimization should be a central focus for every designer.

A second general goal is to avoid the use of hazardous materials. This topic has been discussed earlier with respect to the extraction or manufacture of materials and their dissemination during industrial processes. The goal is equally important in the product recycling arena, where the presence of such materials is a deterrent to detailed disassembly, eventual reuse, or, if necessary, safe incineration and energy recovery. Where hazardous materials must be utilized in a design they should be easily identifiable, and the components that contain them readily separable.

Another general recommendation for the designer is not to join dissimilar materials in ways that make separation difficult. A simple example of a product not designed for recycling is the glass container for liquids whose top twists off while leaving a metal ring affixed; small cutting pliers are required for the conscientious housekeeper to properly sort the materials if the local recycling facility is unable to quickly and cheaply do so. More complex variations on this theme are metal coatings applied to plastic films, plastic molded over metal or over a dissimilar plastic, and the "upscale" automobile dashboard, which is a complex mixture of metal, wood, and plastic. Any time a designer uses dissimilar materials together, she or he should picture whether and how they can eventually be easily separated, an important concept since labor costs tend to be a significant barrier to recycling.

When planning for product end of life, two complementary types of recycling should be considered: *closed-loop* and *open-loop*. As seen in Figure 14.4, closed-loop

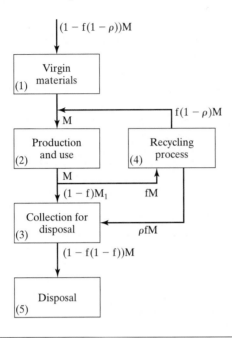

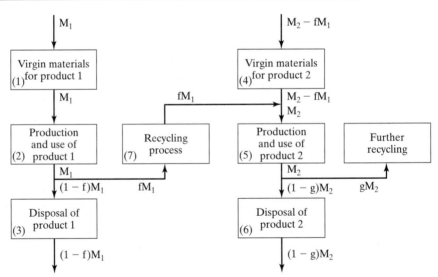

Figure 14.4

Closed-loop (top) and open-loop (bottom) recycling of materials. In the diagrams, M refers to mass flows, *f* and *g* to the fractions of the flows delivered to the recycling process, and ρ to the fraction of those flows rejected as unsuitable for recycling. (Adapted from B.W. Vigon, D.A. Tolle, B.W. Cornaby, H.C. Latham, C.L. Harrison, T.L. Boguski, R.G. Hunt, and J.D. Sellers, *Life-Cycle Assessment: Inventory Guidelines and Principles*, EPA/600/R-92/036, U.S. Environmental Protection Agency, Cincinnati, OH, 1992.)

recycling involves reuse of the materials to make the same product over again (sometimes called *horizontal recycling*), while open-loop recycling reuses materials to produce different products (sometimes called *cascade recycling*). (Typical examples could be aluminum cans to aluminum cans in the first instance, office paper to brown paper bags in the second.) The mode of recycling will depend on the materials and products involved, but closed-loop should generally be preferred.

14.3   REMANUFACTURING

Most products designed for long life do not wear out all at once: A mechanical part may fail, an essential fluid may leak out or become contaminated, or a critical component may become obsolete. As we noted in Figure 14.1, recycling should preferentially take place as high up the embodied resource chain as possible. An efficient way to accomplish this goal is by remanufacturing.

Remanufacturing involves the reuse of nonfunctional products by retaining serviceable parts, refurbishing usable parts, and introducing replacement components (either identical or upgraded). Such a process is frequently cost effective and almost always environmentally responsible. It requires close relationships between customer and supplier, frequently on a lease contract basis; these relationships are often competitive advantages in any case. Remanufacture requires thoughtful design, because the process is often made possible or impossible by the degree to which products can be readily disassembled and readily modified.

A general concept of Design for Recycling is to make designs modular. If a designer can anticipate that a certain portion of a design is likely to evolve or to need repair or replacement, while other portions of the product probably will not, the portion likely to change can be designed in modular fashion so that it can be efficiently replaced and recycled. The use of plug-in circuit boards in modern television sets is a good example of this philosophy. Another is NCR's Personal Image Processor, shown in Figure 14.5, which embodies a high level of modularity throughout.

14.4   RECYCLING

14.4.1   Metals

Pure metals are supremely recyclable, and many of them have historically been recycled to a very high degree. Recycling involves the reentry of metal scrap into the refining process, often after a purification step involving the removal of oxides and other corrosion products.

Metals recycling is complicated by the presence in scrap of noncompatible metals. If possible, therefore, the use of metals should emphasize a single metal or metal grouping in order to facilitate eventual recycling. This guideline is especially important if a small amount of a metal is used with a large amount of another metal, such as when steel is plated with cadmium. When the material is recycled, the plated metal is generally difficult and uneconomical to recover and tends to be discarded.

A second example is when steel automotive scrap mixed with copper wire is recycled. Copper impurities in steel substantially degrade its mechanical properties; aluminum wire is preferred if steel and an electrical wire are to be recycled together.

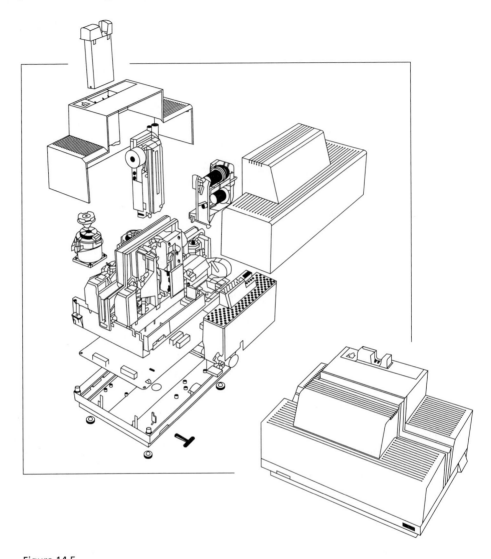

Figure 14.5

The NCR 7731 Personal Image Processor. This optical imaging product features modular design so that it can be adapted or refurbished to fit a customer's evolving needs without the necessity for total product disposal or replacement. Courtesy of A. Hamilton, NCR Canada Ltd.

The desirability of avoiding the mixing of materials streams is neatly illustrated in Figure 14.6, which shows that the selling prices of virgin materials vary approximately logarithmically with their degree of concentration in the matrix from which they are extracted. Also shown on the figure are points for a number of metals that are currently being extracted from residue streams rather than from virgin sources. In most cases, the residue streams are richer (i.e., less diluted) than virgin material matrices, so that efficient recycling operations can be expected to be financially rewarding.

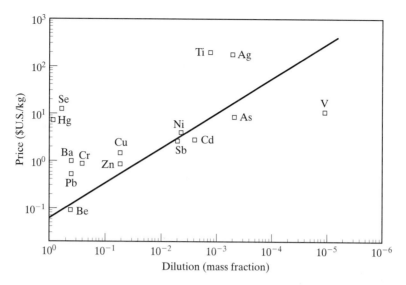

Figure 14.6

The relationship between materials dilution and price. The solid line shows the relationship between the price of virgin materials and their dilution in the matrices from which they are extracted. The individual points are for industrial residues undergoing recycling. (Reproduced with permission from D.T. Allen and N. Behmanesh, Preprint: Wastes as raw materials, in *The Greening of Industrial Ecosystems*, B.R. Allenby and D.J. Richards, eds., Washington, DC: National Academy Press, 1994.)

(Ironically, it is environmental regulations in many cases that make such desirable recycling uneconomical.) In the cases of mercury and cadmium, recycling is now being done because of toxic hazard and not because of economic viability; that viability could be enhanced by efforts to decrease the dilution of those materials in residue streams.

14.4.2 Plastics

Given careful attention to design and materials selection, many of the plastics in industrial use can be recycled. The most basic approach consists of cleaning, melting, and repelletizing. Next higher in complexity are depolymerization techniques that break down polymers to varying degrees. Most ambitious are processes that transform the polymer to its original constituents.

The type of plastic makes a great difference to potential recyclability. Thermoplastics can be ground, melted, and reformulated with relative efficiency. Among the thermoplastics for which recycling facilities now exist are polyethylene terephthalate (PET), polyvinyl chloride (PVC), polystyrene (PS), and the polyolefins (including high-density polyethylene [HDPE], low-density polyethylene [LDPE], and polypropylene [PP]). The utility of recycling these materials is a function of their purity, so the use of paint, flame retardants, and other additives should be minimized or avoided if at all possible. Having plastics of many different colors in a product line limits recyclability options as well. A further consideration is that designs should minimize the potential for plastics to become coated with oil or grease in use, since in-use adulteration also limits the efficacy of recycling.

Recycling is much more difficult for thermoset plastics, the group that includes phenolics, polyesters, epoxides, and silicones. Thermosets form cross-linked chemical bonds as they are created; recycling consists of reduction to lower molecular weight species by pyrolysis or hydrolysis. These processes are endothermic, however, and much of the embedded utility in the thermosets is thereby sacrificed. Incineration for energy recovery is preferable to landfilling, but represents a complete degradation of the material and is generally a recycling process of last resort.

When planning for recyclability of a product containing plastics, the thermal stability of the base resin and the other constituents must be considered, since reprocessing will involve heating of the material. It is usually the case that additives such as adhesives, paints, and coatings that are not removed prior to reprocessing will degrade in molding and extrusion machines. The result is outgassing that inhibits or prevents the reprocessing cycle. Another potential difficulty occurs when mixed polymers are recycled. Degraded mechanical properties of the resulting material often result, but each case must be evaluated individually; a very small polyethylene part on a polycarbonate housing is satisfactory, for example, since a small degree of impurity does not compromise polycarbonate recycling.

No matter how efficiently a plastic can be recycled if the relevant information concerning it is known, the multiplicity of plastics in use often makes it difficult to tell one from another, especially if recycling occurs a decade or more after product manufacture. To alleviate this problem, international standards have been developed for the marking of plastic parts. Although several versions have been promulgated, the most widely accepted is that of the International Organization for Standardization (ISO). A firm rule for physical designers is that no plastic part of significant size should be used without its ISO standard identity being marked upon it.

14.4.3 Forest Products

Of all the raw materials that humankind utilizes, the one that best approaches the cyclization goal of industrial ecology may be forest products: paper, cardboard, wood, and so forth. The raw material itself can be regenerated within a few decades, and the processed material is often utilized successfully for very long periods of time, as in housing timbers and frames. Following use, a significant fraction of forest products is recycled. The most highly developed recycling system is the several-stage process for paper. At each recycling stage, the fibers in the paper become shorter and the acceptable use is more restricted, a normal cycle being from white bond to colored bond to newspaper to grocery bags to toilet paper.

A significant limitation on forest product recycling occurs when the forest product materials are combined with dissimilar materials, such as by adding organic adhesives to wood chips or plastic coatings to hot drink cups, or by using metal-containing preservatives in timber. Most recycling systems find these materials difficult to handle, with the result that adulterated forest products are often landfilled rather than returned to the materials stream. Nonetheless, if coatings or preservatives render a material usable for a much longer time than would otherwise be the case, the overall environmental result may be positive, not negative.

14.5   FASTENING PARTS TOGETHER

The way in which parts are fastened together has a great effect on whether or not a product will be recycled after its useful life, independent of whether its materials were wisely selected. Poor fastening design can be enormously challenging for the recycler: Not only are a wide variety of types often used, but the quantities are staggering—275 for a refrigerator, 1000 for a fork-lift truck, 3500 for an automobile, 1,500,000 for a large airplane. The challenge to the designer is to create a product that is rugged and reliable during service, but easily dismantleable when it becomes obsolete. In general, if a product is designed to be easy to manufacture (i.e., a minimum of fasteners, commonality of fasteners, modular components and few of them, and so forth), much of the design for recycling has probably already been accomplished. Starting with that general advice, designers should be aware of the environmental advantages and disadvantages of different types of fastening techniques.

Screws are perhaps the simplest of the conventional ways to fasten components together. As few as possible should be used, and the number of different sizes and types should be minimized. (If a product is being assembled from modules produced by different suppliers, special attention to specifications is necessary to encourage uniformity of fasteners.) An especially difficult design approach from the recycling standpoint is the use of threaded metal inserts embedded into plastic; these generally require heating to dislodge the inserts, which even then often cannot be separated by the magnetic approaches commonly used in recycling facilities.

In place of more traditional fastening approaches, a variety of quick-release connectors are available. Clips and similar fasteners are generally made of metal, but an increasingly common approach is to use sheets of hook-and-loop fasteners, especially for large panels such as automobile interior roof linings. Hook-and-loop fasteners are available in both normal and industrial strengths, are secure in use, and are readily detached after use. A more precise technique, requiring very accurate machining, is to use parts that snap tightly together, perhaps with break-out inserts, thus avoiding the use of any fasteners at all.

Labels and corporate logos should preferably be embossed rather than affixed on products. If formed of a different material applied or inserted into a product, they should be readily removable.

Fastening techniques to avoid, if possible, are those that make it difficult and time-consuming to disassemble a product. Rivets fall in this class, as do nuts and bolts, and chemical bonds between similar or (worse) dissimilar plastics. Welds between metals are also very difficult to deal with in a recycling facility.

14.6   PLANNING FOR RECYCLABILITY

14.6.1   Design for Disassembly

Designing durable goods for disassembly may sound like something of an oxymoron, but it is being done. An example of Design for Disassembly is the teapot brought to market by Polymer Solutions, Inc., a joint venture of GE and Fitch Richardson Smith, which makes the injection molded parts in the United States and uses British heating elements and switches (Figure 14.7). Because the parts snap together, engineers found

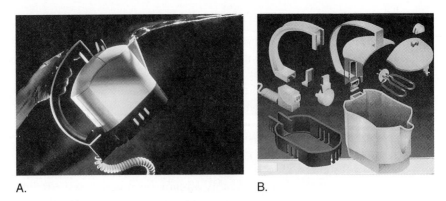

A.   B.

Figure 14.7

The teapot designed for disassembly by Polymer Solutions, Inc. for Great British Kettles Ltd.
(Photographs courtesy of Fitch, Inc., Columbus, OH.)

that the tolerance requirements were much more stringent than older manufacturing methods if leaking was to be prevented. In the short term, the enhanced tooling costs were a negative factor, but in the longer term they provided the capability to fabricate products with much more precise, and thus more desirable, properties.

There are two methods of disassembly. One is *reverse assembly*, in which screws are removed, snap-fit parts are unsnapped, and so forth. The second is *brute force*, in which parts are broken or cut apart. If parts are designed with the goal of rapid and efficient separation of components in mind, either option may be suitable. Otherwise successful design choices can often be disasters at disassembly time: Special tools that were needed to assemble specialized parts may not be available at disassembly, or inserts or coatings may contaminate otherwise usable materials after brute force disassembly. As with many other aspects of DfE, the simple and common is generally to be preferred to the exotic.

DfD is aided by convenient and effective tools. Among the most useful is the reverse fishbone diagram. The standard fishbone diagram, in common use in industrial engineering, is a graphical illustration of the sequence in which the components of a product are assembled from materials or lower level components, and the sequence in which the final product is assembled from the components. The reverse fishbone diagram is a picture of the ideal *dis*assembly process, showing the order of removal and separation of the parts. An example is shown in Figure 14.8. By constructing such a diagram, the engineer can often discover opportunities for increasing product recyclability while retaining other desirable product characteristics.

The demands of design for disassembly can often inspire great ingenuity, such as the effort by BMW to design a sports car with an all-plastic skin. The body panels are designed to be completely disassembled from the metal chassis in 20 minutes; they are of recyclable thermoplastic supplied by GE Plastics Corporation. An unexpected side benefit of this design is that it has proven much easier to repair, since damaged components can readily be removed and replaced.

The design for disassembly scenario can be appreciated by plotting the cost of different disposal options against the number of steps required for product disassem-

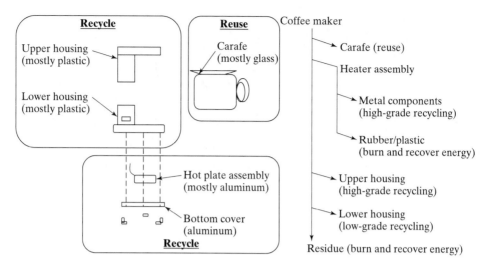

Figure 14.8

The reverse fishbone diagram for a coffee maker. (Courtesy of Kosuke Ishii, Stanford University.)

bly (Figure 14.9). If the product is to be landfilled, the highest cost generally occurs if no disassembly at all is performed, since the volume and difficulty of handling of the product is at the maximum. That cost decreases as some disassembly is performed, but levels off before many steps have occurred. A contrasting behavior results if disassembly and reuse of modules or materials is to be performed. The cost of doing so obviously increases with the number of disassembly steps, and does so rapidly as the remaining modules become smaller and disassembly becomes more difficult and time consuming. The designer can minimize end-of-life cost if the product can be more or less completely disassembled in only a few steps. Conversely, landfilling becomes the financially preferable option if many disassembly steps are required.

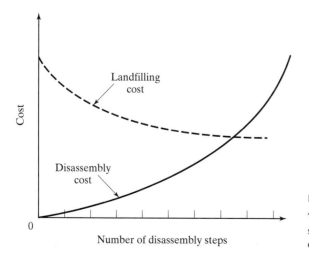

Figure 14.9

The conceptual relationship between the number of steps required for dissassembly of a product and the cost of end-of-life disposition options.

Identification of the materials from which a product has been made and understanding the functions of its modules and components is critical at disassembly time. This is not a particular concern if a firm receives its own products for recycling, but may be a major recycling roadblock in facilities dealing with the products of numerous industrial organizations. To alleviate this difficulty, Sony's technology center in Stuttgart, Germany has proposed that all products incorporate in their design a "green port," i.e., an electronic module that contains retrievable data in tamper-proof form. The module would be made to an industry-wide standard and addressable through a diagnostic connector. It seems likely that some variation of this idea will eventually be implemented, at least for relatively expensive and long-lived products.

14.6.2 Just-in-Case Designs

Among the more unusual topics that enter into design for environment are those concerning products designed in the hope and expectation that their use may seldom or never be required. Spare parts are an obvious example, but entire products can fit the definition as well. Examples include emergency equipment such as air packs or medical supplies, sprinkler systems for fire suppression, and backup safety devices for elevators. The range of technology in such equipment extends to the most sophisticated, as in intercontinental ballistic missiles with computer-guided control systems.

Just as society needs green products in its everyday activities, so too it needs green spares, green rarely used items, and (an intriguing oxymoron) green weapons. In other words, just because an item is designed with a reasonable expectation that it will sit around for a decade or two without being used and then be discarded, but must work the first time if needed, the designer of that item is not excused from the responsibilities of design for environment. Such features as materials selection, modularity, Design for Disassembly, and especially Design for Maintenance need to be given special attention for "just-in-case" products. This topic has received little attention to date, but the enormous inventory of materials and products involved suggests that the time for focused activity is long overdue.

14.6.3 Priorities for Recyclability

As final guidance to the designer, we reproduce a priority list of recycling options developed by the American Electronic Association.

> Usually most preferable
>> Reduce materials content
>> Reuse components/ refurbish assemblies
>> Remanufacture
>> Recycle materials
>> Incinerate for energy (if safe)
> Usually least preferable
>> Dispose of as waste

The impacts of the transportation involved in the recycling operation can be a major consideration when deciding on a recycling strategy. Even otherwise environmentally preferable activities, such as post-consumer product takeback systems, can founder if the environmental costs of transportation are not considered. Germany's packaging takeback system, for example, resulted in high transportation impacts per unit of packaging as a result of recycling of heavy glass bottles, carried out far from the locations where the residue was generated. A case-by-case analysis is generally required to determine the best solution to any individual situation; this analysis must consider the full range of impacts, the value of the recovered materials, and the alternatives to low-grade recycling or none at all. While deciding not to recycle a specific material or type of product seems intuitively wrong, a careful analysis will sometimes show it to be the best solution from an overall standpoint.

Thus, it is not automatically a sound policy to recycle products, and it is not automatically sensible to recycle only in a certain way. It makes sense to recycle only if the energy, environmental, and labor costs are such that recycling is preferable to not recycling. An example is that reusing or recycling beverage containers should be done only if it is easy and inexpensive to collect, transport, and reutilize the materials. Another example is that with old assemblies or equipment not designed for reuse or remanufacture, it is seldom possible to do more than recover the materials, and even that may be difficult and costly. Recycling decisions must be made from a logical perspective, therefore, and often that perspective requires a low-grade disposal of old, non-DfE items. As with other aspects of industrial ecology, tradeoffs are always present, and a comprehensive analysis is often needed to determine the most sensible approach to a specific problem. In particular, performing recycling should not result in a greater environmental impact than not performing recycling.

What to do with Old Tires?

Old tires are a good object lesson in disposal and reuse. The numbers of them are daunting—millions are discarded every year. For decades these tires were dumped in landfills and other less suitable places, but the tendency of tires to "float" in landfills and the feeling that there must be better alternatives are gradually changing that approach.

Retreading tires is useful in lengthening the service life, though it merely delays the inevitable. Upon eventual discarding, a fraction of today's old tires are sent to modern facilities that shred and separate them into three flow streams: small tire chunks, steel shards, and crumbs. The steel is readily recyclable. The crumbs are burned for energy (each tire contains more than eight liters of recoverable petroleum). The chunks see a variety of uses—for running tracks, rubber boots, and rubberized asphalt, to name a few.

Although adequate technology is becoming available for the various types of tire recycling, appropriate economics may not yet be in place. Tire disposal costs have never been internalized; unlike aluminum beverage cans, tire users do not currently pay a recyling deposit upon initial purchase. The result is that recycling happens only if it is profitable, and in many cases it is not. Part of the reason for this situation is that some legislation prohibits the use of tires as feedstock for incinerators (though the petroleum from which tires are made is an approved fuel), or makes it difficult to transport old tires into a city or across a state line to where a recycling facility is located. It is obvious that the recycling industry, economists, and politicians all have roles to play if the issue is to be properly addressed.

The tire design engineer also has a role to play: envisioning the end of life of the tires before they are manufactured. Today's recyclers are dealing with tires designed with no consideration for their eventual disposal. Perhaps a tire's composition can be changed to make it burn more efficiently while releasing few or no toxics. Perhaps a tire can be made so as to be more quickly and easily separated into its components. Perhaps a tire can be reformulated so as to be more readily transformed into a new product. In the case of old tires, the recycling engineers have made substantial progress and the economists and politicians are beginning to think about the situation. The design engineer has yet to play a significant role.

FURTHER READING

Craig, P.P., Energy limits on recycling, *Ecological Economics, 36*, 373–384, 2001.

Field, F.R. III, J.R. Ehrenfeld, D. Roos, and J.P. Clark, *Automobile Recycling Policy: Findings and Recommendations*, Cambridge, MA: Center for Technology, Policy, and Industrial Development, Massachusetts Institute of Technology, 1994.

Guide, V.D.R., Jr., and L.N. van Wassenhove, Closed-loop supply chains, in *A Handbook of Industrial Ecology*, R.U. Ayres and L.W. Ayres, eds., Cheltenham, U.K.: Edward Elgar Publishers, 497–509, 2002.

Henstock, M.E., *Design for Recyclability*, London: Institute of Metals, 1988.

Klausner, M., W.H. Grimm, and C. Hendrickson, Reuse of electric motors in consumer products, *Journal of Industrial Ecology, 2* (2), 89–102, 1998.

Lave, L.B., C. Hendrickson, and F.C. McMichael, Recycling decisions and green design, *Environmental Science & Technology, 28*, 19A–24A, 1994.

Lund, R.T., Remanufacturing, *Technology Review*, Feb/Mar., 19–28, 1984.

EXERCISES

14.1 In a closed-loop recycling system, the mass flow M is 5000 kg/hr, f is 0.7, and ρ is 0.1. Diagram the system and indicate all flow rates on the diagram.

14.2 In an open-loop recycling system, the mass flows M_1 and M_2 are 8000 kg/hr and 6000 kg/hr, f is 0.6, and g is 0.1. Diagram the system and indicate all flow rates on the diagram.

14.3 In the open-loop system of problem 14.2, assume that the recyling process rejects 15% of the material provided to it. Diagram this altered system and indicate on the diagram all the flow rates.

14.4 You are the designer of a table to be used for sorting fruit in a field near a cannery. The table is to have a steel surface and wooden legs. The surface is to be covered with a soft foam top to reduce fruit damage. It is expected that the foam top and the legs will need to be replaced periodically and the cannery owner, who expects to purchase several hundred tables, wants component replacement to be quick and efficient. With the help of your local hardware store (if needed), design the table for optimum disassembly.

CHAPTER 15

An Introduction
to Life-Cycle Assessment

15.1 THE LIFE CYCLE OF INDUSTRIAL PRODUCTS

A central tenet of industrial ecology is that of life-cycle assessment (LCA), introduced briefly in Chapter 8. The essence of LCA is the examination, identification, and evaluation of the relevant environmental implications of a material, process, product, or system across its life span from creation to disposal or, preferably, to recreation in the same or another useful form. The Society of Environmental Toxicology and Chemistry defines the LCA process as follows:

> The life-cycle assessment is an objective process to evaluate the environmental burdens associated with a product, process, or activity by identifying and quantifying energy and material usage and environmental releases, to assess the impact of those energy and material uses and releases on the environment, and to evaluate and implement opportunities to effect environmental improvements. The assessment includes the entire life cycle of the product, process or activity, encompassing extracting and processing raw materials; manufacturing, transportation, and distribution; use/reuse/maintenance; recycling; and final disposal.

The life-stage outline assumes that a corporation is manufacturing a final product for shipment and sale directly to a customer. Often, however, a corporation's products are intermediates—process chemicals, steel screws, brake systems—made for sale to and incorporation in the products of another firm. How does that concept apply in these circumstances?

Picture the process of manufacture as shown in Figure 15.1, an expansion of Figure 8.4. Three different types of manufacture are illustrated: (A) the produc-

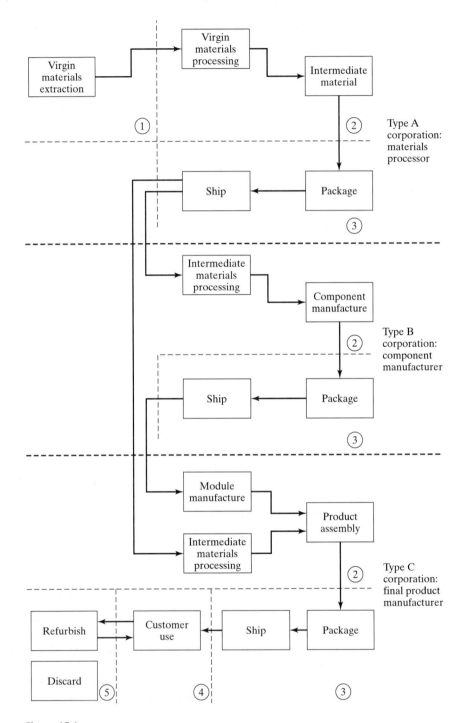

Figure 15.1

The interrelationships of product life stages for corporations of type A (materials processors), type B (component manufacturers), and type C (final product manufacturer).

tion of intermediate materials from raw materials (examples: plastic pellets from petroleum feedstock or rolls of paper from bales of recycled mixed paper); (B) the production of components from intermediate materials (examples: snap fasteners from steel stock or colored fabric from cotton); and (C) the processing of intermediate materials (example: cotton fabric) or the assembly of processed materials (example: plastic housings) into final products (examples: shirts or tape recorders).

Figure 8.4 is for an operation of type C, where the design and manufacturing team had virtually total control over all product life stages except stage 1: Premanufacture. For a corporation whose activities are of types A or B, the perspective changes for some life stages, but not for others:

Stage 1, Premanufacture. Unless a type A corporation is the actual materials extractor, the concept of this life stage is identical for corporations of types A, B, and C.

Stage 2, Manufacture. The concept of this life stage is identical for corporations of types A, B, and C.

Stage 3, Product Delivery. The concept of this life stage is identical for corporations of types A, B, and C.

Stage 4, Product Use. For type A corporations, product use is essentially controlled by the type B or C receiving corporation, though product properties such as intermediate materials purity or composition can influence such factors as by-product manufacture and residue generation. For type B corporations, their products can sometimes have direct influence on the in-use stage of the type C corporation final product, as with energy use by cooling fans or lubricant requirements for bearings.

Stage 5, Refurbishment, Recyling, or Disposal. The properties of intermediate materials manufactured by type A corporations can often determine the potential for recyclability of the final product. For example, a number of plastics are now formulated with the goal of optimizing recyclability. For type B corporations, the approach to the fifth life stage depends on the complexity of the component being manufactured. If it can be termed a component, such as a capacitor, the quantity and diversity of its materials and its structural complexity deserve review. If it can be termed a module, the concerns are the same as those for a manufacturer of a final product—ease of disassembly, potential for refurbishment, and the like.

Thus, type A and B corporations can and should deal with DfE assessments of their products, much as should type C corporations. The considerations of the first three life stages are, in principle, completely under their control. For the last two life stages, the products of type A and B corporations are influenced by the type C corporation with which they deal and, in turn, their products influence the life stage 4 and 5 characteristics of type C products.

15.2   THE LCA FRAMEWORK

A life-cycle assessment can be a large and complex effort, and there are many variations. Nonetheless, there is general agreement on the formal structure of LCA, which contains three stages: *goal and scope definition, inventory analysis*, and *impact analysis*, each stage being followed by *interpretation of results*. The concept is pictured in Figure 15.2. First, the goal and scope of the LCA are defined. An inventory analysis and an impact analysis are then performed. The interpretation of results at each stage guides an analysis of potential improvements (which may feed back to influence any of the stages, so that the entire process is iterative). Finally, the Design for Environment guidance to the technological activity is released.

There is perhaps no more critical step in beginning an LCA evaluation than to define as precisely as possible the evaluation's scope: What materials, processes, or products are to be considered, and how broadly will alternatives be defined? Consider, for example, the question of releases of chlorinated solvents during a typical dry cleaning process. The purpose of the analysis is to reduce environmental impacts. The scope of the analysis, however, must be defined clearly. If it is limited, the scope might encompass only good housekeeping techniques, end-of-pipe controls, administrative procedures, and process changes. Alternative materials—in this case, solvents—might be considered as well. If, however, the scope is defined broadly, it could include alternative service options: Some data indicate that a substantial number of items are sent to dry cleaning establishments not for cleaning per se but simply for pressing. Accordingly, offering an independent pressing service might reduce emissions considerably. One could also take a systems view of the problem: Given what we know about polymers and fibers, why are clothing materials and designs still being provided that require the use of chlorinated solvents for cleaning? Among the considerations that would influence the choice of scope in cases such as the above are: (a) who is perform-

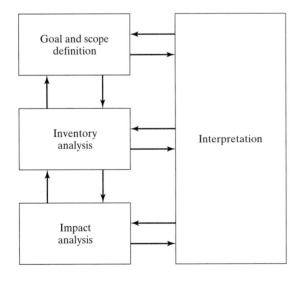

Figure 15.2

Stages in the life cycle assessment of a technological activity. The arrows indicate the basic flow of information. At each stage, results are interpreted, thus providing the possibility of revising the environmental attributes of the activity being assessed. (Adapted from Society of Environmental Toxicology and Chemistry (SETAC), *Guidelines for Life-Cycle Assessment: A Code of Practice*, Pensacola, FL, 1993.)

ing the analysis, and how much control they can exercise over the implementation of options; (b) what resources are available to conduct the study; and (c) what is the most limited scope of analysis that still provides for adequate consideration of the systems aspects of the problem?

The resources that can be applied to the analysis should also be assessed. Most traditional LCA methodologies provide the potential for essentially open-ended data collection—and, therefore, virtually unlimited expenditure of resources. As a general rule, the depth of analysis should be keyed to the degrees of freedom available to make meaningful choices among options, and to the importance of the environmental or technological issues leading to the evaluation. For example, an analysis of using different plastics in the body of a currently marketed portable disk player would probably not require a complex analysis: The degrees of freedom available to a designer in such a situation are already quite limited because of the constraints imposed by the existing design and its market niche. On the other hand, a government regulatory organization contemplating limitations on a material used in large amounts in numerous and diverse manufacturing applications would want to conduct a fairly comprehensive analysis, because the degrees of freedom involved in finding substitutes could be quite numerous and the environmental impacts of substitutes implemented widely throughout the economy could be significant.

The second component of LCA, inventory analysis (sometimes termed LCIA), is by far the best developed. It uses quantitative data to establish the levels and types of energy and materials used in an industrial system and the environmental releases that result, as shown schematically in Figure 15.3. The approach is based on the idea of a

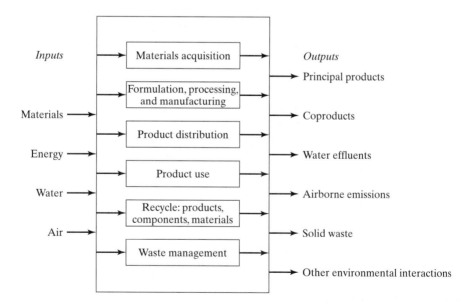

Figure 15.3

The elements of a life cycle inventory analysis. (Adapted from Society of Environmental Toxicology and Chemistry (SETAC), *A Technical Framework for Life-Cycle Assessment*, Washington, DC, 1991.)

family of materials budgets, in which the analyst measures the inputs and outputs of energy and resources. The assessment is done over the entire life cycle.

The third stage in LCA, the impact analysis, involves relating the outputs of the system to the impacts on the external world into which those outputs flow, or, at least to the burdens being placed on the external world. Aspects of this difficult and potentially contentious topic are discussed in the next chapter.

The interpretation of results phase is where the findings from one or more of the three stages are used to draw conclusions and recommendations. The output from this activity is often the explication of needs and opportunities for reducing environmental impacts as a result of industrial activities being performed or contemplated. It follows ideally from the completion of stages one through three, and occurs in two forms: (1) the proactive DfE, and (2) the reactive Pollution Prevention. Less comprehensive but still valuable actions can also be taken by interpreting the results of the scoping and inventory stages.

15.3   GOAL SETTING AND SCOPE DETERMINATION

A common LCA goal is to conduct an assessment of the environmental attributes of a specific product or process and to derive information from that assessment on how to improve environmental performance. If this exercise is conducted early in the design phase, the goal may be to compare two or three alternative designs. If the design is finalized, or the product is in manufacture, or the process is in operation, the goal can probably be no more than to achieve modest changes in environmental attributes at minimal cost and minimal disruption of the existing operation.

It is possible for an assessment target to be much more ambitious than the evaluation of a single product or process. This usually occurs with the evaluation of a system of some sort, the operation of an entire facility or corporation, for example, or of an entire governmental entity. In such a case, it is likely that alternative operational approaches can be studied, but not alternative systems. In addition, a system that makes a logical entity from an LCA viewpoint may involve more than one implementer, so collaborative goal setting may be required. If a goal can be quantified, such as "achieve a 20 percent decrease in overall environmental impact," it is likely to be more useful and the result more easily evaluated than with qualitative goals. Quantification of the goal requires quantification of each assessment step, however, and quantitative goals should be adopted only when one is certain that adequate data and assessment tools are available.

The scope of the assessment is perhaps best established by asking a number of questions: "Why is the study being conducted?" "How will the results be used, and who will use them?" "Do specific environmental issues need to be addressed?" "What level of detail will be needed?" It is useful to recognize that life-cycle assessment is an iterative process, and that the scope may need to be revisited as the LCA proceeds.

15.4   DEFINING BOUNDARIES

The potential complexity of comprehensive LCAs is nowhere better illustrated than by the problem of defining the boundaries of the study. There are many potential issues for discussion in this regard, and no consensus on the best ways of approach. The discussion that follows explores a number of these issues, and concludes with some general recommendations concerning choices of boundaries in LCA.

15.4.1 Life Stage Boundaries

Early attempts to evaluate industrially related environmental impacts focused exclusively on activities within the manufacturing facilities themselves. From the life-cycle perspective, these approaches can be regarded as being restricted to life stage 2, or what today is sometimes termed *gate-to-gate analysis*, i.e., from the factory gate through which materials enter to the gate where products leave. Traditional pollution prevention activities have similar limitations.

The environmental performance of some products during their use became an issue in the late 1980s. Emissions from automobile exhaust began to be widely controlled, and regulations or guidelines on energy use by appliances and office machines began to be issued. Manufacturers were thus encouraged to think about the environmentally related aspects of life stage 4.

In the early 1990s, Germany implemented regulations requiring manufacturers to take back the packaging used for their products: boxes, cushioning foam, plastic wrap, and so on. This encouraged manufacturers to minimize packaging and make it more recyclable, in effect adding life stage 3 to corporate environmental assessment.

Life stage 5 has also begun to come under consideration. Several European countries have "takeback" laws or agreements that compel manufacturers to recover their products when the customer no longer wants them. Some manufacturers are discovering that such recovery, followed by refurbishment and reuse, can be profitable. These actions in turn encourage design and materials selection decisions that optimize the value of recovered products. The result is the incorporation of life stage 5 in environmental planning.

Life stage 1, premanufacture, has proven challenging to address, since that life stage is outside the direct control of the manufacturer. A number of corporations, however, are working with their suppliers on topics such as materials selection and packaging, which influence waste at the manufacturing plant and the hazard characteristics of manufactured products. These topics can readily be incorporated into the life stage 1 component of an LCA.

15.4.2 Level of Detail Boundaries

How much detail should be included in an LCA? An assessor frequently needs to decide whether effort should be expended to characterize the environmental impacts of trace constituents, such as minor additives in a plastic formulation or small brass components in a large steel assembly. With some modern technological products containing hundreds of materials and thousands of parts, this is far from a trivial decision. One way it is sometimes approached is by the *5% rule*: If a material or component comprises less than 5% by weight of the product, it is neglected in the LCA. A common amendment to this rule is to include any component with particularly severe environmental impacts. For example, the lead-acid battery in an automobile weighs less than 5% of the vehicle, but the toxicity of lead makes the battery's inclusion reasonable. Potential items for inclusion are mercury relays, chrome-plated parts, and radioactive materials.

15.4.3 The Natural Ecosystem Boundary

In a number of industrial processes, nature's processes interact with those of the technological society. Consider the generation of electric power by burning wood, as shown in Figure 15.4. The components that are obviously industrial are the harvesting of the wood (using fuel oils for cutting and transport) and the combustion process itself. The natural components (perhaps aided by human action) are the formation of the wood biomass (i.e., the growing of trees) and the biodegradation of the harvesting wastes. Some LCA assessors would choose to draw the assessment boundary around only the industrial components of the process, while others would include the natural components as well. The latter is more complete, but the inclusion of the natural components is likely to make quantification considerably more difficult.

A second natural ecosystem issue that arises when choosing LCA boundaries is that of biological degradation. When industrial materials are discarded, as into a landfill, biodegradation produces such outflows as methane from paper, chlorofluorocarbons from blown foam packaging, and mobilized copper, iron, and zinc from bulk metals. LCA approaches to these complications have included incorporating these flows in the inventory, excluding landfill outflows completely, or including those flows for a specific time period only. Flows from landfills are generally difficult to estimate, so one is faced with a tradeoff between comprehensiveness and tractability.

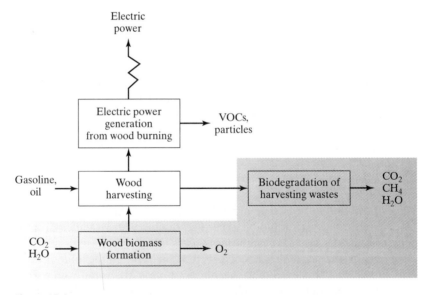

Figure 15.4

Electric power generation by wood burning. The boundary for the LCA might be chosen to include the shaded areas or to exclude them. The implications of this choice are discussed in the text. (Adapted from L.-G. Lindfors, et al., *Technical Report No. 4*, Tema Nord 1995. 502. Copyright 1994 by the Nordic Council of Ministers.)

A third example of the natural/industrial boundary issue is the process of making paper from wood biomass, as shown in Figure 15.5. Here the assessor has several possible levels of inventory detail to choose from. The basic analysis is essentially a restriction of the inventory to life stage 2. The energy envelope incorporates some of the external flows related to the production of energy. The extended envelope includes all life-cycle stages and flows directly connected with the industrial system. The comprehensive envelope adds the natural processes of biomass formation and the degradation of materials in a landfill. None of these options is inherently correct or incorrect, but the choice that is made could determine the LCA results.

15.4.4   Boundaries in Space and Time

A characteristic of environmental impacts is that their effects can occur over a very wide range of spatial and temporal scales. The emission of large soot particles affects a local area, those of oxides of nitrogen generate acid rain over hundreds of kilometers, and those of carbon dioxide influence the planetary energy budget. Similarly, emissions causing photochemical smog have a temporal influence of only a day or two, the disruption of an ecosystem several decades, and the stimulation of global climate change several centuries. LCA boundaries may be placed at short times and small distances, long times and planetary distances, or somewhere in between. The choice of any of these boundary options in space and time may be appropriate depending on the scope of the LCA.

15.4.5   Choosing Boundaries

It should be apparent that the choice of LCA boundaries can have enormous influence on the time scale, cost, results, meaningfulness, and tractability of the LCA. The best guidance that can be given is that the boundaries should be consistent with the goals of the exercise. An LCA for a portable radio would be unlikely to have goals that encompass impacts related to energy extraction, for example, both because the product is not large and because its energy impacts will doubtless be very modest. A national study focusing on flows of a particular raw material might have a much more comprehensive goal, however, and boundaries would be drawn more broadly. The goals of the LCA thus define much of the LCA scope, as well as the depth of the inventory and impact analyses.

15.5   APPROACHES TO DATA ACQUISITION

Once the scope of the LCA assessment has been established, the analyst proceeds to the acquisition of the necessary data. The process is begun by constructing, in cooperation with the design and manufacturing team, a DfE inventory flow diagram. The aim is to list, at least qualitatively but preferably quantitatively, all inputs and outputs of materials and energy throughout all life stages. Figure 9.1 shows an example of such a diagram for the manufacture of a desktop telephone in which the housing is molded in the plant from precolored resin, the electronics boards are constructed from components furnished by suppliers, and those parts and others (microphone, electronic jacks, batteries, etc.) are assembled into the final product. The diagram indicates a number of

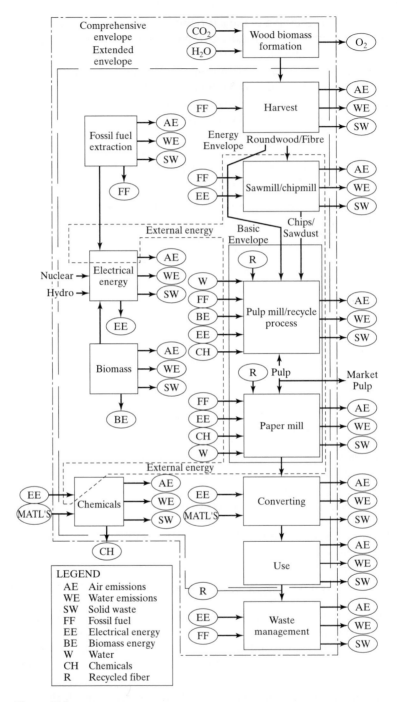

Figure 15.5

A simplified quantitative inventory flow diagram for the manufacture of paper. Four levels of possible detail are shown. (Adapted from a diagram provided by Martin Hocking, University of British Columbia.)

material and energy byproducts (the latter being mostly unused heat). Once the inventory flow diagram is constructed, in as much detail as possible, the actual inventory analysis can begin.

Some of the information needed for an inventory analysis is straightforward, such as the amounts of specific materials needed for a given design or the amount of cooling water needed by a particular manufacturing process. Quantitative data obviously have advantages: They are widely utilized in high technology cultures, they offer powerful means of manipulating and ordering information, and they simplify choosing among options. However, the state of information in the environmental sciences may not permit the secure quantification of environmental and social impacts because of fundamental data and methodological deficiencies. The result of inappropriate quantification might be that those concerns that cannot be quantified would simply be ignored—thereby undercutting the systemic approach inherent in the LCA concept.

In some cases, the information that is available may not be quantitative, but still useful. Accordingly, the analyst should be willing to approach data needs with a broad perspective. Qualitative information, whether it applies to materials selection, processes, components, or complex products, can often be as valuable as quantitative data. A qualitative approach can be somewhat controversial, especially among engineers and business planners who are biased toward quantitative systems, but its utility is great enough so that its incorporation in some form is generally worthwhile.

In order to maximize efficiency and innovation and avoid prejudgment of normative issues, an LCA information system should be nonprescriptive. It should provide information that can be used by individual designers and decision makers given the particular constraints and opportunities they face, but should not, at early stages of the analysis, arbitrarily exclude possible design options. In some cases, the use of highly toxic materials might be a legitimate design choice—and an environmentally preferable choice from among the alternatives—where the process designer can adopt appropriate engineering controls. In others, a process choice involving the use of substantial amounts of lead might require only modest amounts of energy use and thus be responsible for modest amounts of CO_2 emissions. The alternative might be less lead, more energy use, and more CO_2 emissions. If the toxic lead can be well contained, the first option may be preferable. Designing products and processes inherently requires balancing such considerations and constraints, and the necessary tradeoffs can only be made on a case-by-case basis during the product realization process.

In the ideal case, it should be possible to mathematically combine LCIA data at different hierarchical levels, e.g., to combine LCIA information for copper wire with that for PVC plastic to get an LCIA result for plastic-insulated copper wire. In practice, however, differences in scope, time scale, etc. generally require that every LCIA start from the beginning. This obvious deficiency in the methodology emphasizes that LCIAs are works in progress and not finished tools.

LCA information should provide not only relevant data but, if possible, also the degree of uncertainty associated with that data. This approach is particularly important in the environmental area, where uncertainty, especially about risks, potential costs, and potential natural system responses to forcing, is endemic. Often the relatively simple ordinal indicators "high reliability," "moderate reliability," and "low reliability" will be of substantial use to those actually making design decisions. An example of this approach is given in Appendix A, where alternative solder systems are evaluated in a matrix-based LCA.

Surfactants

Surfactants are constituents of detergents that aid in releasing soil from clothing, linens, and other items. Various surfactants can be used, and each is made from various sources of raw materials. To evaluate whether some options for surfactant sourcing and production are preferable to others, Procter & Gamble Company, in cooperation with Franklin Associates, performed an extensive life-cycle inventory study.

To begin the study, production flow diagrams from raw materials to products were developed. An example using palm fruit as the principal raw material is shown in Figure 15.6. Crude oil and natural gas are also needed to synthesize the desired product, a family of alcohol ethoxylates (AE). For each flow of materials, the mass requirements were measured and emissions (not shown on the diagram) were determined. Similar assessments were made for the production of surfactants manufactured from petrochemicals, palm kernel oil, and tallow. A selection of the results is shown in Table 15.1.

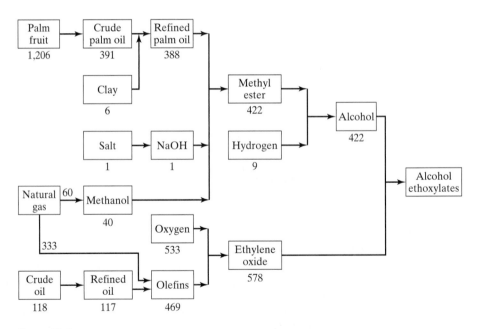

Figure 15.6

The flow of materials for the production of detergent surfactants from palm oil. Mass requirements (kg) are expressed on the basis of 1000 kg alcohol ethoxylate produced. (Reproduced with permission from C.A. Pittinger, J.S. Sellers, D.C. Janzen, D.G. Koch, T.M. Rothgeb, and M.L. Hunnicutt, Environmental life-cycle inventory of detergent-grade surfactant sourcing and production, *Journal of the American Oil Chemists' Society, 70*, 1–15. Copyright 1993 by the American Oil Chemists' Society.)

TABLE 15.1   Raw Material Requirements, Energy Requirements, and Net Life-Cycle Emission for AE Production

	Feedstock		
Flow stream	Petrochemical	Palm oil	Tallow
Organic raw materials*	935	899	891
Water use*	40	49	415
Energy use**			
Raw materials	50	26	27
Transport	3	5	7
Processing	39	37	40
Atmospheric emissions[†]			
Particles	2.2	9.0	8.0
Hydrocarbons	39.2	33.2	34.8
Aqueous emissions[†]			
Dissolved solids	5.6	5.3	4.3
Oil	0.065	0.12	0.30
Land-applied emissions[†]	76	111	139

*kg per 1000 kg of raw materials

**GJ of energy per 1000 kg of raw materials

[†] kg of emissions per 1000 kg of surfactant

Abstracted from C.A. Pittinger, J.S. Sellers, D.C. Janzen, D.G. Koch, T.M. Rothgeb, and M.L. Hunnicutt, Environmental life-cycle inventory of detergent-grade surfactant sourcing and production, *Journal of the American Oil Chemists' Society, 70*, 1–15, 1993.

The results show more similarities than differences, but the differences are worth comment. One is the much larger water consumption in the case of tallow, attributable to the crop irrigation required to provide feed for beef cattle (beef cattle are the primary source of tallow). The energy requirements for petrochemical recovery are higher than those for palm oil or tallow. Particulates and land-applied emissions from the petrochemical feed stock are significantly lower than for the other options. The conclusion of the study was that benefits from one process appeared offset by liabilities in another, and environmental concerns did not support any fundamental shifts in the worldwide mix of feedstocks used for surfactant manufacture.

The implications of studies such as this are much larger than which feedstock is preferable. Indeed, they have potential application to all industrial operations, both those dealing with components entering the manufacturing operation and with products leaving the manufacturing cycle. In practice, they lend encouragement to pollution prevention. However, the feedstock assessment and the spectrum of pollution prevention activities do not face the really crucial issue in implementing LCA in an industrial setting: Assigning relative impact values to each of the material flow and energy use comparisons made in this analysis. Without such assignments, one is implicitly assuming that the emission of a kilogram of wastewater from a facility is no more and no less important than is the emission of a kilogram of ozone-depleting gas or a kilogram of toxic solid waste. Obviously, all environmental impacts are not equal, and attempting to evaluate them on a relative basis brings regulatory, legal, environmental, corporate, and social factors into play. This crucial LCA step is that of impact analysis, the subject of the next chapter.

FURTHER READING

Introduction and Overviews

Curran, M.A., *Environmental Life-Cycle Assessment*, New York: McGraw-Hill, 1996.

Guinee, J., et al., *Handbook on Life Cycle Assessment—Operational Guide to the ISO Standards*, Dordrecht, the Netherlands: Klewer Academic Publishers, 2002.

Lindfors, L.-G. et al., *Nordic Guidelines for Life Cycle Assessment*. Copenhagen, Denmark: Nordic Council of Ministers, Report Nord 20, 1995.

U.S. Environmental Protection Agency, *LCA101,* Cincinnati, OH, 2000. Available on *http://www.epa.gov/ORD/NRMRL/lcaccess/lca101.htm*, accessed Jan. 9, 2002.

Inventory Case Study

Pittinger, C.A., J.S. Sellers, D.C. Janzen, D.G. Koch, T.M. Rothgeb, and M.L. Hunnicutt, Environmental life-cycle inventory of detergent-grade surfactant sourcing and production, *Journal of the American Oil Chemists' Society, 70,* 1–15, 1993.

EXERCISES

15.1 Choose a common but sophisticated household appliance such as a refrigerator, a television set, or a washing machine. Describe the life stages of the appliance, including identifying who is primarily responsible for the environmental concerns at each life stage.

15.2 You are the LCA analyst for a papermaking company and are asked to do an LCA for a new type of paper to be used for printing currency. Define and describe the scope of your assessment.

15.3 Repeat exercise 15.2 for the situation in which you work for a forest products company that supplies wood fiber for the paper.

15.4 Choose one of the following products: a bar of soap, a bicycle, a car wash, an ocean cargo ship. For each, construct a materials and energy flow diagram of the type of Figure 9.1, but with minimal detail in the manufacturing stage and enhanced detail in the product use stage. Suggest appropriate actions by the design engineer for each of those items.

CHAPTER 16

The LCA Impact and Interpretation Stages

16.1   LCA IMPACT ANALYSIS

The previous chapter discussed the component of LCA termed inventory analysis. Quantitative information on materials and energy flows is acquired at that stage in some cases, qualitative information in others. The data presentations in the previous chapter made it obvious that some aspects of life cycle analysis had the potential to be more problematical than others, but the approach begged the question of priorities. One could easily foresee a situation where alternative designs for a product or process each had similar materials use rates, but used different materials. How does the analyst make a rational, defensible decision among such alternatives? The answer is that (1) the influences of the activities revealed by the LCA inventory analysis on specific environmental properties must be accurately assessed, and (2) the relative seriousness of changes in the affected environmental properties must be given a priority ranking. Together, these steps constitute LCA's impact assessment.

Assessing environmental influences is a complicated procedure, but it can, in principle at least, be performed by employing stressors, which are items identified in the inventory analysis that have the potential to produce changes in environmental properties (e.g., the generation of carbon dioxide as a result of energy use). The relationships among stressors and the environment are developed by the environmental science community, and are not always available with the degree of detail and precision needed in the LCA. Ideally, however, the needed stressor relationships will be established and available for use. By combining LCA inventory results with these relationships, a manufacturing process might be found, for example, to have a minimal impact on local water quality, a modest impact on regional smog, and a substantial impact on stratospheric ozone depletion. The procedure that results in these conclusions is structured in several steps, as follows:

Classification. Classification begins with the raw data on flows of materials and energy from the inventory analysis. Given those data, the classification step consists of identifying environmental concerns suggested by the inventory analysis flows. For example, emissions from an industrial process using a petroleum feedstock may be known to include methane, butene, and formaldehyde. Classification assigns the first to global warming, the second to smog formation, the third to human toxicity.

Characterization. Characterization is the process of combining different stressor-impact relationships into a common framework. An example is the use of ozone depletion potentials, in which the effects of a molecule of one substance on stratospheric ozone are compared quantitatively with those of another.

Localization. Localization is the operation of prescribing environmental impacts occurring in different geographical regions that posses different characteristics. For example, the process of localization attempts to compare the emission of 1 kg of moderately toxic material into a pristine ecosystem with the impact of the same emission into a highly polluted ecosystem. Two considerations are involved. The first is the relationship of emissions from the product or process being assessed relative to all similar emissions in the region. The second is the degree to which the region possesses assimilative capacity for the emittant. Examples of approaches to this step are provided later in this chapter. (Localization has sometimes been titled normalization, but the word localization is more descriptive. It is also less confusing, because the common mathematical process of normalizing data to different scales is sometimes used in LCA.)

Valuation. Valuation is the process of assigning weighting factors to the different impact categories based on their perceived relative importance as set by social consensus. For example, an assessor or an international standards organization might choose to regard ozone depletion impacts as twice as important as loss of visibility and apply weighting factors to the normalized impacts accordingly.

Although it has been much more common for environmental scientists to evaluate impacts individually rather to attempt to rank them in priority order, several different approaches to prioritization have been made, and the rank-ordering of environmental problems that emerges from these efforts is discussed below. The purpose of the present chapter is not to present or defend specific ordering of impacts, but to illustrate the process.

16.2   INDUSTRIAL PRIORITIZATION: THE IVL/VOLVO EPS SYSTEM

An initial attempt to establish a more structural basis for assessing the environmental responsibility of an individual manufacturing process was that of Roger Sheldon of the Delft (Netherlands) Institute of Technology, who proposed in the context of the synthesis of organic chemicals the atom utilization concept (AU), calculated by dividing the molecular weight of the desired product by that of the sum total of all products and residues produced. Enlarging on this concept, and realizing that the nature of the residue is important as well as its amount, he proposed the environmental quotient (*EQ*), given by

$$EQ = AU \cdot U \tag{16.1}$$

where *U* is the "unfriendliness quotient," a measure of toxicity.

Sheldon offers no advice on how to assign unfriendliness quotients, but this difficult problem has not gone unaddressed. To begin to deal with something like it in a formal way, the Swedish Environmental Institute (IVL) and the Volvo Car Corporation have developed an analytic tool designated the Environment Priority Strategies for Product Design (EPS) system. The goal of the EPS system is to allow product designers to select components and subassemblies that minimize environmental impact. Analytically, the EPS system is quite straightforward, though detailed. An environmental index is assigned to each type of material used in automobile manufacture. Different stages of the index account for the environmental impact of this material during product manufacture, use, and disposal. The three life-cycle stage components are summed to obtain the overall index for a material, expressed in environmental load units (ELUs) per kilogram of material used (ELU/kg). The units may sometimes vary, however. For example, the index for a paint used on the car's exterior is expressed in ELU/m².

When calculating the components of the environmental index, the following factors are included:

Scope: A measure of the general environmental impact

Distribution: The size or composition of the affected environmental area

Frequency or intensity: Extent of the impact in the affected environmental area

Durability: Persistence of the impact

Contribution: Significance of impact from 1 kg of material in relation to total effect

Remediability: Cost to remediate impact from 1 kg of material

These factors are calculated by a team of environmental scientists, ecologists, and materials specialists so as to obtain environmental indices for every applicable raw material and energy source, together with their associated pollutant emissions. A selection of the results is given in Table 16.1, where a few features are of particular interest. One is the very high values for platinum and rhodium in the raw materials listings, a result of the high extraction energy required for these two metals. The use of CFC-11 is given a high environmental index as well because of its effects on stratospheric ozone and global warming. Finally, the assumption is made that the metals are emitted in a mobilizable form. To the extent that an inert form is emitted, the environmental index may need to be revised.

The environmental indices are multiplied by the magnitudes of material used to obtain environmental load units for processes and finished products. The entire procedure is schematized in Figure 16.1. Note that it includes the effects of materials extraction, emissions, and the impacts of manufacturing, and follows the product in its various aspects through the entire life cycle.

TABLE 16.1   A Selection of Environmental Indices (ELU/kg)*

Raw materials		Emissions—air		Emissions—water	
Co	256	CO_2	0.108	BOD	0.002
Cr	84.9	CO	0.331	Phosphorus	0.055
Fe	0.96	NO_x	2.13		
Mn	5.64	N_2O	38.3		
Mo	2120	SO_x	3.27		
Ni	160	CFC-11	541		
Pb	175	HCFC-22	194		
Pt	7,430,000	CH_4	2.72		
Rh	4,950,000				
Sn	1190				
V	56				

*From a compilation in B. Steen, *A Systematic Approach to Environmental Priority Strategies in Product Development (EPS). Version 2000—General System Characteristics*, CPM Report 1999:5, Chalmers University, Göteborg, Sweden, 1999.

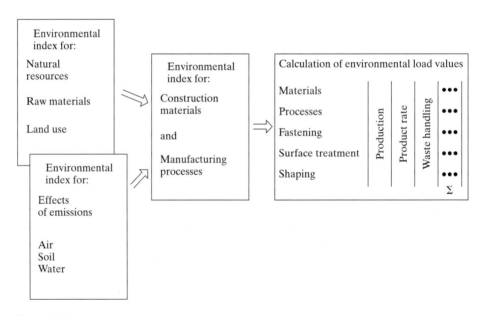

Figure 16.1

An overview of the EPS system, showing how the calculation of summed environmental load values proceeds. (Reproduced with permission from B. Steen, and S. Ryding, *The EPA Enviro-Accounting Method: An Application of Environmental Accounting Principles for Evaluation and Valuation of Environmental Impact in Product Design.* Copyright 1992 by Swedish Environmental Research Institute [IVL].)

GMT—composite
Material consumption: 4.0 kg
(0.3 kg scrap)
Component weight: 3.7 kg

Galvanized steel
Material consumption: 9.0 kg
(3.0 kg scrap)
Component weight: 6.0 kg
Painted area: 0.6 m^2

Figure 16.2

Design options for automotive front ends. (Courtesy of I. Horkeby, Volvo Car Corporation.)

As an example of the use of the EPS system, consider the problem of choosing the more environmentally responsible material to use in fabricating the front end of an automobile. As shown in Figure 16.2, two options are customarily available: galvanized steel and polymer composite (glass-mat thermoplastic, or GMT). The front ends are assumed to be of comparable durability, though differing durabilities could potentially be incorporated into EPS.

Based on the amount of each material required, the environmental indices are used to calculate environmental load values at each stage of the product life cycle. Table 16.2 illustrates the total life cycle ELUs for the two front ends. All environmental impacts, from the energy required to produce a material to the energy recovered from incineration or reuse at the end of product life, are incorporated into the ELU calculation. To put the table in the LCA perspective, the kg columns are the results of LCA stage one, the ELU/kg columns are the environmental indices, and the ELU columns are the results of LCA stage two, the latter being given by

$$ELU = \Sigma_i \Sigma_s (ELU/kg)_i M_{i,s} \qquad (16.2)$$

where i indicates the type of material, s the life stage, and M is the mass of material i at life stage s. There are several features of interest in the results. One is that the steel product has a larger materials impact during manufacturing, but is so conveniently reusable that its end of life ELU is lower than that of the composite. However, the steel front end is heavier than the composite unit, and that factor results in much higher environmental loads during product use. The overall result is one that was not intuitively obvious: that the polymer composite front end is the better choice in terms of environmental impacts during manufacture, the steel unit the better choice in terms of recyclability, and the polymer composite unit the better overall choice because of lower impacts during product use. Attempting to make the decision by analyzing only part of the product life cycle would result in an incompletely guided and potentially incorrect decision.

The EPS system and similar LCA methods and software are currently being refined and implemented by a number of organizations. Many of the software packages include extensive data bases and are quite easy to use. Perhaps their greatest weakness tends to be the need to quantify data of uncertain validity and to compare unlike risks, in the process making assumptions that gloss over serious value and equity issues.

TABLE 16.2   Calculation of Environmental Load Values for Automobile Front Ends

Materials and processes	Production			Product use			Waste						Total ELU
							Incineration			Reuse			
	ELU/kg	kg	ELU	ELU/kg	kg	ELU	ELU/kg	kg	ELU	ELU/kg	kg	ELU	
GMT—composite													
Production													
GMT material	0.58	4.0	2.32										2.32
Reused production scrap	−0.58	0.3	−0.17										−0.17
Compression molding	0.03	4.0	0.12										0.12
Product use													
Petrol				0.82	29.6	24.27							24.27
Recycling													
GMT material							−0.21	3.7	−0.78				−0.78
Total			2.27			24.27			−0.78				25.76
Galvanized steel													
Production													
Steel material	0.98	9.0	8.82										8.82
Steel stamping	0.06	9.0	0.54										0.54
Reused production scrap	−0.92	3.0	−2.76										−2.76
Spot welding (spots)	0.004	48	0.19										0.19
Painting (m²)	0.01	0.6	0.02										0.02
Product use													
Petrol				0.82	48.0	39.36							39.36
Recycling													
Steel material										−0.92	6.0	−5.52	−5.52
Total			6.81			39.36						−5.52	40.65

Women's Shoes LCA

The production of leather footwear and its subsequent use and end of life stages form the basis of a life-cycle assessment, designed to show the environmental impacts of various stages of the life cycle. Most of the processes of interest refer to the raising of animals and the acquisition and treatment of the hides, but textiles and paper must also be taken into account (Figure 16.3).

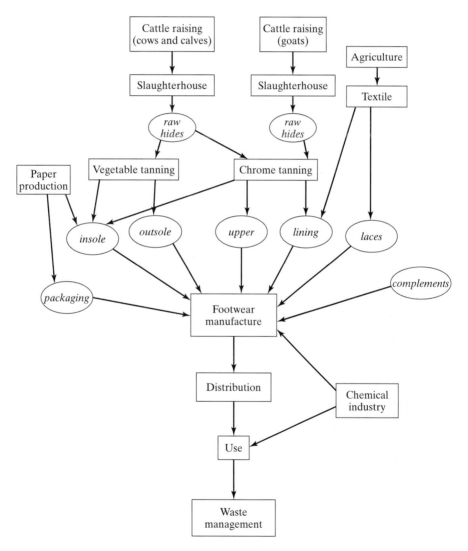

Figure 16.3

A life-cycle process diagram for women's footwear. (Reproduced with permission from Milà, et al., 1998—see Further Reading.)

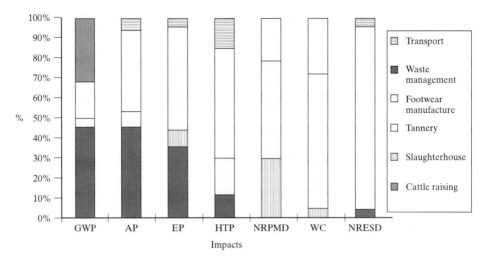

Figure 16.4

Contributions of different life-cycle stages in women's footwear to major environmental concerns. GWP = global warming potential; AP = air pollution; EP = eutrophication potential; HTP = human toxicity potential; NRPMD = nonrenewable primary materials depletion; WC = water consumption; NRESD = nonrenewable energy sources depletion. (Reproduced with permission from Milà, et al., 1998—see Further Reading.)

The life-cycle stages were defined as (1) cattle raising, (2) slaughterhouse, (3) tanning, (4) footwear manufacture, (5) waste management, and (6) transportation. We will not present the inventory results here; they are available in the reference given below. During impact assessment, however, the input/output list items were classified and their contributions to a small number of impacts characterized. The results, expressed as percentages of the toal impacts, are shown in Figure 16.4. Localization and valuation were not performed as part of this process.

The agricultural phase of the life cycle turned out to be important for ecologically related impacts: global climate change, acidification potential, and eutrophication potential. In the case of water consumption, the tannery stage is most important. The tannery stage is also highly significant for eutrophication potential and the depletion of nonrenewable materials. Footwear manufacture is the largest energy-consuming stage, and its impacts are especially significant for energy-related metrics: air pollution, human toxics potential, and fossil-fuel depletion. Thus, two life stages that were not thought particularly significant in environmental terms, agriculture and footwear manufacturing, were identified by the LCA as deserving enhanced attention.

(Details of this study are given in Milà et al.—see Further Reading.)

16.3   INTERPRETATION ANALYSIS

16.3.1   Explicit and Implied Recommendations

The first step in the LCA improvement stage is to use the information flowing from the LCA inventory and impact stages to develop a set of recommendations relating to the activity under study. The intention is to produce environmental benefits or, at least, minimize environmental liabilities. In some circumstances, this stage is almost trivial, as in the Volvo study to choose the environmentally preferable front end. In addition to the obvious utility of choosing a design option on environmental grounds, however, LCA studies often reveal longer term opportunities for environmental improvement. The Volvo study, for example, suggests that once the material is selected for use in manufacturing bumpers, its overall environmental load could be further reduced by any or all of the following:

- Reduce the weight of the composite bumper while maintaining the necessary physical characteristics—strength, manufacturability, etc. (Addressing this recommendation will lead to the testing of alternative materials, alternative designs, or both.)
- Reduce the production-related impacts of the composite material. (Addressing this recommendation will involve working with manufacturing and process engineers, and with suppliers.)
- Reduce the impacts attributable to the compression molding process.
- Develop alternative polymeric-based materials that can be recovered and reused, rather than merely incinerated when the vehicle on which the bumper is mounted is scrapped.

In practice, it is unlikely that potential improvements like this will be considered only after a design approach is chosen and specific materials are selected. A more common procedure is to evaluate potential improvements in parallel with assessments of design options.

Products more complex than components such as automobile bumpers and the like tend to generate longer lists of recommendations. For example, here are selections from a list resulting from the LCA of a telecommunications product:

Manufacturing

- Rewrite specifications for equipment frames to encourage or mandate the use of some recycled material in their manufacture.
- Work with suppliers to minimize the diversity of packaging material entering the facility, so that recycling of solid waste may be optimized.
- Use nitrogen inerting on wave-solder machines to reduce solder dross buildup.
- Minimize the diversity of materials in outgoing equipment packaging, and develop labels to indicate appropriate recycling procedures to the customer.
- Develop reusable shipping containers that satisfy physical and electrostatic protective criteria and are ultimately recyclable.

Design

- Eliminate the use of chromate as a metal preservative in favor of removable organic coatings.
- Review specifications and requirements with the goal of using as few different plastics as possible and of using thermoplastics instead of thermosets.
- Mark all plastic parts, using ISO standards.

Product Management

- Implement a customer information on-line service, to contain not only the operator's manual but also instructions on recycling of parts, components, and packaging during service life and of the entire unit at end of life.
- Develop and implement a strategy for the recovery of used batteries from the field.

In developing a list of recommendations based on LCA results, it is important for the assessor to be inclusive, and to range widely. Recommendations that subsequently prove to be infeasible for one reason or another will be identified and discarded at the prioritization step, the second activity in improvement analysis. Some items, such as the marking of plastic parts, will not require the procedure of a full LCA to indicate their desirability, but would normally be at least implied by LCA results if not explicitly called out. Both more obvious and less obvious recommendations should be considered.

It is worth noting that some recommendations are very specific (i.e., avoid the use of chromate), while others are much more diffuse (i.e., minimize the diversity of packaging materials). Both types are important to include. The highly specific recommendations are easier to generate, and their accomplishment is more easily measured. The diffuse recommendation may be more difficult to deal with, but may in some cases be very important; their inclusion is crucial to a successful implementation of the LCA improvement stage.

16.3.2 Prioritization Tables

The environmental performance of an assessed product can usually be substantially improved by adopting the bulk of the recommendations made in the assessment report. Complete implementation may not be possible, however, and in any case the recommended actions cannot be accomplished simultaneously. Prioritization is thus useful, and in order to prioritize the recommendations one should consider more than just environmentally related characteristics. Some researchers have proposed that the LCA recommendations be prioritized on the basis of how much environmental benefit will result. This procedure does not take into account, however, the fact that industrial decision making incorporates many factors in addition to environmental ones. A first attempt to address that situation was made as part of a Dutch program in "integrated substance chain management," in which priority was given to recommendations that had the most favorable benefit–cost ratio. The costs, however, were established on a society-wide basis, as were the benefits, and the overall accuracy of the technique is therefore difficult for an industrial actor to assess or incorporate. In addition, many other factors of importance to industrial decision making were not addressed. Thus, actions suggested as a result of an LCA process are properly regarded as a subset of possible actions, both environmental and nonenvironmental.

A broadly tractable prioritization approach is to discard quantification and deal with the "binning" of recommendations, i.e., dividing them into a small number of categories on the basis of expert information, much as is done in Pugh Matrix and House of Quality approaches. For example, one can rank each recommendation on a "+/−"

scale (++ being the most desirable score and −− being the least desirable score) across the following product constraints:

- *Technical feasibility:* Rates the technical facility of implementing a particular recommendation; ++ means the recommendation presents no technical challenges and is therefore very easy to implement.
- *Environmental improvement:* Judges to what extent implementation of a recommendation will respond to an important environmental concern, the situation being evaluated on both a scientific and social basis; ++ means implementation will strongly support desirable environmental initiatives.
- *Economic benefit:* Rates the net financial impact for an organization of implementing a particular recommendation; ++ means the product will cost less if the recommendation is incorporated. Here the total life-cycle cost to the manufacturer is considered. For example, some parts may cost more due to DfE constraints but will also yield a higher residual value when an item of leased equipment is returned to the manufacturer for recycling.
- *CVA impact:* Accounts for the Customer-Perceived Value Added by implementing a particular recommendation; ++ means the DfE attribute has a very high perceived value.
- *Production management:* Estimates the production schedule impact or other manufacturing management influence resulting from implementing a particular recommendation; ++ means adoption of the recommendation would reduce the amount of time required to develop and/or manufacture the product; +/− means it would have no significance.

An example of prioritization of the recommendations listed above is given in Table 16.3. The individual scores were assigned by the LCA assessor and the recommendations were then sorted in order of decreasing overall value to the manufacturing organization in each of the three categories: Manufacturing, Design, and Management.

16.4   PRIORITIZATION DIAGRAMS

16.4.1   The Action–Agent Prioritization Diagram

Although the prioritization table is helpful in developing additional supporting information relative to LCA recommendations, its extensiveness may make the most significant information difficult to extract readily, particularly if the number of recommendations is larger than shown here. An alternate display of the information is with a prioritization diagram, as shown in Figure 16.5. The first step in constructing the diagram is to normalize the assessment sum of Table 16.3 by reducing each sum by 10; the philosophy is that the maximum score is 20, and a score at or below 10 reflects neutral or negative overall impacts and thus can be regarded as pertaining to a recommendation that would produce little net benefit. The practical effect of the adjustment is to make it easier to distinguish between and choose among the more highly rated recommendations. The adjusted prioritization sums are plotted in three groups, each group

TABLE 16.3   Prioritization Table for DfE Recommendations

Recommendations	Life stage	Technical feasibility	Environmental sensitivity	Economic impact	CVA impact	Production management	Total
Manufacturing							
Recycled metal specs.	L1.1	++	++	+/-	+	+/-	15
Packaging diversity—inflow	L2.1	++	+	+/-	+/-	+/-	13
Packaging diversity—outflow	L3.1	++	+	+/-	+	+/-	14
Reusable shipping containers	L3.2	++	+	+/-	+	+/-	15
Solder bath N$_2$ inerting	L2.2	++	++	-	+/-	-	12
Design							
Avoid chromate	L1.2(5)	+	+	+/-	+/-	+/-	12
Less plastic diversity	L5.1	+/-	+	+/-	+	-	11
Mark plastic parts	L5.2	++	++	+/-	+	+/-	15
Management							
On-line information	L4.1	++	+	-	+	-	12
Battery recovery	L4.2	++	++	-	++	+/-	15

Symbol	Value	Points
++	Very good/high	4
+	Good/high	3
+/-	Moderate, average	2
-	Little/bad	1
- -	Very little/bad	0

208

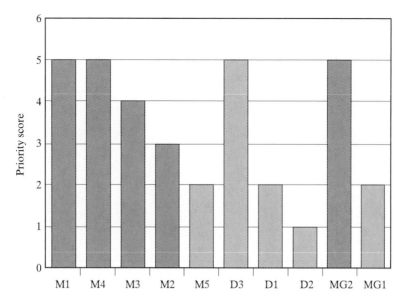

Figure 16.5

An action-agent prioritization diagram of the recommendations from the streamlined life-cycle assessment of a telecommunications product. The designations on the *x*-axis refer to the recommendations (in the order given in Table 16.3) for Manufacturing, Design, and Management. On the *y*-axis, higher numbers indicate greater priority.

representing recommendations that would need to be carried out by specific action agents: manufacturing engineers, design engineers, or management personnel.

The highest priority recommendations are quickly distinguished from those of lower priority in Figure 16.5. In the manufacturing area two actions have the highest priority rating: (1) Specify that major metal parts contain recycled content, and (2) Use reusable shipping containers for modules and components. Several other actions listed in the table are rated high (though not highest) in priority; accomplishing these would also be well justified. The economic impact for all these actions is small to negligible. In the design area, the recommendation that stands out is to mark the major plastic parts with ISO symbols (as discussed in Chapter 14). For management, one priority action is also identified: the development of a program to efficiently take back discharged batteries from the field.

16.4.2   The Life-Stage Prioritization Diagram

An alternative display of the prioritized recommendations is provided by a life-stage diagram. As with the action-agent diagram, the basic information is taken from Table 16.3 and normalized. The recommendations are then divided into five groups, one for each life stage: (1) premanufacture, (2) manufacturer, (3) product delivery, (4) product use, and (5) end of life. If a recommendation pertains to more then one life stage, it is included in each life stage group to which it pertains. The result for the telecommunications product example is shown in Figure 16.6.

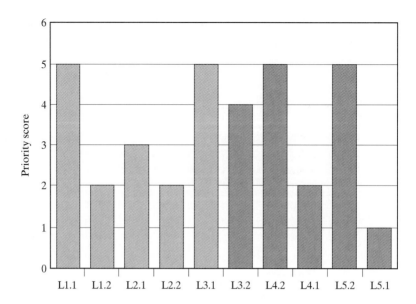

Figure 16.6

A life-stage prioritization diagram of the recommendations from the streamlined life-cycle assessment of a telecommunications product. The first digit under each bar refers to the life stage; the second is a recommendation identification number (see Table 16.3).

The life-stage diagram provides a different perspective on the recommendations, one that varies in time and space rather than in the action agents. The environmental aspects of the manufacturing stage, for example, are seen as relatively benign, as the priority scores of the applicable recommendations are low. In contrast, the end-of-life stage has a recommendation with a higher priority score. Attention is also indicated for the product use stage. The latter two stages are under the direct control of product designers. The premanufacture stage merits activity that requires the participation of the procurement organization in working with suppliers.

16.5   DISCUSSION

Two types of LCA exist: retrospective and prospective. The former is performed subsequent to product design and manufacture. Its goal is to serve as a learning tool to develop generic environmentally superior design approaches. It is most effective when the design team expects to be addressing incremental design changes in similar products, so that most of the learning from the retrospective LCA is directly applicable. Time constraints do not play a major role in retrospective assessments.

Prospective LCAs are performed as part of the design process leading to new products. The aim is to identify aspects of the design (or alternative possible designs) that might be changed to improve the environmental profile of the product. The usefulness of a prospective LCA depends in large measure in the stage of the design process where it can be employed. Time constraints are often severe.

The approach to the LCA interpretation stage basically consists of two parts: (1) deriving a set of conclusions and/or recommendations from earlier LCA stages, and (2) then prioritizing those items from the standpoint of both environmental and nonenvironmental factors. Tables and graphical displays are used as tools for communicating the results.

Prioritization of recommendations is an essential step in a prospective LCA. Quantification is not crucial, since the identification of concerns and the generation of an approximate indication of relative importance appears to be more important than determination of the precision of the assessments. In the experience of the authors, the use within corporations of tools such as those discussed above has led to product environmental improvements in a way that a less structured approach might not have been able to accomplish.

It is interesting to examine how this product improvement process differs as a function of whether those using the tools are industrial firms, policy makers, or interested third parties. The first step, deriving recommendations, could in principle provide identical results no matter which type of agency is performing the assessment. (In practice, insofar as the recommendations deal with production processes or product characteristics in which the firms are naturally well versed, the most useful recommendations will probably come from the firms themselves.) The second step, prioritization, seems likely to be much less uniform, as it involves information on the preferences of external actors as well as on internal and external costs; some of this information is unlikely to be available to nonindustrial parties. From a pragmatic standpoint, therefore, prioritization is probably useful primarily within the organization whose operations are under assessment.

Although the examples in this chapter have treated LCA for products, similar approaches are effective for processes, facilities, service industries, and infrastructure. Life stage diagrams and streamlined LCA approaches to these other applications have been derived and, in some cases, implemented.

There is little question that LCA is generally effective at raising environmentally related issues for product, process, and service designers to address, and in separating the more important of those issues from the less important. It is less certain that LCA can reliably enable the user to say with certainty that Product A is environmentally preferable to Product B, not withstanding Volvo's satisfaction with the EPS approach. The most utilitarian way to use LCAs appears to be to perform retrospective analyses to frame the design approaches for new products, processes, and services, and then to perform prospective LCAs to identify and address specific design choices of environmental interest. The result will invariably be a product more environmentally superior than would have been the case had the LCA approach not been taken.

FURTHER READING

Impact Assessment

Barnthouse, L. et al., *Life-Cycle Impact Assessment: The State of the Art*, Pensacola, FL: Society of Environmental Toxicology and Chemistry, 1997.

Guinée, J., et al., *Handbook on Life Cycle Assessment—Operational Guide to the ISO Standards*, Dordrecht, The Netherlands: Kluwer Academic Publishers, 2002.

Schenck, R.C., Land use and biodiversity indicators for life cycle impact assessment, *International Journal of Life Cycle Assessment 6*, 114–117, 2001.

Steen, B., *A Systematic Approach to Environmental Priority Strategies in Product Development (EPS). Version 2000—General System Characteristics*, CPM Report 1999:4, and *A Systematic Approach to Environmental Priority Strategies in Product Development (EPS). Version 2000—General System Characteristics*, CPM Report 1999:5, Chalmers University, Göteborg, Sweden, 1999.

Evaluation and Uncertainty

Ehrenfeld, J., The importance of LCAs—Warts and all, *Journal of Industrial Ecology, 1* (2), 41–49, 1997.

Field, F.R. III, J.A. Isaacs, and J.P. Clark, Life cycle analysis and its role in product and process development, *International Journal of Environmentally Conscious Design & Manufacturing, 2* (2), 13–20, 1993.

Finnveden, G., Valuation methods within LCA—Where are the values? *International Journal of Life Cycle Assessment, 2*, 163–169, 1997.

Steen, B., On uncertainty and sensitivity of LCA-based priority setting, *Journal of Cleaner Production, 5*, 255–262, 1997.

LCA Case Studies

Jönsson, A., T. Björklund, and A.-M. Tillman, LCA of concrete and steel building frames, *International Journal of Life Cycle Assessment, 3*, 216–224, 1998.

Lippiatt, B.C., and A.S. Boyles, Using BEES to select cost-effective green products, *International Journal of Life Cycle Assessment, 6*, 76–80, 2001.

Milà, L., X. Domènech, J. Rieradevall, P. Fullana, and R. Puig, Application of life cycle assessment to footwear, *International Journal of Life Cycle Assessment, 3*, 203–208, 1998.

Rafenberg, C., and E. Mayer, Life cycle analysis of the newspaper Le Monde, *International Journal of Life Cycle Assessment, 3*, 131–144, 1998.

EXERCISES

16.1 Using the list of risks in Chapter 3, create your own risk prioritization list. Explain and defend your choices, in each case differentiating between scientific and technical assessments on the one hand and values and ethical judgments on the other.

16.2 **(a)** Suppose that global warming is thought less likely to occur than had previously been assumed, and that as a result the ELU/kg for product use is lowered to 0.6. What effect does this have on the comparative ratings of the two front ends of Figure 16.2?

(b) A new high-strength honeycomb steel has been developed and is being considered for use in automobile front ends. Rather than the steel front end weighing 6.0 kg, a satisfactory front end weighing only 4.0 kg can be formed from 6.0 kg of the new steel. Since the new steel's improved properties, which are due to added trace alloying elements, have negligible effects on processing or recycling, the same ELU/kg assessments apply to those stages (Table 16.2). Compute the ELU values for the new front end and compare them with the two options in the table.

(c) Assume that the global warming revision of part (a) occurs as well as the availability of the new honeycomb steel front end. What effect do these two changes taken together have on the relative impact results?

16.3 Suppose that the shortage of petroleum (used as a feedstock for the manufacture of plastic composites) became so great that the ELU/kg of the composite materials was set to 1.90 and that the honeycomb steel front end was available.

  (a) What effect do these two changes taken together have on the relative impact results?

  (b) What are the messages to designers implied in the analyses in the earlier parts of this exercise?

CHAPTER 17

Streamlining the LCA Process

17.1   THE ASSESSMENT CONTINUUM

If no limitations to time, expense, data availability, and analytical approach existed, a comprehensive LCA as described in chapters 15 and 16 would provide the ideal advice for improving environmental performance. In practice, however, these limitations are always present. As a consequence, although very extensive LCAs have been performed, a complete, quantitative LCA has never been accomplished, nor is it ever likely to be. There are many compromises of necessity, among which have often been the use of averages rather than specified local values for energy costs, landfill rates, and the like, the omission of analyses of catalysts, additives, and other small (but potentially significant) amounts of material, neglect of capital equipment such as chemical processing hardware, and the failure to include materials flows and impacts related to supplier operations. As a consequence, detailed LCAs cannot be regarded as providing rigorous quantitative results, but rather as providing a framework upon which more efficient and useful methods of assessments can be developed.

The question of data availability as it relates to product design and manufacture deserves added discussion. Experts agree that roughly 80% of the environmental costs of a product are determined at the design stage, and that modifications at later stages of product development will have only very modest effects. The ideal time to conduct an LCA analysis is thus early in the design phase. At that point, however, the characteristics of the product tend to be quite fluid—materials may not have been selected, no manufacturing facility may have been built, no packaging approach may have been determined, etc. Hence, there is often no possible way to complete a quantitative LCA at the precise time when one would be most useful.

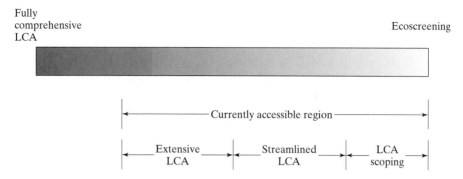

Figure 17.1

The LCA/SLCA continuum.

Techniques that purposely adopt some sort of simplifying approach to life-cycle assessment are termed *streamlined life-cycle assessments* (SLCAs). As shown in Figure 17.1, the family of assessment techniques forms a continuum of effort, the degree of detail and expense generally decreasing as one moves from the left extreme toward the right. The region termed *extensive LCA* is that of detailed, quantitative LCAs such as the IVL/Volvo front end analysis discussed in Chapter 16. The *scoping* or *ecoprofile* regions are those that are purposely sketchy, done to ensure that no truly disastrous design choices have been made or to determine whether additional assessment is needed. Somewhere within the SLCA region is the ideal point—where the assessment is complete and rigorous enough to be a definite guide to industry and an aid to the environment, yet not so detailed as to be difficult or impossible to perform.

17.2   PRESERVING PERSPECTIVE

If streamlining is to be a universal characteristic of LCAs, how can one tell if an SLCA has not streamlined away the LCA's legitimacy? In a survey of ways in which practitioners from academia, government, industry, and consulting firms have attempted to abridge the LCA, Keith Weitz of North Carolina's Research Triangle Institute and his coworkers identified the eight approaches discussed below:

1. *Screen the product with an "inviolates" list.* This approach treats some activities or choices as so obviously incorrect from an environmental standpoint that no design or plan to which they apply should be allowed to go forward. Examples of inviolates are the use of mercury switches in a product or of CFCs in manufacturing. While an inviolates list is useful as an assessment tool, limiting an assessment to the use of such a list obviously has the potential to overlook many life stages and environmental concerns.

2. *Limit or eliminate life-cycle stages.* Some studies limit the LCA to practices occurring within an industrial facility. This "gate-to-gate" approach amounts to a version of pollution prevention. While meritorious, it clearly does not satisfy the criterion of treating the entire life cycle. A second common approach is to limit

or eliminate only upstream stages (resource extraction, for example). This approach is more defendable than gate-to-gate, especially if the evaluation of upstream stages is limited rather than eliminated.

3. *Include only selected environmental impacts.* Some studies limit the LCA to impacts of highest perceived importance or those that can be readily quantified. Such choices tend to be responsive to public pressure rather than to environmental science, and to be anthropocentric rather than balanced.

4. *Include only selected inventory parameters.* This is a variation of the approach immediately above, because if only selected impacts are of interest, only the inventory data needed to evaluate those impacts will be gathered.

5. *Limit consideration to constituents above threshold weight or volume values.* An assessment may be limited only to major constituents or modules. This limitation overlooks small but potent constituents (it would fail as a tool for an SLCA of medical radioisotope equipment, for example), but may sometimes be justifiable from the standpoint of efficiency and tractability. It obviously applies only to quantitative assessments.

6. *Limit or eliminate impact analysis.* Impact analysis is a major component of LCA, and eliminating it clearly abridges the process. The result is that the overall assessment can rely only on a "less is better" philosophy. While pursuing such an approach will probably result in some useful actions, the result provides absolutely no connection between the knowledge base of environmental science and the recommendations made by the abridged LCA.

7. *Use surrogate data.* It is sometimes possible to use data on a similar material, module, or process when the specific data desired for an assessment are not available. The use of surrogate data is often contentious, and has many of the same limits in usefulness as qualitative data.

8. *Use qualitative rather than quantitative information.* Quantitative data are often difficult to acquire, or may not even exist. However, qualitative data can often be sufficient to reveal the potential for environmental impacts at different life stages. The qualitative approach makes it difficult or impossible, however, to compare one product with another or with a new design.

One additional approach not mentioned by Weitz and colleagues is:

9. *Eliminate interpretations or recommendations.* In some studies, inventory and impact results are provided in detailed reports, with the recipient left to devise actions that should be taken in response to the report. If an SLCA is to be useful, however, specific recommendations should be provided by the assessment team, and a method for implementing those recommendations developed.

17.3   THE SLCA MATRIX

A number of SLCA approaches have been developed in recent years. Invariably, they adopt a matrix approach in which the several life stages are evaluated for their potential impacts on a number of environmentally related concerns. An ideal assessment system for environmentally responsible products should have the following character-istics: It should lend itself to direct comparisons among rated products, be usable and consistent across different assessment teams, encompass all stages of product life cy-cles and all relevant environmental concerns, and be simple enough to permit rela-tively quick and inexpensive assessments to be made. Clearly, it must explicitly treat the five life-cycle stages in a typical complex manufactured product, as was illustrated in Figure 8.4.

The assessment system recommended here was developed by the authors in 1993 at AT&T. It has as its central feature a 5 × 5 matrix, the Environmentally Responsi-ble Product Assessment Matrix. One dimension of the matrix is the life-cycle stage; the other is environmental concern (Table 17.1). In use, the assessor studies the product design, manufacture, packaging, in-use environment, and likely disposal scenario and assigns to each element of the matrix an appropriate value. There is no a priori reason why the matrix element values must be continuous. Expert systems of various kinds often use data that are quantized: The values may be either binary (as in problem/no problem decision systems) or ordinal (as in a 1–10 severity ranking system). In the ap-proach we recommend, the assessor assigns an integer rating from 0 (highest impact, a very negative evaluation) to 4 (lowest impact, an exemplary evaluation). In essence, what the assessor is doing is providing a figure of merit to represent the estimated re-sult of the more formal LCA inventory analysis and impact analysis stages. She or he is guided in this task by experience, a design and manufacturing survey, appropriate checklists, and other information. The process is purposely qualitative and utilitarian, but does provide a numerical end point against which to measure improvement.

Although the assignment of integer ratings seems quite subjective, experiments have been performed in which comparative assessments of products are made by sev-eral different industrial and environmental engineers. When provided with checklists and protocols as guidance, overall product ratings differ by less than about 15% among groups of several assessors.

Once an evaluation has been made for each matrix element, the overall Environ-mentally Responsible Product Rating (R_{ERP}) is computed as the sum of the matrix ele-ment values:

$$R_{ERP} = \Sigma_i \Sigma_j M_{i,j} \tag{17.1}$$

TABLE 17.1 The Environmentally Responsible Product Assessment Matrix*

Life stage	Environmental concern				
	Materials choice	Energy use	Solid residues	Liquid residues	Gaseous residues
Resource extraction	1,1	1,2	1,3	1,4	1,5
Product manufacture	2,1	2,2	2,3	2,4	2,5
Product delivery	3,1	3,2	3,3	3,4	3,5
Product use	4,1	4,2	4,3	4,4	4,5
Refurbishment, recycling, disposal	5,1	5,2	5,3	5,4	5,5

*The numerical entries in the table are matrix element indices.

TABLE 17.2   Examples of Product Inventory Concerns

| Life stage | Environmental concern | | | | |
	Materials choice	Energy use	Solid residues	Liquid residues	Gaseous residues
Resource extraction	Use of only virgin materials	Extraction from ore	Slag production	Mine drainage	SO_2 from smelting
Product manufacture	Use of only virgin materials	Inefficient motors	Sprue, runner disposal	Toxic chemicals	CFC use
Product delivery	Toxic printing ink use	Energy loss in packing	Polystyrene packaging	Toxic printing ink use	CFC foams
Product use	Intentionally dissipated metals	Resistive heating	Solid consumables	Liquid consumables	Combustion emissions
Recycling, disposal	Use of toxic organics	Energy loss in recycling	Nonrecyclable solids	Nonrecyclable liquids	HCl from incineration

Since there are 25 matrix elements, a maximum product rating is 100.

Designers who have never performed a product audit may wonder about the relevance of some of the life-stage–environmental concern pairs. To aid in perspective, Table 17.2 provides examples for each matrix element. The basis for some of these examples is that the industrial process is responsible (implicitly, if not explicitly) for the embedded impacts of the processing of raw materials that are used and for the projected impacts as the products are used, recycled, or discarded.

17.4   TARGET PLOTS

The matrix displays provide a useful overall assessment of a design, but a more succinct display of DfE design attributes is provided by target plots, as shown in Figure 17.2. To construct the plots, the value of each element of the matrix is plotted at a specific angle. (For a 25-element matrix, the angle spacing is $360/25 = 14.4°$.) A good product or process shows up as a series of dots bunched toward the center, as would occur on a rifle target in which each shot was aimed accurately. The plot makes it easy to single out points far removed from the bull's-eye and to mark their topics out for special attention by the design team. Furthermore, target plots for alternative designs of the same product permit quick comparisons of environmental responsibility. The product design team can then select among design options, and can consult the checklists and protocols for information on improving individual matrix element ratings.

17.5   ASSESSING GENERIC AUTOMOBILES OF YESTERDAY AND TODAY

The automobile and its manufacture provide a widely known and widely studied example of how SLCA is accomplished in practice. Automobiles have both manufacturing and in-use impacts on the environment, in contrast to many other products such as furniture or interior hardware. The greatest impacts result from the combustion of gasoline and the release of tailpipe emissions during the driving cycle. However, there

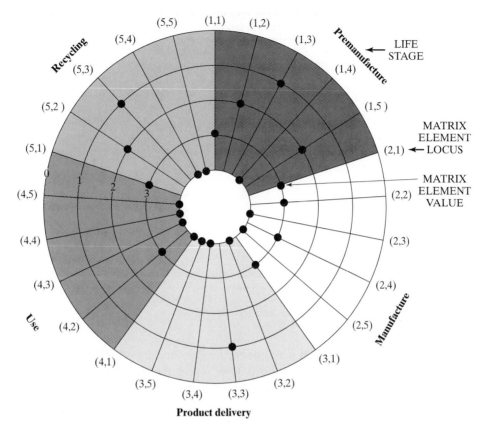

Figure 17.2

The features of the target plot. The data are for demonstration purposes and do not represent an actual product.

are other aspects of the product that affect the environment, such as the dissipative use of oil and other lubricants, the discarding of tires and other spent parts, and the ultimate retirement of the vehicle. To assess these factors, environmentally responsible product assessments have been performed on generic automobiles of the 1950s and 1990s. Some of the relevant characteristics of the vehicles are given in Table 17.3. In overview, the 1950s vehicle was substantially heavier, less fuel efficient, prone to greater dissipation of working fluids and exhaust gas pollutants, and had components such as tires that were less durable.

Premanufacturing, the first life stage, treats impacts on the environment as a consequence of the actions needed to extract materials from their natural reservoirs, transport them to processing facilities, purify or separate them by such operations as ore smelting and petroleum refining, and transport them to the manufacturing facility. Where components are sourced from outside suppliers, this life stage also incorporates assessment of the impacts arising from component manufacture. The ratings assigned to this life stage of generic vehicles from each epoch are given in Table 17.4, where the

TABLE 17.3 Characteristics of Generic Automobiles

Characteristic	ca. 1950s automobile	ca. 1990s automobile
Materials (kg)		
Plastics	0	101
Aluminum	0	68
Copper	25	22
Lead	23	15
Zinc	25	10
Iron	220	207
Steels	1290	793
Glass	54	38
Rubber	85	61
Fluids	96	81
Other	83	38
Total weight (kg)	1901	1434
Fuel efficiency (mi/gal)	15	27
Exhaust catalyst	No	Yes
Air conditioning	CFC-12*	HFC-134a

*Air conditioning entered the automobile market in the late 1950s on top-of-the-line vehicles.
Estimates from *Ward's Automobile Yearbook.*

two numbers in parentheses refer to the matrix element indices. The higher (that is, more favorable) ratings for the 1990 vehicle are mainly due to improvements in the environmental aspects of mining and smelting technologies, improved efficiency of the equipment and machinery used, and the increased use of recycled material.

The second life stage is product manufacture (see Table 17.5). The basic automobile manufacturing process has changed little over the years, but much has been done to improve its environmental responsibility. One potentially high-impact area is the paint shop, where various chemicals may be used to clean the parts and volatile organic emissions can be generated during the painting process. There is now greater

TABLE 17.4 Premanufacturing Ratings

Element designation	Element value and explanation
1950s auto	
Materials choice (1,1)	2 (Few toxics are used, but most materials are virgin)
Energy use (1,2)	2 (Virgin material shipping is energy intensive)
Solid residue (1,3)	3 (Iron and copper ore mining generates substantial solid waste)
Liquid residue (1,4)	3 (Resource extraction generates moderate amounts of liquid waste)
Gas residue (1,5)	2 (Ore smelting generates significant amounts of gaseous waste)
1990s auto	
Materials choice (1,1)	3 (Few toxics are used; much recycled material is used)
Energy use (1,2)	3 (Virgin material shipping is energy intensive)
Solid residue (1,3)	3 (Metal mining generates solid waste)
Liquid residue (1,4)	3 (Resource extraction generates moderate amounts of liquid waste)
Gas residue (1,5)	3 (Ore processing generates moderate amounts of gaseous waste)

TABLE 17.5   Product Manufacture Ratings

Element designation	Element value and explanation
1950s auto	
Materials choice (2,1)	0 (CFCs used for metal parts cleaning)
Energy use (2,2)	1 (Energy use during manufacture is high)
Solid residue (2,3)	2 (Lots of metal scrap and packaging scrap produced)
Liquid residue (2,4)	2 (Substantial liquid residues from cleaning and painting)
Gas residue (2,5)	1 (Volatile hydrocarbons emitted from paint shop)
1990s auto	
Materials choice (2,1)	3 (Good materials choices, except for lead solder waste)
Energy use (2,2)	2 (Energy use during manufacture is fairly high)
Solid residue (2,3)	3 (Some metal scrap and packaging scrap produced)
Liquid residue (2,4)	3 (Some liquid residues from cleaning and painting)
Gas residue (2,5)	3 (Small amounts of volatile hydrocarbons emitted)

emphasis on treatment and recovery of waste water from the paint shop, and the switch from low-solids to high-solids paint has done much to reduce the amount of material emitted. With respect to material fabrication, there is currently better utilization of material (partially due to better analytical techniques for designing component parts) and a greater emphasis on reusing scraps and trimmings from the various fabrication processes. Finally, the productivity of the entire manufacturing process has been improved, substantially less energy and time being required to produce each automobile.

The environmental concerns at the third life stage, product delivery, include the manufacture of the packaging material, its transport to the manufacturing facility, residues generated during the packaging process, transportation of the finished and packaged product to the customer, and (where applicable) product installation (see Table 17.6). This aspect of the automobile's life cycle is benign relative to the vast ma-

TABLE 17.6   Product Delivery Ratings

Element designation	Element value and explanation
1950s auto	
Materials choice (3,1)	3 (Sparse, recyclable materials used during packaging and shipping)
Energy use (3,2)	2 (Over-the-road truck shipping is energy intensive)
Solid residue (3,3)	3 (Small amounts of packaging during shipment could be further minimized)
Liquid residue (3,4)	4 (Negligible amounts of liquids are generated by packaging and shipping)
Gas residue (3,5)	2 (substantial fluxes of greenhouse gases are produced during shipment)
1990s auto	
Materials choice (3,1)	3 (Sparse, recyclable materials used during packaging and shipping)
Energy use (3,2)	3 (Long-distance land and and sea shipping is energy intensive)
Solid residue (3,3)	3 (Small amounts of packaging during shipment could be further minimized)
Liquid residue (3,4)	4 (Negligible amounts of liquids are generated by packaging and shipping)
Gas residue (3,5)	3 (Moderate fluxes of greenhouse gases are produced during shipment)

TABLE 17.7   Customer Use Ratings

Element designation	Element value and explanation
1950s auto	
Materials choice (4,1)	1 (Petroleum is a resource in limited supply)
Energy use (4,2)	0 (Fossil fuel energy use is very large)
Solid residue (4,3)	1 (Significant residues of tires, defective or obsolete parts)
Liquid residue (4,4)	1 (Fluid systems are very leaky)
Gas residue (4,5)	0 (No exhaust gas scrubbing; high emissions)
1990s auto	
Materials choice (4,1)	1 (Petroleum is a resource in limited supply)
Energy use (4,2)	2 (Fossil fuel energy use is large)
Solid residue (4,3)	2 (Modest residues of tires, defective or obsolete parts)
Liquid residue (4,4)	3 (Fluid systems are somewhat dissipative)
Gas residue (4,5)	2 (CO_2, lead [in some locales])

jority of products sold today, since automobiles are delivered with negligible packaging material. Nonetheless, some environmental burden is associated with the transport of a large, heavy product. The slightly higher rating for the 1990s automobile is due mainly to the better design of auto carriers (more vehicles per load) and the increase in fuel efficiency of the transporters.

The fourth life stage, product use, includes impacts from consumables (if any) or maintenance materials (if any) that are expended during customer use (see Table 17.7). Significant progress has been made in automobile efficiency and reliability, but automobile use continues to have a very high negative impact on the environment. The increase in fuel efficiency and more effective conditioning of exhaust gases accounts for the 1990s automobile achieving higher ratings, but clearly there is still room for improvement.

The fifth life stage assessment includes impacts during product refurbishment and as a consequence of the eventual discarding of modules or components deemed impossible or too costly to recycle (see Table 17.8). Most modern automobiles are re-

TABLE 17.8   Refurbishment/Recycling/Disposal Ratings

Element designation	Element value and explanation
1950s auto	
Materials choice (5,1)	3 (most materials used are recyclable)
Energy use (5,2)	2 (moderate energy use required to disassemble and recycle materials)
Solid residue (5,3)	2 (a number of components are difficult to recycle)
Liquid residue (5,4)	3 (liquid residues from recycling are minimal)
Gas residue (5,5)	1 (recycling commonly involves open burning of residues)
1990s auto	
Materials choice (5,1)	3 (most materials recyclable, but sodium azide presents difficulty)
Energy use (5,2)	2 (moderate energy use required to disassemble and recycle materials)
Solid residue (5,3)	3 (some components are difficult to recycle)
Liquid residue (5,4)	3 (liquid residues from recycling are minimal)
Gas residue (5,5)	2 (recycling involves some open burning of residues)

cycled (some 95% of those discarded enter the recycling system in most countries), and from these approximately 75% by weight is recovered as used parts or returned to the secondary metals market. Improvements in recovery technology have made it easier and more profitable to separate the automobile into its component materials.

In contrast to the 1950s, at least two aspects of modern automobile design and construction are worse from the standpoint of their environmental implications. One is the increased diversity of materials used, mainly the increased use of plastics. The second aspect is the increased use of welding in the manufacturing process. In the vehicles of the 1950s, a body-on-frame construction was used. This approach was later switched to a unibody construction technique in which the body panels are integrated with the chassis. Unibody construction requires about four times as much welding as does body-on-frame construction, plus substantially increased use of adhesives. The result is a vehicle that is stronger, safer and uses less structural material, but is much less easy to disassemble.

The completed matrices for the generic 1950s and 1990s automobile are illustrated in Table 17.9. Examine first the values for the 1950s vehicle so far as life stages are concerned. The column at the far right of the table shows moderate environmental stewardship during resource extraction, packaging and shipping, and refurbishment/ recycling/ disposal. The ratings during manufacturing are poor, and during customer use are abysmal. The overall rating of 46 is far below what might be desired. In contrast, the overall rating for the 1990s vehicle is 68, much better than that of the earlier

TABLE 17.9 Environmentally Responsible Product Assessments for the Generic 1950s and 1990s Automobiles

Life stage	Environmental concern					
	Materials choice	Energy use	Solid residues	Liquid residues	Gaseous residues	Total
Resource extraction						
1950s	2	2	3	3	2	12/20
1990s	3	3	3	3	3	15/20
Product manufacture						
1950s	0	1	2	2	1	6/20
1990s	3	2	3	3	3	14/20
Product delivery						
1950s	3	2	3	4	2	14/20
1990s	3	3	3	4	3	16/20
Product use						
1950s	1	0	1	1	0	3/20
1990s	1	2	2	3	2	10/20
Recycling, disposal						
1950s	3	2	2	3	1	11/20
1990s	3	2	3	3	2	13/20
Total						
1950s	9/20	7/20	11/20	13/20	6/20	46/100
1990s	13/20	12/20	14/20	16/20	13/20	68/100

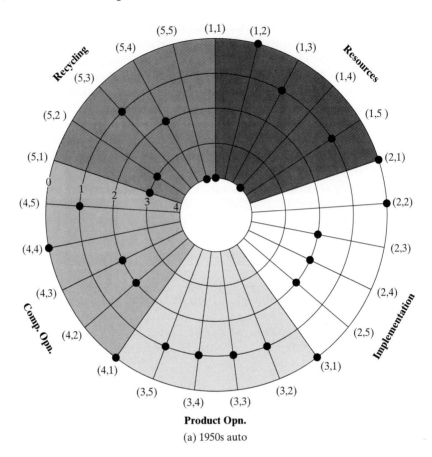

Figure 17.3

Comparative target plots for the display of the environmental impacts of the generic automobile of (a) the 1950s and of (b) the 1990s.

vehicle but still leaving plenty of room for improvement. A more succinct display of DfE design attributes is provided by the target plots of Figure 17.3.

17.6 SLCA ASSETS AND LIABILITIES

As suggested above, when the LCA concept is streamlined in some way, part of the legitimacy of a comprehensive LCA might be thought to be sacrificed. What is it that is really lost? Conversely, what are the gains? Several of each can be listed and discussed. SLCAs are potentially superior to LCAs in the following ways:

- SLCAs are much more efficient and less costly, typically taking a few days of effort rather than several months.
- SLCAs complement LCAs by evaluating design attributes such as ease of disassembly that are inherently qualitative.

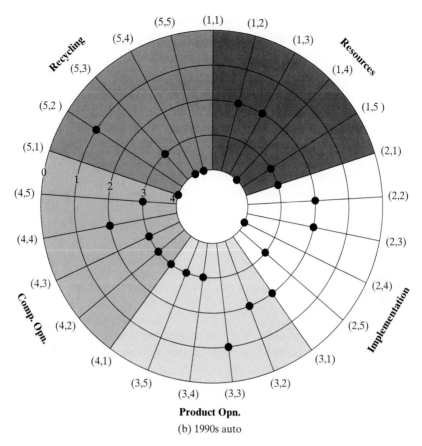

Figure 17.3

Continued

- Many SLCAs are usable in the early stages of design, when opportunities for change are great but quantitative information sparse.

Because of the above attributes, SLCAs are much more likely to be carried out routinely and thus to be applied to a wide variety of products and industrial activities. SLCAs are potentially inferior to LCAs in the following ways:

- SLCAs have little or no ability to track overall materials flows. Within a corporation, for example, SLCAs on all products might well indicate whether a particular material was used, but not whether its use in a particular product was a significant fraction of total corporate usage.
- SLCAs have minimal ability to compare completely dissimilar approaches to fulfilling a need.

- SLCAs have minimal ability to track improvements over time, e.g., to reliably determine whether a product is environmentally superior to its predecessor.

17.7 DISCUSSION

Streamlined product assessment systems can be readily adapted to the manufacture of a variety of products. In cases where one corporation's product is another corporation's feedstock, as is the case with producers of plastic pellets eventually used for auto body panels, for example, assessments can be (and have been) done on an intercorporate basis.

Unlike classical inventory analysis and perhaps impact analysis, streamlined life-cycle assessment is less quantifiable and less thorough. It is also inestimably more practical and utilitarian; it is far better to conduct a number of streamlined LCAs than to conduct one or two comprehensive LCAs. This is particularly true because LCAs and SLCAs are about equally successful at identifying actions that will improve environmental performance, because the options of designers are generally constrained to a fairly high degree, and because the results of full-scale LCAs remain contentious. An SLCA survey of the modest depth advocated here, performed by an objective professional or a cross-functional group of professionals, will succeed in identifying perhaps 80–90% of the useful design actions that could be taken, and will identify them early in the design process, while consuming relatively small levels of time and money. Consequently, the assessment has a good chance of being carried out and its recommendations of being implemented.

A perspective on the SLCA assets and liabilities harks back to a statement earlier in this chapter, that a complete LCA has never yet been carried out and probably never will be. That being the case, it seems only reasonable to proceed with the use of SLCAs while recognizing their limitations. The results of SLCAs are often regarded as approximately correct; if they come even close to that characterization, carrying out the assessments and implementing their recommendations will be of much value.

FURTHER READING

Field F., J. Ehrenfeld, D. Roos, and J. Clark, *Automobile Recycling Policy: Findings and Recommendations*, prepared for the Automotive Industry Board of Governors, Cambridge, MA: Massachusetts Institute of Technology, February, 1994.

Graedel, T.E., B. Allenby, and P. Comrie, Matrix approaches to abridged life cycle assessment, *Environmental Science & Technology, 29*, 134A–139A, 1995.

Henstock, M.E., *Design for Recyclability*, London: The Institute of Metals, 1988.

Hunt, R.G., T.K. Boguski, K. Weitz, and A. Sharma, Case studies examining LCA streamlining techniques, *International Journal of Life Cycle Assessment, 3*, 36–42, 1998.

Noyes, R., ed., *Pollution Prevention Technology Handbook*, Park Ridge, NJ: Noyes, 1993.

Weitz, K.A., M. Malkin, and J.N. Baskir, *Streamlining Life-Cycle Assessment: Conference and Workshop Summary Report*, Research Triangle Park, NC: Research Triangle Institute, 1995.

U.S. Congress, Office of Technology Assessment, *Green Products by Design: Choices for a Cleaner Environment*, Report OTA-E-541, Washington, DC: U.S. Government Printing Office, 1992.

EXERCISES

17.1 Product inventories have been prepared for two different designs of a high speed widget. The matrices are reproduced below, the figure on the left side of each matrix element referring to Design 1, that to the right to Design 2. Select the better product from a DfE viewpoint. What features of each design would you address if improvement were needed?

Life stage	MC	EU	SR	LR	GR
PM	1/1	4/3	4/3	2/2	3/2
M	2/1	1/2	1/2	2/1	2/4
PD	3/2	1/1	2/3	1/1	1/1
PU	1/2	1/2	1/3	1/1	1/3
RD	2/1	2/2	2/1	1/2	1/2

17.2 Select a product of moderate complexity, such as a toaster, a desktop telephone, or an overhead projector. Conduct an SLCA on the product. Prepare a report that summarizes your findings, comment on where it was difficult to assign ratings because of lack of information, and propose design changes that would improve the environmental responsibility of the product.

Using the Corporate Industrial Ecology Toolbox

18.1 STAGES AND SCALES IN INDUSTRIAL ENVIRONMENTAL MANAGEMENT

Activities by industry that have environmental considerations as their focus cover a very wide range of time scales and organizational entities. An attempt to indicate the locations of common activities in temporal and organizational space appears in Figure 18.1. At the lowest scales, the manufacturing operation, Pollution Prevention activities are appropriate. Environmental Engineering approaches deal with concerns that could not be completely avoided by pollution prevention. Considerations of the entire project lifecycle call into play approaches with wider perspectives: Design for Environment and Manufacture for Environment. When all of the technological system is considered, the full sweep of industrial ecology is required. The largest scales, those of all of society and of the span of civilization, require addressing the concept of Sustainable Development. The boundaries between these concepts are fuzzy, and definitions of the approaches overlap, but Figure 18.1 is useful in approximating their scope and focus.

18.2 THE FIRST STAGE: REGULATORY COMPLIANCE

For most corporations, environmental thinking began with the passage of regulations requiring certain levels of environmental performance. As with other regulations pertaining to the conduct of business—financial, worker safety, commitments to customers—complying with the applicable regulations is a necessary activity. In the environmental area, most regulations focus on the health of workers and permitted levels of emissions to air, water, and soil.

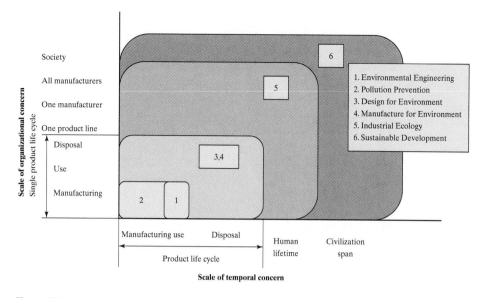

Figure 18.1

Organizational and temporal scales for approaches to industrial environmental management. (Adapted from B. Bras, Incorporating environmental issues in product realization, *United Nations Industry and Environment, 20* (1–2), 7–13, 1997.)

A conscientious business approaches regulatory compliance by maintaining an inventory of its current and projected emissions, and monitoring any other activities that might be influenced by local, regional, or national regulations. This intent is supported by a program of regulatory compliance audits, in which levels of emissions and the corporate status of other relevant activities are matched against applicable regulations to ensure that no violations exist, or might exist at some time in the future. In the area of corporate environmental superiority, the ongoing process of ensuring regulatory compliance is the first step.

18.3   THE SECOND STAGE: POLLUTION PREVENTION

Pollution prevention is directed to the task of reviewing current products and processes and minimizing their environmental impact, primarily through simple, straightforward approaches such as better housekeeping and material substitution. The typical time scale for pollution prevention actions and their effects is a year or two. Typical actions include leak prevention, energy conservation, and packaging improvements. No significant changes are made to products or processes, but rather the way in which they are manufactured is optimized. Almost by definition, pollution prevention is concerned only with activities within the factory gates.

As described in Chapter 9, P^2 generally begins with the construction of one or more process characterization flow charts. Once the pattern of resource flows through the facility is established, materials balances are used to deduce for chosen materials the locations of major inputs, outputs, and losses. Facility-wide improvements can be made by completing waste audits, energy audits, and water audits, and making improvements based on the audit results.

P^2 actions are generally not mandated by regulators or customers. Rather, they are performed because they are financially rewarding as well as good general practice. P^2 is simple housekeeping carried to its highest level.

18.4   THE THIRD STAGE: DESIGN FOR ENVIRONMENT

Design for Environment is the process by which a full spectrum of environmental considerations is taken into account as a routine step in the product or process design sequence. Its components, described in previous chapters, are generally not required by regulations and may or may not provide the immediate financial incentives of P^2. More and more often, however, customers are demanding products that can be demonstrated to be environmentally superior. DfE is the design and development structure that provides access to those markets.

The simplest DfE tool is the Inviolates List, employed early in the design process to avoid making intolerable choices at the component, material, or systems level. Design guidance handbooks can also help the designer avoid historic mistakes and communicate best practice approaches. The next simplest tool is environmental features checklists, which suggest to the designer approaches that are likely to make the prod-

---

Guidelines for Environmentally Superior Product Design

A number of environmental features checklists have been prepared to offer general DfE guidance to product designers. Here is one of the best.

- Make it durable.
- Make it easy to repair.
- Design it so that it can be remanufactured.
- Design it so that it can be reused.
- Use recycled materials to make it.
- Use commonly recyclable materials.
- Make it simple to separate the recyclable components of a product from the nonrecyclable components.
- Eliminate the toxic and problematic components of a product or make them easy to replace or remove before disposal.
- Make products more energy and resource efficient.
- Make products manufacturable using environmentally superior processes.
- Work toward designing source reduction-inducing products (i.e., products that eliminate the need for subsequent waste).
- Adjust product design to reduce packaging.

Adapted from P. Kaldjian, Ecological design: Source reduction, recycling, and the LCA, *Innovation*, 11–13, Special Issue, 1992.

uct more environmentally responsible. Once designs are complete, or nearly so, life-cycle assessments and streamlined life-cycle assessments can be used to conduct overall design evaluations with the goal of environmentally superior products in mind.

18.5 ENVIRONMENTAL OPPORTUNITIES AT THE PRP GATES

In Chapter 8 we presented the concept of the product realization process (PRP), and of the gates that must be successfully passed if a product is to be approved for manufacturing and sale. Environmental considerations have often been omitted in gate reviews, largely because tools for bringing environmental information into the process have not been formalized. The relevant information is therefore not presented even if available within the corporation. However, relevant environmental information can, in principle, be provided at each gate if corporate knowledge-sharing is practiced. And if that information is part of the gate decision, a better overall result is likely.

To illustrate these points, Table 18.1 lists the product and process information available at each gate for a typical manufactured product. The items in this table serve as a guide to the environmental information that can be brought to bear at each gate review.

Gate 1: From Concept to Preliminary Design The environmentally related questions at the first gate are basic, and are designed to discourage product concepts that involve highly disfavorable environmental attributes. Typical questions include: Are forbidden substances included in the product? Will forbidden or highly regulated substances be required in manufacture? Table 18.1 demonstrates that these questions can be addressed at gate 1 only for the principal material or materials and for the principal process or processes.

The appropriate environmental tool at gate 1 is thus a list of inviolates: product or process attributes that the corporation has decided will not be permitted. A typical list for high-technology product manufacture, compiled from those of several corporations, is given in Table 18.2. Each corporation can be expected to have its own list of inviolates, forbidding or discouraging the use not only of illegal materials, processes, or practices but also those which involve potential liabilities the corporation would rather not assume even though current regulation is not an issue.

Gate 2: From Preliminary Design to Mature Design Gate 2 (and subsequent gates) brings another look at the inviolates list of Table 18.2, since some environmentally related issues on that list that could not be addressed at gate 1 can now be reviewed. The availability of a reasonably complete design at this stage allows use of additional environmentally related tools and approaches as well. Hence, a new activity at gate 2 is a review of the environmental aspects of product and process design approaches, together with associated guidance provided by the corporation. This guidance is sometimes informal, in which case it is hard to evaluate compliance at the gate review. Some corporations have systematized this process, however. Lucent Technologies, for example, publishes an internal Designer's Companion, which is a series of case studies of fortunate and unfortunate design choices from the past, some environmental, some not, and guidance in accomplishing the fortunate and avoiding the un-

TABLE 18.1 Information Known at Product Development Gates

Product	Process
Gate 1	
Principal material(s)	Key manufacturing processes (with technology and chemicals)
Critical electrical characteristics	
Critical mechanical characteristics	
Size	
Gate 2	
Major components	Principal manufacturing processes (with technology and chemicals)
Preliminary electrical design	
Preliminary mechanical design	
Preliminary visual appearance	
Gate 3	
All components	All manufacturing processes (with technology and chemicals)
Final electrical design	
Final mechanical design	Process energy consumption
Final visual appearance	
Mold designs	
Gate 4	
Final materials list (constituents and quantities)	All by-product streams
Recyclability	All residue streams
Packaging	Outside supplier interactions
Gate 5	
Marketing	Shipping

fortunate. The result is a manual of design preferences that can be reviewed as part of the gate 2 approval process.

Gate 3: From Mature Design to Development Environmental information at gate 3 can be derived from detailed guidelines and checklists. (Table 18.3 illustrates an excerpt from such a checklist.) In a number of corporations, these tools are now available as part of the CAD (computer-aided design) process, which automatically renders them part of the designer's product development activities. The gate 3 review can thus evaluate the degree to which a product design incorporates recommended product attributes from the checklists. At this gate there is also enough information available to permit a semi-quantitative or streamlined life-cycle assessment (SLCA) to be performed. In such an assessment, the entire range of potential environmental impacts is evaluated for each product life stage. The combination of the checklists and the SLCA allow for corrections of environmentally disfavorable attributes before the product design is finalized.

Gate 4: From Development to Manufacture With product and process information now finalized, either an enhanced SLCA or a fully comprehensive life-cycle assessment can be performed. Most items of environmental concern will have been identified by gate 4, but product delivery implications can be addressed in de-

TABLE 18.2   Typical Product and Process Environmental Inviolates

Gate 1 inviolates

- Chlorofluorocarbons and hydrogenated halocarbons that are restricted by the Montreal Protocol may not be used in any manufacturing process.
- Radioactive substances may not be used in any product.

Gate 2 inviolates

- Mercury switches may not be used in any manufacturing process.
- Plastics must not contain additives (colorants, stabilizers, etc.) formulated with the following metals: Ag, As, Ba, Cd, Cr, Hg, Pb, Se.

Gate 3 inviolates

- Cadmium-plated metal components may not be used in any product.
- Plastic components may not be used without an appropriate International Organization for Standardization symbol.
- Plastics may not contain polybrominated flame retardants.

Gate 4 inviolates

- Recycled stock must be used for all packaging material and descriptive literature.

Gate 5 inviolates

- Products may not be advertised as environmentally superior to competing products (but their positive environmental attributes should be pointed out).

tail for the first time, and the overall results can be made quantitative to the degree desired.

Gate 5: From Manufacture to Sales and Use   From an environmental standpoint, one asks at gate 5 whether environmental issues have been properly reviewed at previous gates, whether the product delivery and marketing activities will meet environmental goals, and whether provisions need to be made for end-of-life activities such as product takeback or battery recycling. As soon as the manufacturing process begins, even in the pilot plant stage, audits for energy, water, and waste can begin. These reviews, and those of earlier stages, are aided by the existence of corporate environmental management protocols, as will be discussed in Chapter 19.

Thus, a wealth of environmental information is available within corporations to aid in the decision making steps of the Product Realization Process. In many cases, however, corporations have not implemented procedures to make that information available and integrate it into their decision making. The PRP format provides an important and convenient way to accomplish that integration.

Although gate passage as described above is a discrete and reproducible sequence of actions, the use of the environmental information tools that have been described is less firm. In different corporations, in different circumstances, and with different gate

TABLE 18.3 A sample checklist for recycling-oriented design of construction and joining techniques for personal computers

1. Are assemblies made of incompatible materials separable or joined by means of separation aids? (Mandatory requirement)

   Joints between the housing and chassis and between the chassis and electronic assemblies are important. If the assemblies and materials are to be reused or recycled, these parts need to be disassembled easily. If the components contain toxic substances, separation has to be done quickly and safely.

2. Are components containing hazardous substances easy to find and simple to remove? (Mandatory requirement)

   The minimum target for recycling is the removal of toxic substances such as batteries and capacitors. They need to be easily separable.

3. Are joints that have to be opened or released easy to find? (Suggested requirement)

   Connections that have to be opened or released in the dismantling of the product should be easily and quickly identifiable. If they are concealed, appropriate advisory labels should be affixed to the product.

4. Can the product be dismantled using only universal tools? (Mandatory requirement)

   The term universal tools refers to everyday, commercially available tools.

5. Have necessary points of application and working space taken into account the need for space for dismantling tools? (Mandatory requirement)

   If tools are needed to engage release mechanisms, adequate working space has to be provided.

Adapted from a checklist developed by R. Steinhilper, Fraunhofer Institut für Produktionstechnik und Automatisierung, Stuttgart, Germany.

review teams, some of the tools may be used at different stages or over several stages, or tool variations may be found useful. The way in which an individual corporation proceeds will be a function of the details of its environmental management plan. The important factor is not that environmental information be used only in a certain way, but that a mechanism is in place to guarantee the use of environmental information at PRP gate reviews. When that process is accomplished there is great potential for benefits to the environment and, increasingly, to the responsible corporations themselves.

18.6 THE INDUSTRIAL ECOLOGY MECHANIC AND THE TOOLBOX

The industrial ecology tools available to the corporation, and the environmental management stage to which they commonly apply, are collected in Table 18.4. New tools are continually being developed, of course, and the approaches and goals of existing tools have some degree of overlap. The list is thus not necessarily comprehensive, but is more than sufficient to demonstrate that the industrial ecology mechanic has a good-sized toolbox on which to call.

The relationship of the tools to the Product Realization Process is shown in Figure 18.2. At each of the PRP gates, industrial ecology tools can be brought to bear on designs or on design implementation. These tools, used skillfully, will provide significant benefits to the environment and, increasingly, to the responsible corporations themselves.

TABLE 18.4   The Industrial Ecology Corporate Toolbox

Stage	Tool
Regulatory Compliance	Emission Inventory*
	Regulatory Compliance Audit*
Pollution Prevention	Process Characterization Flow Chart
	Materials Balance
	Waste Audit*
	Energy Audit*
	Water Audit*
Design for Environment	Inviolates List*
	Environmental Features Checklist
	Design Guidance Handbook
	Life-Cycle Assessment
	Streamlined Life-Cycle Assessment*
	Reverse Fishbone Diagram*

*Tools potentially useful as well within the service sectors.

18.7 INDUSTRIAL ECOLOGY TOOLS FOR THE SERVICE SECTOR

Industrial ecology began as a response to perceived problems in the manufacturing sector, and the material in chapters 8–18 has largely addressed issues related to product design and manufacturing processes. As we will discuss in Chapter 21, however, a large fraction of economic activity and related environmental risk occurs in the service sectors. The analytic approaches for the industrial ecology of these sectors are considerably more primitive than is the case for manufacturing.

On the toolbox list of Table 18.4, we have indicated those tools that are potentially useful without substantial modification for service sector industrial ecology. Regulatory compliance is certainly a consideration for many service sector firms—hospitals, for example. Emissions to air, water, and soil are concerns for many as well,

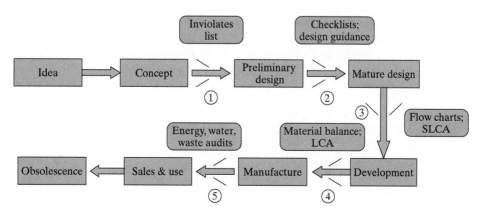

Figure 18.2

Industrial ecology tools to be employed at successive stages of the Product Realization Process.

such as dry cleaners and auto body repair shops. In the pollution prevention category, audits of energy, water, and waste are certainly performed by many. Some service industry firms have developed an inviolates list, which relates not to their manufacturing but to constituents in products they buy—the avoidance of lead solder in electronic devices, for example.

Perhaps the most ambitious application of industrial ecology tools to the service sectors thus far has been efforts by Elizabeth Bennett and Thomas Graedel of Yale University to evaluate specific service sectors or products using streamlined life-cycle analysis. These studies suggest that the greatest potential environmental gains that can be achieved by service organizations are associated with environmentally sensitive design of the buildings in which the services are located and the ways in which those buildings and contents are provisioned, not with service-related operations per se. These are potentially useful conclusions, but clearly the industrial ecology toolkit for the service sector and the insights that will flow from its use remain works in progress as of this writing.

FURTHER READING

Bennett, E.B., and T.E. Graedel, "Conditioned air": Evaluating an environmentally preferable service, *Environmental Science & Technology, 34*, 541–545, 2000.

Graedel, T.E., *Streamlined Life-Cycle Assessment*, Chapter 10, Upper Saddle River, NJ: Prentice Hall, 1998.

Thurston, D.L., Environmental design trade-offs, *Journal of Engineering Design, 5*, 25–36, 1994.

Van Berkel, R., E. Willems, and M. Lafleur, Development of an industrial ecology toolbox for the introduction of industrial ecology in enterprises, I, *Journal of Cleaner Production, 5*, 11–25, 1997.

EXERCISES

18.1 You are part of the design team for a new toaster. What items do you anticipate being raised for discussion at each of the PRP gates?

18.2 Is it useful to develop a Service Realization Process similar to the PRP of Figure 18.2? Test this idea by choosing a service and determining what questions might be asked at each of the gates. Comment on the usefulness of this approach.

PART IV   Corporate Industrial Ecology

C H A P T E R   1 9

Managing Industrial Ecology in the Corporation

19.1   OVERVIEW

The private corporation is an important entity in industrial ecology, whether one is performing mass flow analyses or exploring means by which environmentally preferable products can be developed and deployed. Private firms dominate the development and deployment of technology and are centers of global economic activity and, increasingly, of global governance. They are thus critical for progress towards a more sustainable economy, both in the environmental and social sense. Some knowledge of relevant corporate activities and the ways in which corporations view and manage change is therefore quite useful to the industrial ecologist.

Like any large institution, firms are more complex, multidimensional, and even internally contradictory than most people realize. Factions, politics, cultural and operational differences, internal conflict, dysfunctional behavior: All are endemic to the firm. Managing change in such circumstances is feasible, but challenging. It requires a more sophisticated view of the firm than some tend to adopt: Environmentalists, especially, tend to view companies as in some sense morally suspect. There appear to be two reasons for this: (1) a deep distrust of technology, based on the environmental impacts bequeathed in the past by our technological civilization; and (2) a suspicion of the profit motive and the concomitant amorality of private firms. An element of skepticism may be inevitable when dealing with powerful institutions of any kind. It becomes counterproductive, however, if it results in opposition to all technological evolution and an adversarial approach to private firms and their role in achieving enhanced economic, social and environmental efficiency. A substantially improved environment requires the cooperation of corporations, not their dissolution.

19.2 ENVIRONMENT AS STRATEGIC FOR THE FIRM

The environmental program of a firm is not necessarily equivalent to an industrial ecology program. In the past, environmental issues for the firm arose from bad or careless practice, and the governmental response—remediation and compliance—was driven by regulation. Moreover, the focus of environmental activity within and outside the firm was primarily on the manufacturing process and generally characterized by end-of-pipe controls on emissions. Regulations were concerned with single media, individual substances, specific sites, or particular process emissions. There was little recognition (or incentive to recognize) that the impacts arising from the firm's activities were fundamentally linked with regional and global natural, technological, and economic systems. Industrial ecology, however, requires that recognition, and encourages a fundamental integration of these issues at the level of the firm.

The principal incentives to action also shift in this process. Although well crafted regulation (of which there is too little) remains important, market demand, which tends to be far more important in the cultural model of businesspeople, becomes increasingly powerful. Sophisticated customers that drive much commercial procurement, such as governments in Japan, Europe, and the United States, increasingly demand environmentally preferable products and services, even though defining such offerings remains difficult. Ecolabeling schemes, such as the Energy Star for energy-efficient electronic products or the German Blue Angel ecolabeling scheme, impose product design and operation requirements, a far cry from end-of-pipe controls on manufacturing facilities.

It is important to recognize the difference in kind, not just degree, between traditional environmental regulation and requirements such as those imposed by product ecolabels. The Blue Angel requirements for personal computers, for example, include modular design of computer systems, customer-replaceable subassemblies, avoidance of bonding between incompatible materials such as plastics and metals, and post-consumer product takeback. Implementing takeback requirements alone requires development of a reverse logistics system to get the products back to a central location where they can be disassembled. This establishes a new relationship with suppliers as they become responsible for the appropriate initial design of their components or subassemblies, it develops a new corporate capability (disassembly and management of end-of-life products and material streams), and it influences the business planning process.

It is obvious that none of these requirements has anything to do with traditional environmental approaches, but rather involves the strategic activities that are the core of any manufacturing company. Failure to perform these functions effectively does not, as with traditional environmental regulation, simply expose the firm to liability. Rather, failure to perform in this new mode of operation affects the firm's ability to market its product, and the cost structure of each product. Failure to design products, logistics systems, and business and marketing plans for post-consumer takeback, efficient disassembly, and recycling does not simply raise a firm's overhead. Instead, it prices a firm's products out of the market, regardless of how efficiently the product can be made initially. Environmental capabilities thus change from being a way to control liabilities to becoming a potential source of sustainable competitive advantage.

When environmental issues are regarded as overhead, they are only peripheral concerns of corporate management. As they become more strategic, however, environmental considerations become one of the routine objectives and constraints that the firm must manage, whether in product design, manufacturing, or business planning. For example, the environmental specialists in an electronics manufacturing firm may identify the use of lead solder in printed wiring board connections as a concern, but the complexity and technological constraints of bonding technology may make substitution of a less toxic alternative for all solder uses impractical, at least for a time. This is because the environmental considerations are part of the design equation, but not the only part: economic, competitive, technological, and other considerations are also balanced in the process. The evolution of environmental issues from overhead to strategic may appear to devalue environmental issues because they no longer stand apart. However, including them in the strategic decisions of the firm provides a far higher level of environmental efficiency over the longer term.

This shift in environment from overhead to strategic for the firm entails substantial organizational change. The environment, health, and safety (EHS) group in most firms has traditionally had almost nothing to do with any of the strategic or substantive decision-making processes; in organizational terms, it has weak or nonexistent coupling to other functional units within the company. Once the firm begins to understand the implications of the shift, however, EHS couplings to corporate operations are greatly strengthened. Complying with Blue Angel requirements clearly is a challenge to R&D organizations, business planning and product management operations, and other operational elements; it cannot be done by any single organization—and, in particular, it cannot be done by the EHS group alone.

19.3   IMPLEMENTING INDUSTRIAL ECOLOGY IN THE CORPORATION

The kinds of organizational restructuring required to implement the principles of industrial ecology in existing firms involve changes in both the internal context—within the firm's existing divisions and organizational entities—and externally, as the firm seeks to achieve new relationships with customers and suppliers. While it is true that more environmentally and economically efficient technologies are an important support for this evolution, it is also true in many firms that the most difficult barriers are cultural, not technological. This can be particularly true of the EHS group. Traditional environmental personnel are highly specialized, with precise and well defined responsibilities, chosen more for detailed knowledge than for business or strategic aptitude. These skills do not prepare them to play a leading role when environmental issues become strategic for the firm. Under these conditions, it is ironic but not surprising that some firms find their environmental organizations to be among the most opposed to this shift in the corporate treatment of environmental issues.

Implementing industrial ecology in the firm involves the types of activities and programs usually associated with any culture change in complex organizations:

1. Individual champions, who come from appropriate organizations within the firm and are willing to take the associated risks, must be identified and supported. Thus, green accounting issues would be the province of a financial organization

champion, while Design for Environment programs might be championed by someone from the product research and development group.

2. Barriers to change, especially those which arise from corporate culture and informal patterns of behavior, must be identified and reduced.

3. The least threatening method of introducing new techniques, tools, and systems should be identified and used. The best changes are often those that are never recognized by those that implement them.

4. Strong rationales for new activities, defined in terms of the target audience's interests and cultural models, must be developed. For example, if a company manufactures personal computers, the Blue Angel requirements can be used with the design team as examples of customer demand patterns; for a service firm, the potential to create a market for services that replace environmentally problematic products can be a driver for behavior change.

The evolution of environmental issues from overhead to strategic requires that environmental expertise be diffused throughout the firm, becoming part of the accounting process, the strategic and business planning process, the research and development process, the product design process, and the marketing process. As this process occurs, environmental considerations may come to look less and less like environmental issues and more like financial, strategic, design, or material selection issues. Such a translation of environmental concerns into other corporate "languages" is highly desirable; it facilitates the desired improvement in environmental performance.

19.3.1 Environmental Management Systems

Perhaps the most familiar component of an industrial ecology implementation program at the firm is reliance on an environmental management system (EMS), defined in ISO 14001 as:

> . . . that part of the overall management system which includes organizational structure, planning activities, responsibilities, practices, procedures, processes and resources for developing, implementing, achieving, reviewing and maintaining the environmental policy.

EMSs range from the relatively simple to the quite complex, as illustrated by ISO 14000 itself, a set of environmental requirements now under development by the International Organization for Standards through a negotiating process involving industrial, government, and public stakeholders. When implemented, ISO 14000 (Table 19.1) creates an environmental standards system that facilitates corporate management of environmental issues. ISO section 14001 describes the basic structure of an EMS, while other sections deal primarily with products and ecolabeling. The focus on products, and implicitly on manufacturing, reveals ISO 14000 as rooted in the older concept of environment as overhead, although it ultimately encourages a transition to more industrial-ecology–based thinking.

EMSs are performance aids, but not always ideal or necessary instruments. Most transnational firms, because of their size and the generally high quality of management

TABLE 19.1   ISO 14000 Environmental Standards System

Standards area	Series	Description
Management systems	14001	Describes basic elements of an environmental management system (EMS)
	14004	Guidance document explaining and defining key environmental concepts
Evaluation and auditing	14010	Guidelines on general environmental auditing principles
	14011	Guidelines on audit procedures, including audits of EMSs
	14012	Guidelines on environmental audit or qualifications
	14013	Guidelines on managing internal audits
	14014	Guidelines on initial reviews
	14015	Guidelines on site assessments
	14031	Definition of, and guidance on, environmental performance evaluation
Environmental labeling	14020	General principles
	14021	Terms and definitions for self-declared environmental claims
	14022	Symbols
	14023	Testing and verification
	14024	Criteria for product evaluation and label awards
	14040	General principles and guidelines
Environmental assessment of products	14041	Guidelines for life cycle assessment impact assessment
Environmental aspects in product standards	14061	Guidance for writers of product standards

that has enabled them to become transnational in the first place, have long had their own internal environmental management systems in place. Because these internal management systems are tailored to the firm and its operations, they tend to be more effective, and certainly more efficient, than the rather bureaucratic and generalized public EMSs. Moreover, although public EMSs purport to apply to all types of firms, they tend to be more easily implemented by manufacturing firms than others. That is, the desirable generality of a uniform structure is paid for by the inefficiency of application to the specific firm. This criticism is strengthened by the fact that certification to EMS standards, especially ISO 14000, requires extensive audits by outside consultants. While the audit validates the process of certification, it also generates significant and unproductive costs. This circumstance does not necessarily mean that EMSs have little value: In particular, they can be excellent guides for smaller manufacturing firms in developed and developing countries that may have little experience in management systems. But even in such cases, unless customers are demanding actual certification, it is unclear that the extra costs of obtaining certification, as opposed to implementing an internal EMS to enhance environmental management, are justified.

A more subtle issue for industrial ecologists is that EMSs tend to embed assumptions appropriate to managing environment as overhead, not as strategic for the firm. In many cases, the significant gains in environmental and social efficiency in the firm are not realized through more efficient implementation of "environmental policy," but through the diffusion of environmental concerns throughout the firm's activities. In this sense, EMSs generally help firms do environment as overhead more efficiently, but are seldom useful in supporting the shift of environment from overhead to strategic within the firm.

19.3.2 Tactical Organizational Structures

Traditional EHS organizations will likely be ineffectual if given the complete mission of implementing industrial ecology principles, because the expertise of many other segments of the organization is inherently involved. An EHS organization purporting to establish product design criteria will have little credibility with product design teams, for example, although their input, properly presented, can help the team generate a preferable design. Similarly, an EHS organization that issues green accounting standards to the firm will have a difficult time achieving credibility with business and accounting managers, even though they can provide the data that support the accounting system.

Accordingly, a corporate entity that is not perceived as part of the core EHS group is frequently more effective in supporting the evolution of environmental issues to strategic within the firm.

19.3.3 Training Programs

Many medium-sized and small firms have little internal capability to comply with existing command-and-control environmental regulations, much less develop and use industrial ecology methodologies such as Design for Environment. Thus, it is often important to establish training programs that not only present specific tools appropriate to the sector, but the broader context within which such practices can be shown to make business sense. While large firms in many cases will do this for themselves, there is a clear role for government support in sectors characterized by many small companies (printing, for example). Choosing the entity involved in conducting the training is important: If it is offered by a government environmental agency, for example, smaller firms will be concerned about the perceived integration of training and compliance functions, and may tend to question the technological expertise of the training. It is better to have an organization perceived as independent and technologically sophisticated—local universities and community colleges, for example—involved in delivering the training.

19.3.4 Technical Support

Integrating environmental considerations into all aspects of a firm's business is very difficult unless appropriate technical support is provided: information on environmentally and economically preferable technologies, information on life cycle and DfE methodologies, information on relevant trade or marketing requirements. In large firms, this capability can be developed internally, perhaps in conjunction with the training program. In other cases, however, offering such services through governmental or educational means is worth consideration. For example, if a country has a series of publicly funded technology centers, these can be upgraded to support DfE, LCA, and other methodologies, and to provide information on environmentally and economically efficient materials, processes, and product design choices. Alternatively, academic institutions can be funded to provide such services. If done properly, this approach has the advantage of training students as well.

19.3.5   The Triple Bottom Line

Regardless of jurisdiction, private firms are legally established with the goal of earning money for their shareholders. The bottom line of the accounts is where all of the costs and revenues for the firm are added up and the total profit or loss of the operation determined. Recently, stakeholders and firms have begun to experiment with an approach known as the Triple Bottom Line, or TBL, wherein the firm attempts to satisfy performance measures along not just economic dimensions, but environmental and social dimensions as well. From this perspective, environmental issues and industrial ecology become part of the broader framework within which the firm manages its activities (Figure 19.1).

Although most large firms have always had a socially responsible element to them—in part to assure continued public support for their operation—the suggestion that firms should be measured on social and environmental performance as part of their core activity is new, and represents a potential shift in the definition of the private firm. It is too early to tell whether approaches such as the TBL will be useful or suc-

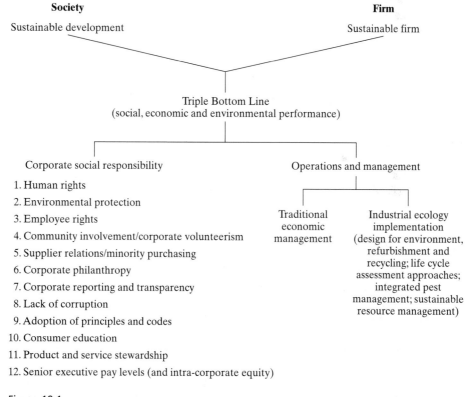

Figure 19.1

The Triple Bottom Line seeks to have firms satisfy not just economic performance measures, but environmental and social measures as well. This forms a context within which corporate industrial ecology activity will increasingly be conducted—and judged—in the future.

cessful, either from the viewpoint of the firm, or society as a whole. While a number of leading firms are beginning to develop annual reports based on the concept, and a new nongovernmental organization called the Global Reporting Initiative has been formed to support such activities, there are still substantial issues to be resolved. Perhaps the most important is the cultural relativism of the concept of social responsibility. There are few mutually accepted objective criteria in such cases, and firms can find themselves in the middle of highly charged and ambiguous debates. For the industrial ecologist, the important point is the increased coupling of relatively objective industrial ecology studies and the perhaps ambiguous and difficult context within which the results of such studies may be applied.

FURTHER READING

Allenby, B.R., *Industrial Ecology: Policy Framework and Implementation*, Upper Saddle River, NJ: Prentice Hall, 1999.

Castells, M. and P. Hall, *Technopoles of the World: The Making of 21st Century Industrial Complexes*, London, Routledge, 1994.

Finster, M., P. Eagan, and D. Hussey, Linking industrial ecology with business strategy, *Journal of Industrial Ecology, 5*(3) 107–125, 2002.

Frankl, P., Life cycle assessment as a managing tool, in *A Handbook of Industrial Ecology*, R.U. Ayres and L.W. Ayres, eds., Cheltenham, U.K.: Edward Elgar Publishers, 530–541, 2002.

Kanholm, J., *ISO Requirements*, Pasadena, CA: AQA Co., 1998.

Schmidheiny, S. (and the Business Council for Sustainable Development), *Changing Course*, Cambridge, MA: MIT Press, 1992.

EXERCISES

19.1 You are the newly assigned environmental director for a major electronics manufacturer. You have been given the assignment of implementing DfE in the firm within a year.
  (a) Develop an organizational structure for carrying out your task.
  (b) Develop a plan, including actions, milestones, and metrics where appropriate, for completing your assignment.
  (c) Does your plan change when you learn that the company expects to be exporting 90% of its production to Europe within five years?

19.2 The president of your company, a business-to-consumer Internet firm that sells CDs and books all around the world, has just called you into her office and asked you to make the company Triple Bottom Line compatible.
  (a) What are the specific issues a company such as yours should consider in evaluating its TBL status?
  (b) What kinds of industrial ecology research might you want to initiate to get a better idea of your TBL status?

19.3 You are the new Environmental Minister of a country that has just decided to become sustainable in one generation (25 years). You are sophisticated enough to know that you must rely principally on market mechanisms, and the evolution and diffusion of new, environmentally preferable, technologies by private firms. Using your insight into private firms, what policies would you implement to encourage firms to become environmentally superior? What metrics of corporate performance might be relevant?

CHAPTER 20

Indicators and Metrics

20.1 THE IMPORTANCE OF INDICATORS AND METRICS

In both government and industrial operations, "what gets measured gets managed." Especially in a complex and ideologically charged area like the environment, metrics and indicators assure transparency and provide incentives for accomplishment. It is almost impossible to imagine any environmental governance system that does not rely on some system of structured evaluation to validate and guide its function. *Metrics* and *indicators* are the tools of structured evaluation, serving as measures of progress and incentives for performance. A metric is a quantitative measure of performance relative to a defined criterion. An indicator is a nonquantitative measure of environmental state, such as the existence or nonexistence of an endangered species. Properly chosen, they are key elements of environmental improvement programs no matter what the organization, or spatial scale, or time scale.

In practice, the task of establishing reasonable measures to monitor the evolution of human and natural systems poses significant challenges to governments, nongovernmental organizations, private firms, and societies. In particular, the higher level, more integrated metrics and indicators contain high normative content; although appearing quantitative on the surface, they may, in fact, have significant cultural, ethical, and economic implications (Table 20.1). Nonetheless, the generation of meaningful policies and an integrated approach to environmental perturbations requires establishing some set of indicators and metrics by which one can evaluate progress toward agreed-upon goals. This chapter introduces the topic and provides a status report on this evolving area.

TABLE 20.1   Metric System Goals

Goal	Normative content of associated metrics
Ethics	High
Interspecific	
Intraspecific	
Intergenerational	
Intragenerational	
Ideology	High
Communitarianism	
Egalitarianism	
Environmentalism	
Stewardship	Medium
Achievement of sustainable world	
Sustainable for how long?	
Which world?	
Eco-efficiency	Low
Reduce impacts (go in right direction)	
Contribute to sustainability	

20.2   METRIC SYSTEMS DEVELOPMENT

Metrics are appropriate for a variety of spatial scales and organizational entities. Although it is interesting to use global metrics for state of the world assessments, it is only where the scale matches the organizational entity that metrics really possess utility. A national government may be concerned about a global problem like sustainability, for example, but can make progress only by utilizing metrics that stimulate actions at the national level. National air and water quality metrics have been in existence for several decades in many countries, and sustainability has inspired new national metrics development. In 1987, for example, the Dutch began work on metrics as part of their effort to formulate the National Environmental Policy Plan of 1989, with a goal of making the country sustainable in a generation (box, Chapter 7). Obviously this required some metrics by which progress towards that goal could be measured. It was eventually determined that the metrics should meet four requirements:

1. They must bear some relation to underlying causal relationships within the system being studied, and must aggregate as much information as possible into a meaningful composite measure. This measure should have intuitive appeal and be easy to understand.

2. They must accurately reflect a trend with an appropriate time scale (e.g., a metric for global climate change should not be evaluated over a time scale of months), and should where applicable display both medium-term and long-term effects.

3. They must link to existing policy objectives and related activities, and ultimately link to the achievement of sustainability.

4. They must be clear and understandable by the public, both to engender support for the policies that are associated with them, and to achieve over time the culture change that will be a prerequisite to a sustainable global economy.

The "themes" for which indicators were developed in the Netherlands included:

- climate change
- acidification of the environment
- eutrophication
- dispersion of toxics
- solid waste accumulation
- disturbance of local environments
- dehydration of soils
- resource depletion

Each theme had a quantitative indicator (a metric) and a quantitative target developed for it. In some cases, the metric consists of a number of underlying constituents, which in turn might be based on aggregated data. For example, the climate change theme has an overall "pressure" metric made up of components based on emissions of carbon dioxide, methane, nitrous oxide, CFCs, and halons. Figure 20.1 shows the Dutch calculation for their climate change theme, expressed in CO_2 equivalents, as well as targets for reductions. The result is a clear and concise statement of progress toward the defined social goal.

Developing such indicators requires understanding not only scientific data, but the context within which such indicators work, including the economic, technical, pol-

Figure 20.1

Progress in the CO_2 equivalents metric, a component of the Dutch sustainability metrics system. (After Adriaanse, A., *Environmental Policy Performance Indicators: A Study on the Development of Indicators for Environmental Policy in the Netherlands,* Sdu Uitgeverij Köninginnegracht, 1993.)

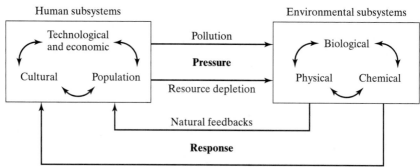

Figure 20.2

The pressure–state–response framework. A good metrics set will include measures to track progress on each of the framework components. (Based on A. Hammond, A. Adriaanse, E. Rodenburg, D. Bryant, and R. Woodward, *Environmental Indicators*, Washington, DC: World Resources Institute, 1995.)

icy, and cultural dimensions. These factors can be pictured as a Pressure–State–Response Framework (Figure 20.2), which can then be used to develop a broader matrix of sustainability metrics as suggested in Table 20.2. Metrics might be adopted to measure pressure on certain issues (as the Dutch did for CO_2 emissions), state (as is done increasingly to examine species existence or abundance), and response (as in the level of investment for water treatment).

20.3   INDUSTRY-LEVEL METRICS

In recent years it has become common for corporations to issue annual environmental reports. A centerpiece of most of these reports is the presentation of the corporation's environmental performance, expressed as a selection of metrics results. Unlike corporate financial reports, for which widely agreed-upon standards and formats are prescribed, the metrics in corporate environmental reports are self-selected. Some of the metrics commonly used are given in Table 20.3, though few reports include a set this extensive. Because metrics development is ongoing, the entries in Table 20.3 should be regarded as examples or templates, to be modified as appropriate for the particular sector, firm, facility or operation under consideration.

Corporate metrics are generally normalized, as in "water use per unit of sales." This is done so that the perceived environmental performance is not unrealistically influenced by, for example, a major production increase. Where after-sale product performance measures are included, the metrics are generally normalized on a per product basis. Normalization is completely reasonable at the corporate level, but may make less sense when applied to society as a whole. As population and affluence increase, maintaining environmental performance on a per capita or per product basis, while commendable, may nonetheless result in overall unsustainable performance on the global systems level.

TABLE 20.2   Matrix of Potential Government-Level Environmental Indicators

Issues	Pressure	State	Response
Climate change	Greenhouse gas emissions	Concentrations	Energy intensity
Ozone depletion	Halocarbon emissions	Chlorine concentrations	Montreal Protocol
Eutrophication	Nitrogen, phosphorous emissions	Nitrogen, phosphorous concentrations; biological oxygen demand	Water treatment investments and costs
Acidification	SO_x, NO_x, NH_3 emissions	Deposition, concentrations	Investments; control agreements
Toxic contamination	Heavy metal, persistent organic compounds POC emissions	POC, heavy metal concentrations	Recovery of hazardous waste; control and investments/costs
Urban quality of life	Volatile organic compounds (VOC), NO_x, SO_x, emissions	VOC, NO_x, SO_x concentrations	Expenditures; transportation policy
Biodiversity	Land conversion; land fragmentation	Species abundance	Protected areas
Waste	Waste generation by sectors and communities	Soil/groundwater quality	Material collection rate; recycling investments and costs
Water resources	Demand and use intensity by sector and communities	Demand/supply ratio by sector; water quality	Expenditures; water pricing; savings policy
Forest resources	Use intensity	Area of degraded forest	Protected area of forest; prevalence of sustainable logging practices
Fish resources	Fish catches	Sustainable stocks	Quotas and economic rationalization
Soil degradation	Land use changes	Top soil loss; soil degradation	Rehabilitation/protection
Oceans/coastal zones	Emissions; oil spills; depositions; eutrophication	Water quality; biological impacts	Coastal zone management; ocean protection
Environmental index	Pressure index	State index	Response index

A characteristic of Table 20.3 and other current corporate metrics sets is that they are oriented toward the product manufacturing industries. But the bulk of industrial activity is not, in fact, in product manufacture, but in services (e.g., telecommunications, retailing, food service), infrastructure (e.g., water, transportation), or resources (e.g., mining, forestry, fisheries). In these areas, metric development is needed just as badly as in manufacture, but is largely embryonic at present.

Early corporate environmental reports dealt largely with traditional topics (air emissions, water emissions) and with a few unregulated but easily measured parameters such as water use. This approach is now evolving into corporate sustainability reports designed around the Triple Bottom Line. Social metrics are problematic at this point, and not well defined with regard to environmental metrics for two principal reasons. First, unlike environmental issues, social issues do not have any underpinning in an objective science such as environmental science or ecology. Second, social issues have a much higher cultural content than environmental issues. Thus, they tend to be much more ideological and highly contentious in practice. For example, an American

TABLE 20.3   Typical Corporate Environmental Metrics

Materials
   Quantity of material used per unit of sales
   Proportion of recycled or reused material

Resource consumption in processing
   Energy used per unit of sales
   Water used during unit of sales

Residuals and emissions
   Quantity of residuals generated per unit of sales
   Quantity of hazardous residuals generated per unit of sales
   Quantity of emissions contributing to specific air, water, or soil perturbations per unit of sales

Operations
   Quantity of packaging residuals generated per unit of product
   Quantity of residuals generated during use per unit of product
   Quantity of energy consumption during use per unit of product

pressure group might argue that all foreign workers be paid the same wages as U.S. workers, on the grounds that to do otherwise amounts to exploitation of workers in developing countries. The developing country, on the other hand, might argue that this position is only a disguised form of protectionism, and (given the lower productivity of its workers) simply a way to ensure that it cannot use its comparative advantage, cheaper labor, to develop itself. Finally, the role that private firms should take in forcing social change beyond their activities as economic actors is still open to question. While there is a strong desire that firms not act in socially irresponsible ways, that is different from empowering firms to act proactively on social rather than economic issues.

Industrial environmental metrics, although less difficult and normative than social metrics, can be problematic as well as utilitarian. A natural approach is to develop metrics that are broadly applicable across sectors or industry groups and thus facilitate cross-firm comparisons. Others, however, prefer metrics that are meaningful for the particular activities and operations of a firm or sector. An example is a metric dealing with the sustainable use of land—highly appropriate for a forest products company, but not very significant for a manufacturer of portable radios. Some green investment groups apply the same general measures (e.g., releases of SO_2) to telecommunications firms that they do to chemical companies on the grounds that this encourages transparency and comparability. The use of such metrics for service companies, however, tends to obscure the environmental impacts, positive and negative, that service companies actually have. For example, if a set of metrics only takes in the negative impacts of a telecommunications firm by using metrics that are meaningful primarily in a manufacturing or energy context but does not comprehend the use of the telecommunications infrastructure for travel avoidance and paperless billing, it fails to provide an objective and balanced picture of the firm's real impact on environmental systems.

20.4   METRICS DISPLAYS AND METRICS AGGREGATION

Effective aggregation and display of metrics data can increase their meaningfulness and utility substantially. The danger of such approaches is that they can be simplistic and misleading; because of the potential payoff, however, they are worth exploring.

The simplest level of reporting conveys uniform values and trend information about individual metrics. A major benefit of having a common metric is that the uniform rating scale permits the environmental performance of corporations in the same industrial sector to be directly compared and the environmental performance of a single corporation to be tracked over time. It is not a significant problem that the scales will differ for different sectors. Such financial measures as debt, equity, and dividend payout ratios are routinely expected to be sector dependent, for instance.

An overview of a corporation's environmental performance is provided by displaying metrics in a grouped chart. Suppose that 10 metrics have been selected and that all have been normalized on a scale of 0–10. We can now create a triangular metrics scoring display divided into 10 areas (one for each of the 10 metrics) (Figure 20.3). Within each area we can express the rating score by degrees of a gray scale, or even better by a color:

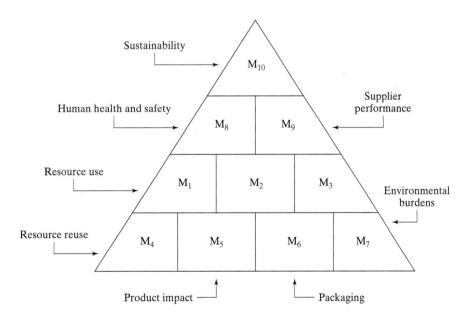

Figure 20.3

A triangular scoring display for a set of 10 corporate metrics. The metrics set is arbitrary; there are many possibilities.

	Red	Orange	Yellow	Cyan	Green
Rating score	0–2	2–4	4–6	6–8	8–10

A colored triangular display shows the metrics in a form that directs the user's attention to areas of highest and lowest performance, and it can be an efficient way to communicate the ensemble of industrial environmental metrics information applicable to an individual facility or a corporation as a whole. Such a display has the potential to become a uniform and readily understood presentation of corporate environmental performance, much as the U.S. Department of Agriculture's chart of daily minimum nutritional requirements efficiently presents the characterization of foods or food products. A final though potentially controversial step might be to aggregate the results from a metrics set into a single indicator of corporate environmental performance.

20.5   HIERARCHICAL METRICS SYSTEMS

Environmentally related goals may be established by individuals, firms, or governments. As the discussion above suggests, rather different approaches result if metrics are chosen and goals are set solely from the perspective of any one of these actors. Environmental metrics and goals assume their most powerful role, however, when they are chosen on a hierarchical basis, with the measures and targets at one level linking clearly to those at levels above and below. Such a framework is the basis for the targets in the Kyoto Protocol, in which national stressor commitments were adopted in the context of a global environmental challenge. The full implementation of any such action plan ultimately involves linking the national goals to goals at local and regional levels and with industry, government, and individuals. This hierarchical concept is pictured in Figure 20.4.

It is clear that a hierarchical metric must be carefully selected, and the metric may be expressed somewhat differently at different levels. A metric directly evaluating sustainability, for example, is not likely to be much help in designing a manufacturing process to clean printed wiring boards in electronics. However, if a system is constructed to link aspects of manufacturing (e.g., lower energy consumption), through intermediate levels (reductions in emissions of carbon dioxide from electricity generation), to sustainability goals (reduced global climate change forcing), one can ensure that initiatives taken at different levels integrate into a coherent structure. In the absence of such a system, linking progress toward an environmental goal across different systems levels will be difficult, if not impossible, to achieve.

A set of environmental metrics, as opposed to an individual metric, can also be formulated as a hierarchical system. Such a system, shown schematically in Figure 20.5, might follow the form of the 10-metric triangle of Figure 20.3, with the individual metrics selections mapping across the multilevel system under examination. This becomes, in effect, the metrics-centered monitoring of progress toward the grand objectives of Chapter 1. The system need not have the three levels indicated in the figure. If monitoring and optimization of national water use or disturbance of local environments were the goal, for example, the global level would be omitted and local and regional levels could be separated or subdivided to allow for more precise evaluation at the user stage.

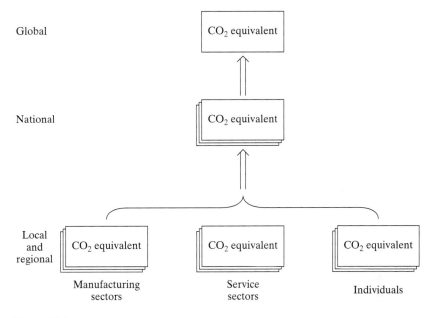

Figure 20.4

A hierarchy of metrics for CO_2 equivalent emissions. The same approach can be used for a hierarchy of indicators such as the existence of sustainable fish stocks.

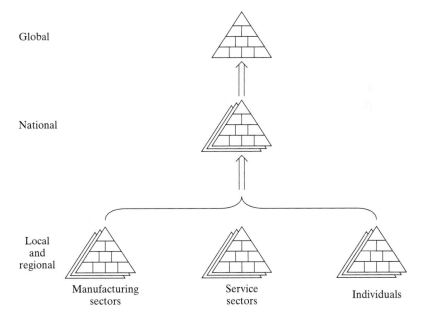

Figure 20.5

A hierarchy of metrics sets. Some of the individual components might resemble those of Figure 20.4, but all of the metrics selections would need to respond to the highest hierarchical level in order for the integrated system to be meaningful.

In either corporate or hierarchical societal metrics, if the metrics are all transformed to a common scale their values can be added to produce a composite assessment. Such highly aggregated approaches are appreciated by the nonexpert as a simple and readily understood summary, although experts will be well aware of the degree to which such approaches tend to obscure issues of data availability and quality, uncertainty, and relative risk. It is hard to imagine, however, that progress toward global sustainability can be made without an aggregated metrics system. As the global community moves toward a more sustainable mode of behavior, it will do so in part by inaugurating and utilizing a negotiated hierarchical metrics system much as the Netherlands has done on a national basis. Corporations should therefore collaborate with other public and private organizations in the development of metrics systems in order to assure that the systems design is realistic, implementable, and hierarchically consistent.

FURTHER READING

Adriaanse, A., *Environmental Policy Performance Indicators: A Study on the Development of Indicators for Environmental Policy in the Netherlands*, Sdu Uitgeverij Köninginnegracht, 1993.

Committee on Industrial Environmental Performance Metrics, *Industrial Environmental Performance Metrics: Challenges and Opportunities*, Washington, DC: National Academy Press, 1999.

Global Reporting Initiative, *Sustainability Reporting Guidelines on Economic, Environmental, and Social Performance*, Boston, 91 pp., 2002.

Hammond, A., A. Adriaanse, E. Rodenburg, D. Bryant, and R. Woodward, *Environmental Indicators: A Systematic Approach to Measuring and Reporting on Environmental Policy Performance in the Context of Sustainable Development*, Washington, DC: World Resources Institute, 1995.

Hardi, P., Trendsetters, followers, and skeptics: The state of sustainable development indicators, *Journal of Industrial Ecology, 4* (4), 149–162, 2001.

Kuik, O. and H. Verbruggen, eds., *In Search of Indicators of Sustainable Development*, Dordrecht: Kluwer Academic Publishers, 1991.

Owens, J.W., Water resources in LCA impact assessment: Considerations for choosing category indicators, *Journal of Industrial Ecology, 5* (2), 37–54, 2001.

Shane, A.M., and T.E. Graedel, Urban environmental sustainability metrics: A preliminary set, *Journal of Environmental Planning and Management, 43* (5), 643–663, 2000.

Wernick, I.J., and J.H. Ausubel, National material metrics for industrial ecology, in *Measures of Environmental Performance and Ecosystem Condition*, P.C. Schulze, ed., Washington, DC: National Academy Press, 157–174, 1999.

EXERCISES

20.1 **(a)** Select the three sustainability indicators you think are most appropriate for: (1) Sweden, (2) the United States, (3) the People's Republic of China, and (4) the Congo.

   (b) Are the indicators you selected different, and if so, why? Justify your selection of each indicator for each country. Consider in your answer the state of development, size, population relative to resources, relationship to the global economy (especially trade), existing state of the environment, and any other factors you think relevant.

(c) Given your answer to part (b), do you think the nation-state is a relevant political level for the establishment of sustainability indicators? Why or why not? Can you think of any alternatives and, if so, what are their benefits and drawbacks?

20.2 You are the operator of a facility producing metal parts, with relevant operations including casting, machining, cleaning, and packaging and shipping. Using Table 20.3, generate a set of metrics by which to evaluate your manufacturing operations. Then generate a set of metrics covering the life cycle of the part. Which set of metrics is less ambiguous, and easier to measure and control? What does this tell you about the nature of industrial environmental metrics?

20.3 You are assigned the task of developing hierarchical metrics for energy, following the framework of Figure 20.4. What precise metrics do you choose for one specific actor (individual, country, etc.) for each of the five types or regions on the figure? Explain how the local and regional metrics aggregate to give a useful picture at larger scales.

C H A P T E R   2 1

Services, Technology, and Environment

Traditional environmental regulation and management has been based on the implicit assumption that manufacturing, rather than economic activity taken as a whole, is the principal source of environmental stress. This perception reflects the visibility of environmental impacts arising from manufacturing, their associated human health impacts, and the often straightforward technological solutions. However, the assumption that regulating and managing manufacturing emissions will create an environmentally and economically sustainable world is simply wrong. It is not that emissions from manufacturing processes do not still require attention, especially in developing countries. But environmental perturbations are increasingly the result of economic activity taken as a whole, and especially in the developed world economies are increasingly dominated by services (Figure 21.1). There is a strong perception that shifting the basis of the consumer economy from selling manufactured artifacts to offering services—called servicizing—is always environmentally preferable, and offers an easy way to move toward a more sustainable economy. While this seems intuitively reasonable, it has not yet been demonstrated by rigorous analysis.

21.1   DEFINING SERVICES

A service can be defined as any commercial activity where the predominant characteristic is not the production of an artifact, or, as one wag put it, anything you buy that can't be dropped on your foot. Manufacturing an automobile is not a service activity, but leasing the automobile is, because what is being transferred in the transaction is not the title and ownership of the product, but its use. From the same perspective, if a pesticide manufacturer provides a chemical formulation to a farmer, it is selling a man-

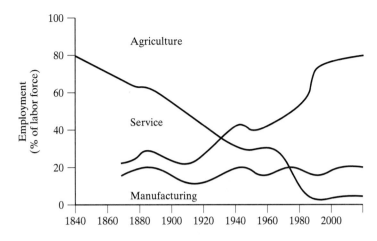

Figure 21.1

Sector employment in the United States from 1840 to 2000. Notice that the shift is not primarily from manufacturing, where employment as a percent of the labor force has remained relatively constant, to service sectors, but rather reflects the long-term mechanization of agriculture, and the concomitant release of people for employment in services. Similar patterns characterize other developed countries.

ufactured product rather than offering a service. If the pesticide manufacturer instead offers the farmer a comprehensive pest control system that includes applying pesticides, giving advice on planting rotations and crop selection, encouraging changes in mulching and plowing practices, employing biological pest controls, and performing real-time monitoring of field conditions with implanted sensors and satellite systems— a so-called integrated pest management (IPM) package—it is offering a service. Between these two extremes lie numerous business models which are some form of hybrid between traditional pesticide manufacturing and IPM.

In most developed economies, the service sector encompasses some 65 to 85% of economic activity. This proportion differs by country—the Japanese economy is more manufacturing oriented than the U.S. economy, for example—and is complicated by definitional issues. There is no universally agreed-upon definition of services, as can be seen by looking at national systems of economic accounts such as the Standard Industrial Classification Codes (SIC Codes) of the United States. Table 21.1 lists the major SIC Code categories: although Division I is identified as services, it is apparent that a number of other Divisions such as E (infrastructure services), F (wholesale trade), G (retail trade), H (financial and real estate) and J involve the provision of services as well. A breakdown of Division I indicates that while it indeed covers a number of services, it is not comprehensive. A number of these services—engineering, repair, accounting, etc.—are business-to-business transactions, rather than business-to-consumer; this is important because the degrees of interaction, expertise, and leverage are quite different.

TABLE 21.1   Standard Industrial Classification (SIC) Codes for the United States

Division	Industry
A	Agriculture, forestry and fishing
B	Mining
C	Construction
D	Manufacturing
E	Transportation, communications, electric, gas, and sanitary services
F	Wholesale trade
G	Retail trade
H	Finance, insurance, and real estate
I	Services
70 Major group	Hotels, rooming houses, camps, and other lodging places
72	Personal services
73	Business services
75	Automotive repair, services, and parking
76	Miscellaneous repair services
78	Motion pictures
79	Amusement and recreation services
80	Health services
81	Legal services
82	Educational services
83	Social services
84	Museums, art galleries, and botanical and zoological gardens
86	Membership organizations
87	Engineering, accounting, research, management, and related services
88	Private households
89	Miscellaneous services
J	Public administration
K	Nonclassifiable establishments

It is useful to divide the components of the service sector into three generic types, since they operate differently and since their environmental evaluations differ as well. These types, termed alpha, beta, and gamma services, are described below.

21.1.1   Type Alpha Services: The Customer Comes to the Service

In one common provisioning of services, alpha services, a service is provided in a fixed location and the customer travels to the service. An example is the dry cleaner, who receives clothing, treats it by one or more chemical and/or physical processes, and returns it. The product involved (the clothing) is owned by the customer, transferred temporarily to the service provider, and then recovered. From a design for environment (DfE) assessment standpoint, the impacts related to the building in which the service is provided, as well as the service itself, are chargeable to the provider. Depending on the philosophy of the assessor, the environmental impacts related to transportation could be chargeable either to the customer or to the provider. Several examples of alpha services are given in Table 21.2.

The environmental responsibility of providing an alpha service is assessed from an LCA or SLCA perspective much as one would perform a facility evaluation. Life

TABLE 21.2 Product and Process Characteristics of Typical Commercial and Industrial Service Facilities

Facility	Product	Process
Alpha services		
Dry cleaner	Clean clothing	Solvent cleaning
Hair salon	Hair maintenance	Chemical and physical treatments
Hospital	Health maintenance	Medical care
Beta services		
Appliance repair	Reconditioned appliance	Part and function maintenance
Grounds care	Property maintenance	Moving, fertilizing, etc.
Package delivery	Transport of packages	Pickup, movement, delivery
Gamma services		
Bank	Financial services	Electronic transactions
Burglar alarms	Building monitoring	Electronic communication

stage 1 (Figure 21.2) treats either facility construction or the modification required of the facility to make it suitable for the intended use. Stage 2 refers to the equipment needed to perform the service—computers, automotive repair hardware, and so forth. Stage 3a is similar to the manufacturing facility matrix, but refers to DfE characteristics of the products being provided rather than those of products being made. Life stages 3b and 4 are evaluated as for manufacturing facilities.

21.1.2 Type Beta Services: The Service Goes to the Customer

A second type of service is termed beta service, and has as its distinguishing characteristic that the provider performs the service at a customer location. Table 21.2 lists some examples, such as appliance repair in the customer's home. Providing and servicing leased photocopy machines is a common commercial example. Another type of beta service is the leasing of a television set. In this case, the material is owned by the service provider, transferred temporarily to the user, and then recovered. From the DfE analyst's standpoint, the inputs related to providing the service, including transportation, are chargeable to the provider, while those related to the facility refer to the facility from which the service originates.

    The environmental responsibility of providing a beta service is assessed similarly to alpha services, given that beta services require a location out of which to do business. The only principal distinction is that stage 3a, performing the service, is accomplished at the customer's location, and thus the environmental impacts related to transportation must be included in the stage 3a assessment.

21.1.3 Type Gamma Services: Remote Provisioning

The advent of modern technology has enabled a new class of services to arise in recent years. Termed gamma services, these are provided without either the customer travelling to the service or the service traveling to the customer (see Table 21.2). Rather, the service is provided by electronic means, such as bank services by telephone. In its more extensive implementations, gamma services permit entire careers to be accomplished

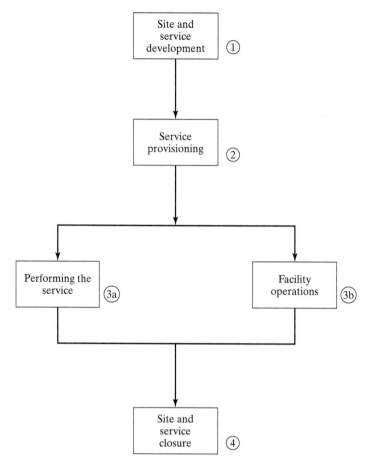

Figure 21.2

The life stages of a service.

with a minimum of physical movement, through computers, telephones, fax machines, modems, and associated software and hardware.

As with other service activities, one wishes to know how to optimize gamma services from an environmental perspective. A characteristic of gamma services is that even though there is no direct physical contact with the customer, accomplishing the service inevitably requires that hardware and supporting personnel be located within a facility of the enterprise supplying the services. For the gamma assessment, therefore, three life stages relate to the facility itself: (1) life stage 1—site selection, development, and infrastructure; (2) life stage 3b—facility operations; and (3) life stage 4—refurbishment, transfer, and closure. The other two life stages refer to the equipment used to provide the service: (4) life stage 2—equipment DfE characteristics; and (5) life stage 3a—equipment operation. In addition, stage 3a includes the environmental implications of customer interactions—energy use, customer service representatives, and so forth.

21.2   THE ENVIRONMENTAL DIMENSIONS OF SERVICES

All services operate on technology platforms: education services require school buildings, e-retailing requires transportation networks, teleworking requires information infrastructures. In fact, services tend to be slightly more capital intensive per employee than manufacturing (Figure 21.3). For example, consider the provisioning of Internet services. Not only is a physical network, with all its operational impacts, required (Figure 21.4), but the artifacts attached to that network, such as computers, have environmental impacts associated with their manufacture, use, and end-of-life stages as well. For example, a simple workstation glass panel is composed of many different oxides, some rare and/or potentially hazardous (Table 21.3).

Evaluating the environmental impacts of the service sector is not trivial, due to the vast variety of economic activities that fall into the service category. For example, operation of a major supermarket is quite different from operating a fast food hamburger chain. And both of these services, oriented toward food, are obviously different than running a hospital, operating a wireless telephone system, or running a bank.

These examples illustrate the difficult issue of boundary definition. In the case of manufacturing operations, or even products, drawing boundaries for purposes of life cycle assessments is relatively straightforward. But understanding what the appropriate boundary is for the case of services is much more complex. Consider, for example, an assessment of a large transoceanic passenger airplane made by Airbus or Boeing. Techniques that would permit life-cycle assessment of the airplane as an artifact can be applied fairly directly. But the airplane as a service entity is much more complex. Airplanes are enablers of higher rates of tourism, which in turn puts increased pressure on

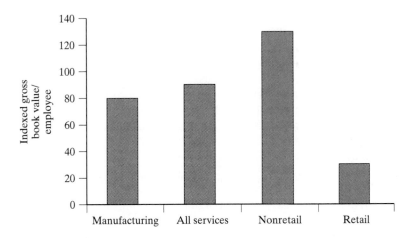

Figure 21.3

The capital intensity per employee of services as opposed to manufacturing activities actually tends to be slightly higher, especially in the nonretail sectors. While such figures should not be taken as quantitatively definitive because of the difficulty of measuring services in most economic systems of accounts, they do caution against the naive view that services are necessarily less resource intensive than manufacturing/ artifact sale systems.

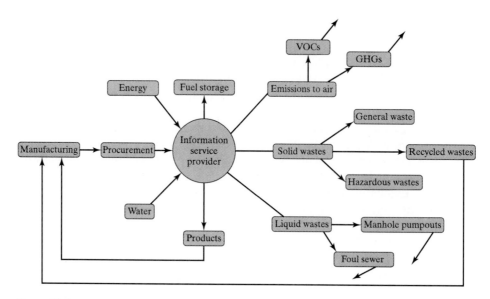

Figure 21.4

The information networks operated by telecommunications companies provide a framework for many environmentally preferable services, such as teleworking, but they have their own environmental impacts associated with their operation. (Adapted from *Annual Report,* London: British Telecom, 1997.)

destination locations, especially if they are ecologically sensitive, requiring lodging, water and energy, and local transport. How much of this activity is appropriately attributable to the airplane—or, conversely, can be ignored when thinking about the airplane? Is the relevant industrial ecology system the artifact, or the tourism industry of which the artifact is a critical part?

Although the environmental impacts of services are substantial, services can also offer the potential for discontinuous improvements in environmental performance. This occurs because a service can potentially provide not just improvements internal to the service firm itself, but also can be deployed across the economy to reduce the environmental impacts of economic activity as a whole. An example is teleworking. Firms

TABLE 21.3 Composition Ranges for Workstation Panel Glass (by percent weight)

SiO_2	60–65
SrO	8–12
Na_2O	7–12
K_2O	8–10
BaO	2–10
PbO	0–3.6
Al_2O_3	.4–2.8
CaO	2.0–3.0
CeO	0–2.5
ZrO_2	0–1.8

that provide teleworking infrastructure can certainly reduce their environmental footprint by using their own service, but this is a relatively small part of the service's environmental potential. The real gains come when teleworking services are rolled out throughout the economy, thereby generating the potential for discontinuous improvements in environmental efficiency far beyond the service itself.

21.3   THE INDUSTRIAL ECOLOGY OF SERVICE FIRMS

With the exception of governmental services, most services are offered by firms that are separate from those that manufacture the supporting technology platforms. From an industrial ecology perspective, these service firms can fulfill five environmentally important roles:

1. Leverage suppliers to improve the environmental efficiency of the supply chain
2. Educate consumers
3. Facilitate environmentally preferable resource consumption and product use
4. Substitute less environmentally intensive services for material and energy use
5. Change the predominant source of perceived quality of life from ownership and consumption of products, to services

21.3.1   Leverage Suppliers

Given the evolution of the service economy, many service firms have grown large in relationship to their suppliers, thereby giving them substantial leverage over the environmental performance of their supply chain. In the retail sector, for example, firms of the scale of Walmart, Office Depot, or Staples can urge or require their suppliers to adopt environmentally appropriate practices. In theoretical terms, what the suppliers are doing by these actions is internalizing externalities. That is, by integrating new requirements based on environmental criteria into their purchasing practices, they bring costs and benefits that were previously not part of the economy into it. In this vein, a number of large multinational firms that retail lumber and wood products have begun requiring environmental certification from their suppliers, partly in order to publicly demonstrate their environmental commitment.

Service firms must be careful how they exercise their leverage. Most service firms are not expert in the technologies or environmental considerations that are embedded in the products that they use as platforms for their services, and are thus ill equipped to make specific recommendations regarding product design. A better course in most cases is to require that suppliers use LCA or DfE methodologies in their design processes, and select the options that enhance the service firm's own environmental performance—in other words, to focus on raising the environmental awareness of their suppliers rather than trying to micro-manage product design. Additionally, service firms can implement structural changes in their supplier relationships that encourage environmentally preferable design. For example, a large telecommunications firm can purchase switches and routers from a supplier on condition that the supplier take the products back when no longer needed, and refurbish or properly dispose of them.

Cooperative Packaging Reductions

The volume of packaging and its environmental implications is a continuing challenge to service sector firms. To address this issue, Sears Roebuck & Co. worked with 2300 of its suppliers to reduce both the amount and type of packaging. The company proposed four goals: (1) reduce the volume and weight of a product's packaging material by at least 10%; (2) increase the level of recycled material in corrugated containers to 25%; (3) increase the use of recycled materials in plastic containers to 20%; and (4) utilize the highest recycled content materials possible in other types of packaging. At the end of the two-year program, packaging volume in Sears stores had been reduced by more than 25% overall, operating costs for handling packaging were reduced, and environmental impacts were decreased.

Adapted from *The Role of the Retailer in Sustainable Production and Consumption,* Geneva: World Business Council on Sustainable Development, 1996.

21.3.2 Educate Consumers

Many service firms are the interface between the consumer and underlying technologies. Retailers can provide environmental and health information on packages and as part of their displays, financial institutions can offer lower credit and technical support to borrowers who use environmentally preferable management systems, and telecommunications firms can encourage their customers to use electronic, rather than paper, billing. Similarly, business to consumer e-commerce firms can put a few sentences in their ordering forms to the effect that selecting ground transportation and one-week delivery times (as opposed to air transportation and two-day delivery times) can minimize energy consumption in package delivery. This preserves consumer choice while educating consumers about some of the environmental implications of their consumption.

21.3.3 Facilitate Environmentally Preferable Resource and Product Use Patterns

When one shifts from offering a product to offering a service based on the product, the new approach frequently changes incentive structures so as to facilitate environmentally preferable resource consumption and product use. This transition can occur even when the reasons for making this transition are wholly economic. A good example is painting operations in automobile manufacturing. The traditional pattern has been that auto manufacturing companies would buy coatings from suppliers and paint the cars themselves. Under the servicized model, the coating supplier manages the materials and operations to produce a painted body for the automobile manufacturer. In other words, rather than simply sell paint, the coatings manufacturer sells the service of painting automobiles. The automobile manufacturers, in turn, pay not by the unit of paint consumed, but on a per painted car basis. From the manufacturers' perspective, the benefits include greater efficiency, lower cost per unit, and higher quality coating operations (it is reported that implementing this service arrangement at Chrysler's Belvedere Neon assembly plant in the United States saved over a million dollars in its first year, a sum that was split between the automobile manufacturer and the coating supplier). From the perspective of the coating supplier, the benefits include not only the resulting profit, but also a competitive advantage: Since the service is more complex than just selling coatings, and requires a closer integration between the auto manufacturer and the coating supplier, the manufacturer is less likely to shift suppliers.

None of these benefits overtly involves environmental considerations, but the environment benefits by the shift in incentives. When a supplier makes its money based solely on the amount of product it ships, it has an incentive to ship as much as possible. Under the old business model, coating suppliers preferred automobile companies to use their products inefficiently. Under the new system, however, the coating supplier directly captures a proportion of the benefits that accrue from using less material per car, from minimizing the production of waste and its toxicity (and thus cost of management), and from developing coatings designed to maximize functionality. From a systems perspective, these environmental efficiencies arise because an inappropriate boundary, drawn between the design and production of the coating and its application in a manufacturing environment, was eliminated.

21.3.4 Substitution of Services for Energy and Material Use

In some instances, one can substitute services for existing activities involving high levels of material and energy use, thus reducing environmental impact per unit quality of life. Perhaps the most familiar examples involve the substitution of information services of various sorts for physical products or energy-intensive activities such as commuting by personal automobile. Thus, teleworking substitutes a home office and information technology for commuting and a commercial office space. Especially where commuting involves use of personal automobiles for long trips under highly congested conditions, the environmental benefits can be substantial.

In many cases, quality of life appears to be enhanced by substitution services. For example, AT&T has a large population of teleworkers—at least 25% of its managers telework at least once a week, including some 10% in virtual offices where they work from home full time—and surveys show that such workers are happier with their jobs and more productive than those that work full time in an office. Not only that, but their families report more satisfaction with the working structure as well. Accordingly, it is not appropriate, at least in some cases, to view servicizing as a substitute for something else; services may in fact be a significant improvement over the status quo.

21.3.5 Services as a Source of Quality of Life

As services become a more important component of the global economy, they necessarily become a more significant source of perceived quality of life. While there are obviously limits to the possibilities of servicizing the economy—in particular, material and energy flows must, by definition, remain an underpinning of any physical system—such a process has significant potential benefits. In particular, as services become the focus of perceived quality of life, it becomes possible to manipulate underlying technology platforms in ways that might not have been possible before. The focus of the consumer shifts from product attribute to service attribute, which in turn can provide product designers with more degrees of freedom in the design process, which may in part lead to more environmentally efficient products. For example, in the now-outdated model of national telephone monopolies, most of the companies supplied telephones and other products on a lease basis. Thus, the customer purchased the use of a telephone, not the telephone itself, and should the telephone become nonoperational it was simply returned for another. The telephones themselves were refurbished and recycled, sometimes for decades.

A product that is designed for a service environment will usually be different than one designed for production and sale. Especially where product takeback is part of the service environment, costs that may have been previously external to the manufacturer (and designer) of the product are now internalized to that firm, and over time affect the way the product is designed. Rather than including just (inexpensive) initial cost, for example, the firm's cost function must expand to include such elements as reverse logistics systems (to get the product back), and the cost of disassembling, refurbishing, or recycling the product, its components, or its materials. The result in most cases of this evolution of product from standalone to service platform is greater environmental efficiency.

FURTHER READING

Chemical Strategies Partnership, *Chemical Management Services: Industry Report 2000*, San Francisco: Chemical Strategies Partnership, 2000.

Davies, T. and D.M. Konisky, *Environmental Implications of the Foodservice and Food Retail Industries*, Discussion Paper 00-11, Washington, DC: Resources for the Future, 2000.

Davies, T. and A.I. Lowe, *Environmental Implications of the Health Care Service Sector*, Discussion Paper 00-01, Washington, DC: Resources for the Future, 2000.

Goedkoop, M.J., C.J.G. van Halen, H.R.M. te Riele, and P.J.M. Rommens, *Product Service Systems: Ecological and Economic Basics*, commissioned by the Dutch Ministries of Environment (VROM) and Economic Affairs (EZ) from PricewaterhouseCoopers N.V./PiMC, 1999.

Graedel, T.E., The life-cycle assessment of services, *Journal of Industrial Ecology, 1 (4)*, 57–70, 1998.

Lifset, R., Moving from products to services, *Journal of Industrial Ecology, 4 (1)*, 1–2, 2000.

Reiskin, E.D., A.L. White, J.K. Johnson, and T.J. Votta, Servicizing the chemical supply chain, *Journal of Industrial Ecology, 3* (2,3), 19–31, 2000.

Rejeski, D., An incomplete picture, *The Environmental Forum, 14* (5), 26–34, 1997.

World Business Council for Sustainable Development, *The Role of the Retailer*, Geneva, 1998.

EXERCISES

21.1 Choose a service facility with which you are familiar, perhaps because you have been employed in helping to provide the service or because you use the service frequently. Classify the service as alpha, beta, or gamma, describe the service provided, and the process by which it is provided.

21.2 What recommendations would you make for improving the environmental attributes of the service facility chosen in Exercise 21.1? How would you prioritize the recommendations you make, and what are the bases for the prioritization?

21.3 A hospital can monitor the status of patients with heart irregularities either by having them visit the outpatient clinic once a week or by giving each patient an electronic monitor and periodically transferring data electronically from home to clinic. Assume the medical care is equivalent. Discuss the environmental advantages and disadvantages of the two options.

PART V   Systems-Level Industrial Ecology

C H A P T E R   2 2

Industrial Ecosystems

22.1   THE ECOSYSTEMS CONCEPT

In Chapter 4 we explored the relevance of biological ecology to technology from the perspective of organisms (manufacturing facilities) and populations (families of products). We demonstrated that tools of biological ecology (BE) could aid industrial ecology (IE) in such areas as studies of resource utilization, production efficiency, and product synergism. The focus throughout was on the individual organism or a group of organisms.

The next hierarchical level to consider is the ecosystem. In BE, an ecosystem consists of the interacting parts of the physical and biological worlds. By analogy, an industrial ecosystem consists of the interacting parts of the technological and nontechnological worlds. The interactions among participants in an ecosystem generally involve the transfer of resources, often in ways or along paths that would not be intuitively anticipated. Such a structure expands upon the notion of the food chain presented in Chapter 4.

Food chains imply a linear flow of resources from one trophic level to the next, as in Figure 22.1a. In such a construct, the interspecies interactions are straightforward. No resource flow systems in BE follow this simple structure, however; they resemble much more the web structure of Figure 22.1b. Here species at one trophic level prey upon several species at the next lower level, and omnivory is common, as in Figure 22.1c. Finally, a fully expressed food web can exhibit all the various features: multiple trophic levels, predation, and omnivory (Figure 22.1d).

Food web analysis has two important goals: to study the flows of resources in ecosystems and to analyze ecosystems for dynamic interactions. Consequently, con-

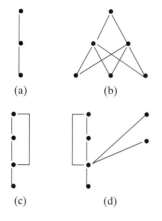

(a)   (b)

(c)   (d)

Figure 22.1

Diagrams of food chains and food webs. Filled circles represent species; lines represent interactions. Higher species are predators of lower species, so resource flow is from bottom to top. Species and rates of interactions vary with time. (a) A food chain in a three-level trophic system. (b) A food web in a three-level trophic system. (c) Omnivorous behavior in a food chain. (d) A food web encompassing multiple trophic levels, predation, and omnivorous feeding behavior.

structing the food web diagram is a prelude to developing the source and sink budget for the resource of interest. In concept, organisms are identified, their trophic levels assigned, and their interactions specified. The food web is then diagrammed, and resource flows and stabilities are analyzed.

In BE, many (but certainly not all) food webs have large numbers of primary producers, fewer consumers, and very few top predators, giving a web of the form in Figure 22.1b. Omnivores may be scarce in these systems, whereas decomposers are abundant. Food web models have provided a potential basis for fruitful analyses of resource flows in both BE and IE. Difficulties arise, however, when one wishes to quantify the resource flows and subject the web structure and stability properties to mathematical analysis. Many of the needed data turn out to be difficult to determine with certainty, particularly with organisms that appear to function at more than one trophic level. This characteristic is not a major complication for resource flow studies, but it seriously complicates stability analysis. The contention that more complex communities are more stable—because disruptions to particular species or flow paths merely shunt energy and resources through other paths rather than putting a roadblock on the entire energy or resource flow—remains a hotly debated topic.

Food web studies in BE are often centered around the question, "Given a perturbation of a certain type and size, how will the ecosystem respond?" The perturbation may be an infestation, the loss of a routine food source, or a severe climatic event. One view is that an ecosystem with modest frequencies and intensities of disturbance may be the most stable, a view that seems in harmony with what we know of industrial ecosystems as well.

Food webs are formed as opportunistic responses to local resource availability and desirability. An example for a biological system is shown in Figure 22.2a. In this diagram, the position of the populations indicates their normal location with respect to the ocean surface. The connecting arrows indicate the flow of resources from one population to another. Diagrams such as this one are often drawn with the lowest trophic level at the bottom and the top predator at the top; were the present system so drawn it would, by the dictates of marine environments, be rather similar. Even without that presentation, it is clear that omnivory is common—the anchovy is fed upon by everything from mackerel to fur seals, for example.

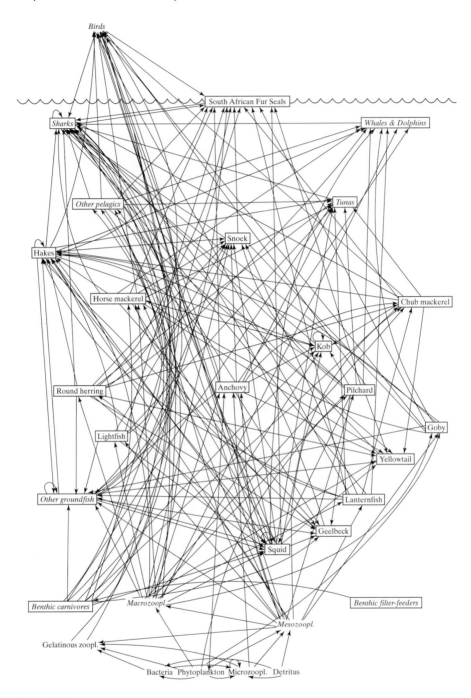

Figure 22.2

(a) A food web for the Benguela marine ecosystem off the southwest coast of South Africa. (Reproduced with permission from P.A. Abrams, B.A. Menge, G.A. Mittelbach, D.A. Spiller, and P. Yodzis, The role of indirect effects in food webs, in *Food Webs*, G.A. Polis and K.O. Winemiller, eds., New York: Chapman & Hall, 371–395, 1996.)

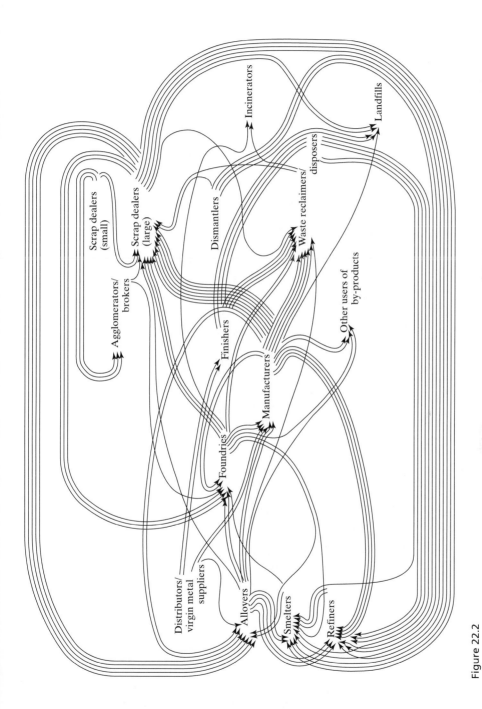

Figure 22.2

(b) A diagrammatic representation of an industrial food web. In this diagram, each line represents one set of transactions between categories of interviewed firms. For example, the transactions of a firm that might send its wastes to four different waste disposal companies appears as only one line. Similarly, a firm that might buy from three different alloyers also receives a single line for these transactions. The magnitudes of the flows are not captured on this diagram. Also, flows to and from the consumers of the industrial goods are not shown in this diagram, nor are fugitive emissions to air, water and land. (Reproduced with permission from A.D. Sagar and R.A. Frosch, A perspective on industrial ecology and its application to a metals-industry ecosystem, *Journal of Cleaner Production, 5*, 39–45, 1997.)

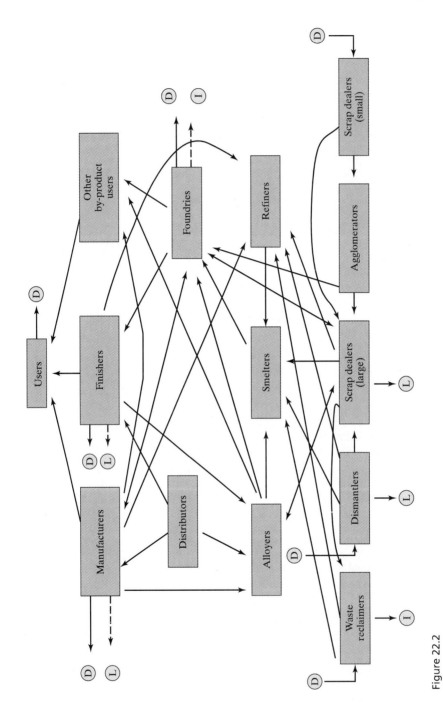

Figure 22.2

(c) The industrial food web of Figure. 22.2c, redrawn as is customary in biological ecology with the lowest trophic level at the bottom and the highest at the top. D = detritus, L = loss to landfill, I = loss to incineration.

Figure 22.2b shows an industrial food web for the flow of copper among industries in the Boston area. Termed a "spaghetti diagram" by its author, it appears so only because of the approach used to create the display; it could alternatively have been drawn to have much the same format as the BE example, as we do in Figure 22.2c.

22.2   INDUSTRIAL SYMBIOSIS

Symbiosis is a recurring situation in BE. It is the intimate association of two species, either for the benefit of one of them (parasitic symbiosis) or for both of them (mutualistic symbiosis). BE symbiosis generally involves a long period of coevolution. The concept can be applied as well to technological systems, the difference being that IE symbiosis may occur opportunistically or can be planned. Planned industrial symbiosis appears to offer the promise of developing industrial ecosystems that are environmentally far superior to unplanned ones.

Marian Chertow of Yale University has designated such systems "eco-industrial parks" (EIPs), and separates them into five types. We describe these types below, each with an example.

1. Type 1 EIP: Through waste exchanges
    In these situations, recovered materials are donated or sold to other organizations. Some of the exchanges are informal or opportunistic, while others are formalized through waste exchange networks. A common example is the automobile scrap yard, which recovers and sells automobile components and prepares the bulk metal body and chassis for recycling. These interactions are essentially unplanned, however, so that the exchange of resources is insufficient to regard Type 1 EIPs as examples of industrial symbiosis.

2. Type 2 EIP: Within a facility, firm, or organization
    In this type of EIP, materials or products are exchanged within the boundaries of a single organization, but among different organizational entities. This is a common approach to the design of petrochemical complexes, for example, where a byproduct from one chemical process serves as the feedstock to another.

3. Type 3 EIP: Among co-located firms in a defined industrial area
    In this type, corporations or other entities located close together, perhaps in an industrial park, organize themselves to exchange energy, water, and materials. An example of such a system is the Monfort Boys Town in Suva, Fiji (Figure 22.3). Here brewery waste inaugurates an EIP that involves mushroom, pig, fish, and vegetable farming.

4. Type 4 EIP: Among nearby firms not co-located
    Type 4 systems are exemplified by that at Kalundborg, Denmark, in which a number of firms within a 3-km radius exchange steam, heat, fly ash, sulfur, and a number of other resources (Figure 22.4). Not designed as an EIP, Kalundborg became one as a consequence of a series of "green twinning" trades, each economically advantageous.

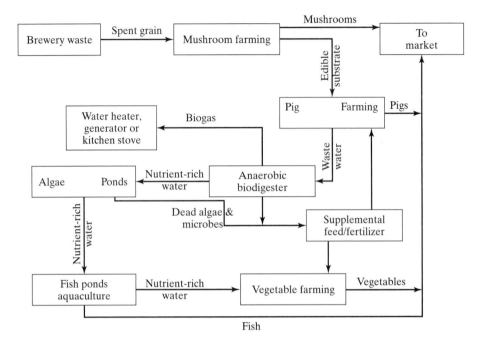

Figure 22.3

The flows of resources in the integrated biosystem of the Monfort Boys' Town in Suva, Fiji. (Courtesy of R. Klee, Yale University.)

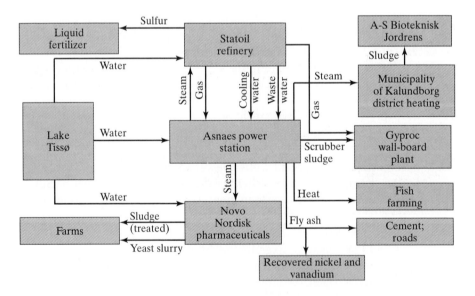

Figure 22.4

Flows of resources in the eco-industrial system at Kalundborg, Denmark. (Courtesy of M. Chertow, Yale University.)

5. Type 5 EIP: Among firms organized across a broader region
The final type consists of exchanges across a broad spatial region. In principle, it can incorporate any or all of the EIP types described above. For a Type 5 EIP to succeed, and none have yet been fully realized anywhere, it would probably require an active management organization to identify additional twinning opportunities and recruit new participants.

As can be seen from this discussion, EIP typology designations are evolutionary, not static. Simple Type 1 systems can become Type 4 or even Type 5 over time, for example. Indeed, EIPs can be consciously developed, often around a key tenant such as a power plant that can readily begin the resource exchange process with a wide variety of potential industrial partners.

22.3   DESIGNING AND DEVELOPING SYMBIOTIC INDUSTRIAL ECOSYSTEMS

How might one design an efficient industrial ecosystem? It seems clear that such a system would need to involve a broad sectoral and spatial distribution of participants, and be flexible and innovative. It is likely that several EIP types (see above) would be included. This concept is sketched in Figure 22.5. The system illustrates Types 3, 4 and 5

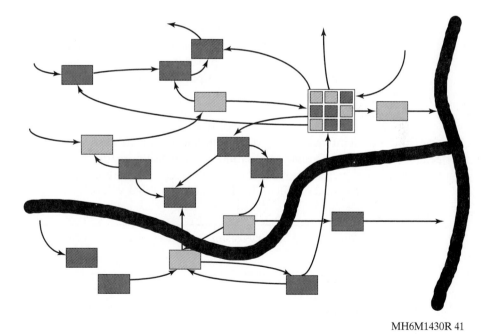

MH6M1430R 41

Figure 22.5

A scenario for a Type 5 eco-industrial system. A central Type 3 system is embedded within, but most resource exchanges occur over longer distances. In this vision, dark squares and rectangles indicate existing firms, light polygons firms that might be added to improve connectance. (Courtesy of David Cobb, Bechtel Corporation.)

components, and it is likely that any such system would have Type 1 and perhaps Type 2 interactions as well.

A key to an effective broad-scale EIP is a high degree of synergy between input and output flows of resources. BE is useful here in providing several statistics for food web analysis. The simplest is species richness (S): the number of different types of organisms contained within the system. The second is connectance (C). C is derived by constructing a community matrix, as shown in Table 22.1 for the web of Figure 22.1b. On the table, a numeral one is entered for each consumer that receives resources from a given producer. If no resource transfer occurs, a zero is entered. C is then calculated by

$$C = 2L/(S[S-1]) \tag{22.1}$$

where L is the number of nonzero interaction coefficients in the community matrix. For the ecosystems of Table 22.1 and Figure 22.1b, $S = 6$ and $C = 16/30 = 0.53$.

Few industrial food webs have been analyzed in this way, but those that have include the ones pictured in Figures 22.2c, 22.3, and 22.4. When those and 16 others have their connectance computed, the median turns out to be very similar to that of biological systems (Figure 22.6). This is a somewhat surprising result, in view of the possible reasons why we might expect $C_{IE} < C_{BE}$:

- Industrial ecosystems are in earlier evolutionary states than biological ecosystems, and connectance generally increases as an ecosystem matures.
- BE organisms all exchange organic nutrients of rather similar chemical form. IE organisms, however, have much broader feeding habits—some desire petrochemicals, some metals, some forest products. In such a system mismatches between output flows and input needs act against high connectance.

Analyses of a large number of industrial food webs may thus start to reveal characteristics not demonstrated by other approaches. In the ecosystem design of Figure 22.5, for example, web analysis may indicate a missing sector or type of industrial activity that could increase connectance. Conversely, such studies might help address a question that remains as difficult for BE as for IE: What will be the response of a par-

TABLE 22.1   Community Matrix for the Ecosystem of Figure 22.1b.

Prey species*	Consuming species					
	1	2	3	4	5	6
1	—	0	0	1	1	0
2	0	—	0	1	1	0
3	0	0	—	1	1	0
4	0	0	0	—	0	1
5	0	0	0	0	—	1
6	0	0	0	0	0	—

*Species at the lowest trophic level are numbered 1, 2, and 3; those at the intermediate level, 4 and 5; and that at the top level, 6.

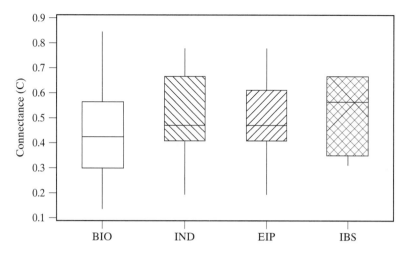

Figure 22.6

Box plots of connectance in different types of food webs. BIO = 113 biological food webs; IND = 19 industrial food webs, consisting of 15 eco-industrial parks (EIP) and 4 integrated biosystems (IBS). (Biological data are from F. Briand and J.E. Cohen, Environmental correlates of food chain length, *Science, 238,* 956–960, 1987; industrial data are from C. Hardy and T.E. Graedel, Industrial ecosystems and food web theory, *Journal of Industrial Ecology, 6,* in press, 2002.)

ticular ecosystem to a particular type of perturbation? These topics provide a rich area for detailed investigation.

22.4   RESOURCE FLOW IN INDUSTRIAL ECOSYSTEMS

The flow of nutrients and energy in individual organisms, both biological and industrial, was described in Chapter 4. In much the same way, flows for entire ecosystems can be studied. The BE version is shown in Figure 22.7a. Here, in a Type II system, nutrients largely recycle but energy flow is strictly linear. The system is stable over time because on time scales of hundreds of million of years the energy source, the Sun, is inexhaustible.

The resource flow diagram for a contemporary Type I industrial ecosystem is shown in Figure 22.7b. Two differences between this system and that of Figure 22.7a are obvious: The IE system as shown is linear in *both* energy and nutrients, and energy is injected at both the extractotroph and the transformotroph levels. The system is unstable over time for two reasons: (1) It will ultimately run out of material in the virgin material pool; and (2) it will ultimately run out of energy, since its energy sources are primarily nonrenewable. To ultimately achieve complete sustainability, industrial ecosystems must solve the first problem by completely recycling nutrients and the second by utilizing only renewable energy; the result would be a diagram of the form shown in Figure 22.7c. Robert Frosch proposes that two of the most important components in making this cycling fully realizable are disassembly technology (to take products apart in selective ways) and "negentropy" technology (to sort mixtures into bins

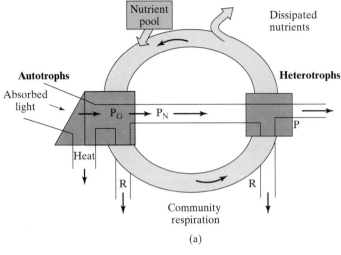

(a)

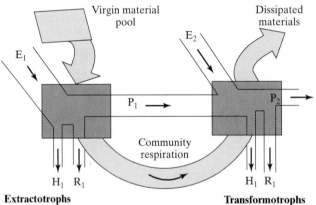

(b)

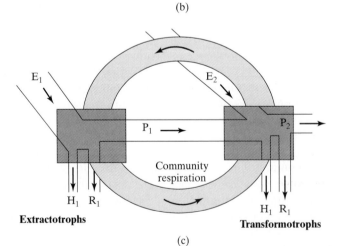

(c)

Figure 22.7

(a) Flows of resources in a Type II biological ecosystem. In such a system, the flow of energy is linear while that of nutrients is cyclic. P_G = gross production, P_N = net production, P = heterotrophic production, R = respiration. The widths of the bands are rough indications of relative flow magnitudes. Reproduced with permission from R. E. Ricklefs, *The Economy of Nature*, 3rd ed., New York: W. H. Freeman, 1993. (b) Flows of resources in a traditional industrial ecosystem. E = energy ingestion, H = heat, R = respiration, P = production. This is the thermodynamic version of the diagram for a Type I industrial ecosystem (see Figure 4.8). (c) Flows of resources in a hypothesized Type III industrial ecosystem. This is the visionary goal for a global technological society, now beginning to create Type II industrial ecosystems of the type shown in Figures 4.9, 22.3, and 22.4.

of things people actually want); both of these components are seriously underdeveloped. As a result, a fully realized Type III industrial system is not a realistic goal at present, but is an attractive target toward which to direct our aim.

Resource flow studies are difficult to perform in industrial ecosystems, and few examples are available. One that provides insight into typical flow behavior, however, was developed by Paul Brunner and colleagues for the region of St. Gallen in northern Switzerland. The overall system flows are shown in Figure 22.8. Water is by far the largest of the input flows, followed by air and construction materials. Water and air also dominate the output flows. It is clear that this is a Type I system, with no recycling significant enough to be shown. Of particular interest is the realization that output flows are less than input flows; thus, St. Gallen's stock of industrial nutrients is growing. This phenomenon has been seen in other IE resource flow studies, and identified as largely an increase in the stock of construction materials in buildings, highways, and other structures. In the short term, this growth of stock may not be problematic. It is clearly an unsustainable trend, however, and at some point stock discard will begin to balance stock buildup.

22.5   PATTERN AND SCALE IN INDUSTRIAL ECOSYSTEMS

It is readily apparent that the natural world is not spatially uniform. Different habitats—river banks, forests, fields—support ecosystems of different richness, connectance, and functionality, even under similar climatic conditions. Realizations such as this form the basis for landscape ecology, a growing field within BE. It appears that landscape ecology may have a role to play within IE as well.

As with the natural world, human technology, population, and environmental impacts are patchy. One obvious example is the existence of cities, and the relationship of urban areas to IE is as yet a poorly-explored phenomenon. It is also one of growing importance, as the world's population is becoming increasingly urbanized (Figure 22.9).

A region that contains large numbers of humans is also a region to which resources flow at high rates. In the more developed world, the per capita flows of resources amount to about 50 kg per day. A large urban conurbation of 10 million inhabitants must therefore deal with inputs of some 500,000 tonnes of "stuff" per day, water and air excluded. The inputs are extremely varied, as are their sources: apples from New Zealand, clothing from Central America, automobiles from Germany, steel from China. Fifty to sixty percent of that flow, mostly food and fossil fuels, is passed through as output relatively quickly. The output is not returned to its sources, however, but tends to be deposited in landfills or exhausted to freshwater reservoirs near the conurbation. Large urban areas are strong attractors, but weak dispersers.

What are the implications of urban resource flows? Historically, their spatial concentration has been largely undesirable, as seen, for example, by the generation of photochemical smog in Los Angeles, although a uniform human population density pattern throughout the state might have been no better for the environment overall, only different. Nonetheless, one can imagine that the re-utilization of resources might be enabled or enhanced in cities, that they could become, in the words of Jane Jacobs, "The largest, most prosperous cities will be the richest, the most easily worked, and the most inexhaustible mines." As an example, consider the contemporaneous use of sil-

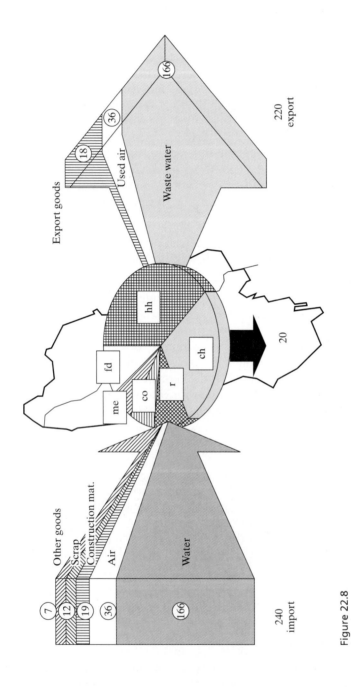

Figure 22.8

Regional flux of goods through the anthroposphere of St. Gallen, Switzerland (tonnes per capita per year). (hh = Private households, ch = Production of chemicals, co = Construction, me = Metal processing, fd = Processing of food and drink, r = Others). (Reproduced by permission from P.H. Brunner, H. Daxbeck, and P. Baccini, Industrial metabolism at the regional and local level: A case-study on a Swiss region, in *Industrial Metabolism: Restructuring for Sustainable Development*, R.U. Ayres and U.E. Simonis, eds., Tokyo, United Nations University Press, 163–193, 1994.)

Level of urbanization (percent)

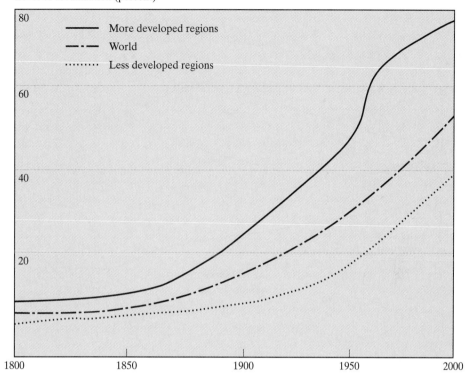

Figure 22.9

Increases in the level of urbanization, 1800–2000. (Reproduced by permission from B.J.L. Berry, Urbanization, in *The Earth as Transformed by Human Action*, B.L. Turner, et al., eds., Cambridge, U.K.: Cambridge University Press, 103–119, 1990.)

ver. About half of it is utilized in photography, and 40% of that amount is used in X-ray imaging. Studies have shown that X-ray films retain much of that silver following film developing. In other words, the X-ray files of dentist's and radiologist's offices are silver mines! Further, modern computer technology permits the information on the films to be scanned and captured electronically. Should our supplies of virgin silver run short for any reason, therefore, urban regions provide alternative sources.

Radiologists' and dentists' offices are not, of course, uniformly distributed in space; they tend to be clustered together in large medical complexes. Figure 22.10 shows a portion of the city of Boston, on which three such complexes are shown. For the urban miner in Boston, it is highly likely that these complexes are the richest veins of silver in the city.

22.6   THE UTILITY OF MIXED ECOLOGICAL APPROACHES

From the discussion in this chapter, and in Chapter 4, there seems little doubt that tools and concepts can be shared between BE and IE. At a minimum, IE can benefit in the following ways:

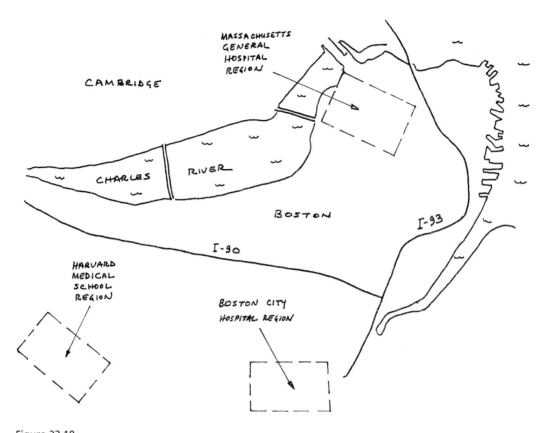

Figure 22.10

An example of a spatial resource pattern in industrial ecology: Silver in medical radiologists' offices in Boston, MA.

- The BE food web approach aids in the understanding of the structure and functioning of industrial ecosystems.
- Food web analysis aids in the identification of missing species or trophic levels in an industrial ecosystem.
- Ecosystem tools aid in conceptualizing and understanding resource use, time delays, potential for resource reuse, and spatial patterns in IE.

IE can also be of service to BE, especially in providing information on ecosystem impacts. As more information becomes available on spatially resolved IE residue flows, and on the chemical nature and integrated magnitude of those flows, the responses of biological systems to technological stresses can be determined with increasing precision and perspective.

FURTHER READING

Allenby, B.R., and W.E. Cooper, Understanding industrial ecology from a biological systems perspective, *Total Quality Environmental Management*, 343–354, Spring, 1994.

Andrews, C.J., Building a micro formulation for industrial ecology, *Journal of Industrial Ecology, 4* (3), 35–51, 2001.

Chertow, M.R., Industrial symbiosis: Literature and taxonomy, *Annual Reviews of Energy and the Environment, 25*, 313–337, 2000.

Chertow, M. (Ed.), *Developing Industrial Ecosystems: Approaches, Cases, and Tools,* Bulletin No. 106, New Haven, CT: Yale School of Forestry & Environmental Studies, 2002.

Ehrenfeld, J., and M. Chertow, Industrial symbiosis: The legacy of Kalundborg, in *A Handbook of Industrial Ecology*, R.U. Ayres and L.W. Ayres, eds., Cheltenham, U.K.: Edward Elgar Publishers, 334–348, 2002.

Forward, G., and A. Mangan, By-product synergy, *The Bridge, 29* (1), 12–15, 1999.

Jacobs, J., *The Economy of Cities*, New York: Random House, 1970.

Keckler, S.E., and D.T. Allen, Material reuse modeling: A case study of water reuse in an industrial park, *Journal of Industrial Ecology, 2* (4), 79–92, 1999.

Sagar, A.D., and R.A. Frosch, A perspective in industrial ecology and its application to a metals-industry system, *Journal of Cleaner Production, 5*, 39–45, 1997.

EXERCISES

22.1 Compute the richness and connectivity of the industrial food web of Figure 22.3.

22.2 Compute the richness and connectivity of the industrial food web of Figure 22.2c.

22.3 Select a local industrial park for study, and identify the principal input and output flows. Determine any existing "green twinning" relationships and propose others for consideration.

22.4 The food web of Figure 22.2c pictures many types of actors at several different trophic levels. For any single actor type (smelters, users, etc.) do input flows equal output flows? Why or why not?

CHAPTER 23

Metabolic and Resources Analyses

23.1   BUDGETS AND CYCLES

Ecology is a concept with cycles at its very heart, and cycles are analyzed by means of materials and energy budgeting. Nearly everyone is familiar with the concept of a household or personal financial budget, whether or not she or he is conscientious about making and sticking to it. An approach very similar to that of financial budgeting is used to fashion budgets in industrial ecology. The situation can be appreciated with the aid of the diagram in Figure 23.1, which shows a tub receiving water from several faucets and having a number of drains of different sizes. When the water is supplied at constant (but probably different) rates by all the faucets and is removed at an equal total rate by drains with (probably) different capacities, the water level remains constant. When the tank is very large, however, and has some wave motion that makes it difficult to tell whether the absolute level is changing, an observer may have difficulty telling whether the system is in balance or not. In that case, he or she may try instead to measure the rate of supply from each of the faucets and the rate of removal in each of the drains over a period of time to see whether the sums are equivalent. A part of this technique involves the determination of the pool size (the quantity of water in the tank) and either the rate of supply or the rate of removal. Determination of changes in the pool size then gives information about rates that are difficult to measure. The process of estimating or measuring the input and output flows and checking the overall balance by measuring the amount present in the reservoir constitutes the budget analysis.

Suppose that the input from one of the sources is increased, i.e., in our analogy, the flow from one of the faucets increases. Will the water level keep increasing? The

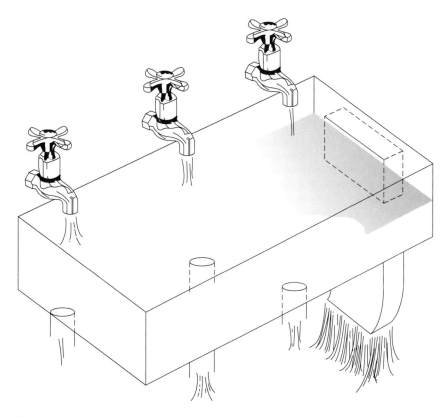

Figure 23.1

A simple conceptual system for budget calculations. The water level in the tub is determined by the water flows in and out, as discussed in the text.

answer depends on whether one of the drains can accommodate the additional supply, as can the trough drain at the right side of the tank in the figure. If no such drain is present, then the water level will indeed increase. Conversely, if the flow into a drain is enhanced for some reason, such as the removal of an obstruction, then the water level will decrease in the absence of a corresponding increase in the supply. Such a process will continue in this manner unless the flows through the drains adjust themselves to this new factor or unless the new factor results in other drains changing their functioning.

All of the circumstances mentioned above occur in budgets devised for various industrial ecology studies, and all budgets involve the same concepts. One is that of the reservoir, in which material is stored. Examples include the shipping department where completed products are prepared for forwarding to customers, or the atmosphere as a whole, where emissions of industrial vapors collect and react. A second concept is that of flux, which is the amount of a specific material entering or leaving a reservoir per unit time. Examples include the rate of evaporation of water from a power plant cooling tower or the rate of transfer of ozone from the stratosphere to the troposphere. Third, we have sources and sinks, which are rates of input and loss of a specific material within a reservoir per unit time. A system of connected reservoirs that transfer and conserve a specific material is termed a cycle.

Industrial ecology budgets have the same three basic components as those for the tank in Figure 23.1: (1) determination of the present level (the concentration of a single material or a group of materials), (2) a measurement or estimate of sources, and (3) a measurement or estimate of sinks. A perfect determination of any two of these three components determines the other. Because any material of interest in an industrial facility or in the environment may have several sources and sinks, each source and sink must generally be studied individually.

It is convenient to define a number of different time scales as part of the budget and cycle process. The first is the *turnover time*, τ_o. This parameter is the ratio between the content $[\beta]$ and the flux of a specific material into (F_i) or out of (F_o) a reservoir that is in steady state:

$$\tau_o = \beta/F_o = \beta/F_i \tag{23.1}$$

The turnover time reflects the spatial or temporal variability of a property within a reservoir, with a small variability indicating a long turnover time and a large variability indicating a short turnover time. If material enters or leaves the reservoir by several paths, then the overall turnover time of the reservoir is related to the individual turnover times of the pathways $\tau_{o,i}$ by

$$\tau_o = \{\beta_i\}/\{\Sigma_i F_i\} = \{\beta_i\}/\{\Sigma_o F_o\} = 1/\{\Sigma_i\{1/\tau_{o,i}\}\} \tag{23.2}$$

A second useful parameter is the *residence time*, τ_r, which is the average time spent in the reservoir by a specific material. If physical rather than chemical processes are involved, then the term *transit time* may be used as an alternative. The average residence time is composed of those of all appropriate substances, weighted with appropriate probability factors. For example, when one evaluates nitrogen flow into and out of the animal reservoir, one finds that some of the nitrogen rapidly flows through the reservoir whereas other nitrogen flows much more slowly, a reflection of animal lifetimes and foraging and excretion characteristics. We can define the average residence time in a situation with a variety of residence times as

$$\tau_r = \int \tau\psi(\tau)d\tau \tag{23.3}$$

where $\psi(\tau)$ indicates the fraction of the constituent having a residence time between τ and $\tau + d\tau$. This probability fraction $\psi(\tau)$ is a function of the reservoir processes. In the case of radioactive decay, for example, it can be shown that $\tau_r = \tau_a$, the time constant for the exponential decay of the radionuclide.

The *age* is the time elapsed since a particle entered a reservoir. The average age of all particles of a specific kind within a reservoir is thus given by

$$\tau_a = \int \tau\Psi(\tau)d\tau \tag{23.4}$$

where $\Psi(\tau)$ is the age probability function.

For a reservoir in steady state, the turnover time τ_o and the average residence time τ_r are equal. They may, however, be significantly different from the average age of the particles in the reservoir, depending on the properties of the functions $\psi(\tau)$ and $\Psi(\tau)$. An obvious example showing that $\tau_r \neq \tau_a$ is the human population: In the developed world, the average age is between 35 and 40 yr, whereas the average residence time (life expectancy) is more than 70 yr.

For a system chosen with a boundary surrounding the entire Earth, as with the nitrogen budget, and assuming no significant loss to interplanetary space, the overall quantity of material will not change with time:

$$\Sigma \beta = \text{Constant} \tag{23.5}$$

If the entire system is in a steady state, the source and sink fluxes into each reservoir exactly balance, so that in each case

$$\beta_r = \text{Constant} \tag{23.6}$$

that is, the contents of each reservoir will not change with time:

$$\Delta_r = \{d(\beta_r)\}/dt = 0 \tag{23.7}$$

In a changing system, however, the source and sink fluxes are not equal and Δ_r is real-valued and expressed as a function of the *response time* τ_e, which is the time needed to reduce the effect of a disturbance from equilibrium in a reservoir to $1/e$ of the initial perturbation value. For example, in many chemical applications the equilibrium of species i in a system is expressed by

$$0 = P_i - \Lambda_i \beta_r \tag{23.8}$$

where P_i is the production rate of species i and $\Lambda_i \beta_r$ its loss rate, β_r being the content of species i in reservoir r. If we now introduce a disturbance from equilibrium β'_r, the change in β_r with time becomes

$$\Delta = d\{(\beta_r)\}/dt = P_i - \Lambda_i(\beta_r + \beta'_r) = -\Lambda_i \beta_r \tag{23.9}$$

so

$$\beta'_r(t) = \beta'_{r0} \exp(-\Lambda_i t) \tag{23.10}$$

In this case,

$$\tau_e = \Lambda_i^{-1} \tag{23.11}$$

The concepts discussed above can be illustrated by examples from nature. When a budget is centered on an organism and encompasses all the significant resource flows in and out, it is termed *metabolic analysis*.

Instead of an organism-centered approach, a budget may be directed toward a specific resource. Depending on the boundaries of the *resource analysis*, it may deal with any spatial scale and with many reservoirs. In Figure 23.2 we reproduce a nitrogen budget for a forest ecosystem. Inputs and outputs cross the boundary of the ecosystem; within it are five reservoirs ((1) above-ground living biomass bound N, (2) below-ground living biomass bound N, (3) forest-floor bound N, (4) mineral-soil bound N, and (5) available soil N) and the flows that connect them. One quickly sees that the largest nitrogen stock is in the mineral soil, and that the largest nitrogen flows are mineralization followed by below-ground biomass uptake. As is the case with many bud-

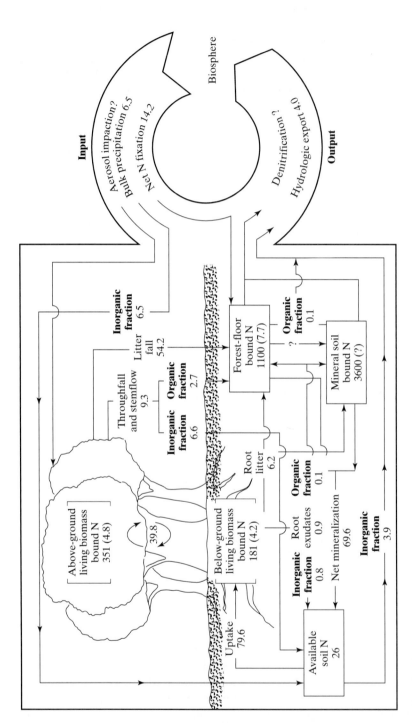

Figure 23.2

An example of a resource analysis budget: the annual nitrogen budget for an undisturbed northern hardwood forest ecosystem at Hubbard Brook, New Hampshire. Bound nitrogen denotes organically bound nitrogen not readily available for plant growth; mineralization is the microbiological transformation of bound to inorganic nitrogen which can be utilized by plants, especially ammonium (NH_4), nitrite (NO_2), and nitrate (NO_3). Nitrogen fixation, the dissociation of atmospheric N_2 into fixed nitrogen (NH_3), is accomplished by denitrifying bacteria which, under anaerobic conditions, convert nitrate to molecular nitrogen and nitrous oxide. The reservoir values in boxes are in kilograms of nitrogen per hectare. The rate of accretion of each pool (numbers in parentheses) and all transfer rates are expressed in kilograms of nitrogen per hectare per year. (Reproduced with permission from F.H. Bormann, G.E. Likens, and J.M. Melillo, Nitrogen budget for an aggrading northern hardwood forest ecosystem, *Science*, *196*, 981–983. Copyright 1977 by AAAS.)

gets, some of the stocks and flows, especially the minor ones or those difficult to measure, remain uncertain.

In industrial ecology, the concepts of budgets and cycles are applied to technological rather than natural systems, or sometimes to the combination of technological and natural systems. For technologically related organisms of various types and degrees of complexity, the results help us evaluate present metabolic needs, and to estimate those that may be required in the future. Similarly, we can study specific resources as they pass through various technological organisms, and thus evaluate resource supply, use, and potential environmental impacts.

23.2   METABOLIC ANALYSES IN INDUSTRIAL ECOLOGY

In organismal metabolic analysis (OMA), a single industrial organism, a person or a factory, for example, is the subject. A classic OMA was performed by Iddo Wernick and Jesse Ausubel of the Rockefeller University, who studied per capita resource flows in 1990 for the United States. Some of the results are given in Table 23.1. Water is such a large flow that it tends to obscure the importance of the others; mine overburden is very large as well. Of the remainder, construction materials and energy-generating fossil fuels are the largest inputs, carbon dioxide from fossil fuel combustion the largest output.

The OMA approach is also very useful for a factory or commercial establishment. By analyzing what enters a facility, what leaves, and to what degree entering material is utilized usefully or is lost, opportunities for substantially improving environmental performance can often be identified.

Flow analysis in industrial ecology can be conducted within and across the boundaries of specific geographic regions; this is Regional Metabolic Analysis (RMA). An example was shown in Figure 22.8 for a region in northeastern Switzerland. The

TABLE 23.1   U.S. Per Capita Resource Flows for 1990

	Flows (kg / day)
Inputs	
Energy materials	23.5
Construction minerals	23.1
Imports	6.9
Agriculture	6.9
Forestry products	2.9
Industrial minerals	2.7
Metals	1.2
Outputs	
Air emissions	19.0
Wastes	6.1
Export	4.5
Dissipation	1.6

Data from I.K. Wernick and J.H. Ausubel, National material flows and the environment, *Annual Review of Energy and the Environment, 25,* 463–492, 1995.

authors of that study chose to include air and water explicitly, and those flows dominate. Of the solid flows, construction materials are the most important. An interesting characteristic of the data is that input and output do not balance; there is clearly an increase of stock within the region. Such an increase is the technological analog of the growth of a biological organism, and obviously cannot be sustained indefinitely; at some point the stock will begin to be part of an exiting flow.

Larger regions are also subjects of RMAs. In fact, since data are often collected on a national basis, a country-level budget is generally easier to assemble than those for smaller regions. Table 23.2 compares RMA results for several countries, as characterized by an international team under the auspices of the World Resources Institute (WRI). The top portion of the table consists of magnitudes of flows to the environment from different sectors. The results are a combination of country size and level of activity—the United States flows are the highest in all categories. Japan's output from the industrial sector is all out of proportion to its size, however, as is that of the Netherlands for agriculture. The differences are markedly decreased when these flows are expressed on a per capita basis, however. Household-related flows are similar, for example, except in Japan, where they are much lower. Construction-related flows are similar as well, except in Austria, where they are much higher. Environmental flows from agriculture in the Netherlands and transport in the United States present obvious possible targets for public policy initiatives.

An important contribution of the study that led to Table 23.2 was the identification and quantification of hidden flows—materials moved or mobilized in the course of providing commodities, but which do not themselves enter the economy. There are two components to these hidden flows: (1) ancillary material removed along with the target material and later separated and discarded, such as the rock matrix containing a

TABLE 23.2   Domestic Processed Outputs to the Environment in Five Countries, 1996

	Austria	Germany	Japan	Netherlands	United States
			National totals (Tg)		
Agriculture	9.5	44.8	30.6	44.6	231.6
Construction	5.4	26.2	25.4	3.0	110.0
Energy supply	2.9	133.1	119.7	12.4	723.9
Industry	11.0	35.3	171.3	18.1	653.0
Household	6.5	56.4	35.2	14.0	280.4
Transport	1.2	57.9	77.8	17.1	668.9
			Per capita averages (Mg)		
Agriculture	1.2	0.6	0.2	2.9	0.9
Construction	0.7	0.3	0.2	0.2	0.4
Energy supply	0.4	1.6	1.0	0.8	2.7
Industry	1.4	0.4	1.4	1.2	2.4
Household	0.8	0.7	0.3	0.9	1.0
Transport	0.7	0.7	0.6	1.1	2.5

Data from E. Mathews, et al., *The Weight of Nations: Material Outflows from Industrial Economies*, Washington, DC: World Resources Institute, 125 pp., 2000.

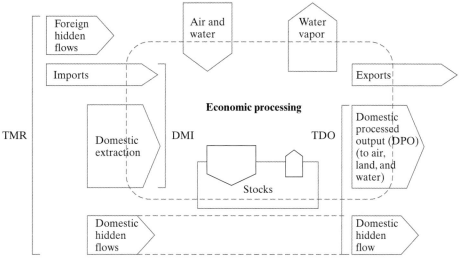

TMR (Total material requirement) = DMI + domestic hidden flows + foreign hidden flows
DMI (direct material input) = domestic extraction + imports
NAS (net additions to stock) = DMI − DPO − exports
TDO (total domestic output) = DPO + domestic hidden flows
DPO (domestic processed output) = DMI − net Additions to stock − exports

Figure 23.3

The materials cycle in a regional metabolic analysis. The terms are defined below the cycle diagram. (Reproduced with permission from E. Mathews, et al., *The Weight of Nations: Material Outflows from Industrial Economies*, Washington, DC: World Resources Institute, 125 pp., 2000.)

metal ore, and (2) excavated and/or disturbed material, such as the soil removed to gain access to an ore body. This inclusion means that all material flows, whether obvious to the user or not, are included in the RMA, as shown in Figure 23.3.

Among the interesting and useful results presented by the WRI-led team for country-level RMAs are the following:

- Industrial economies are becoming more efficient in their use of materials, but waste generation continues to increase.
- One-half to three-quarters of annual resource inputs to industrial economies are returned to the environment as wastes within a year.
- The extraction and use of fossil energy resources dominate output flows in industrial countries.
- On average, about 10 metric tons of materials per capita are added each year to domestic stocks in industrial countries, mostly in buildings and infrastructure.

23.3   RESOURCE ANALYSES IN INDUSTRIAL ECOLOGY

Resource analyses occur at several levels that may be distinguished by the chemical state of the resource. If one follows a resource in either atomic or molecular form, the study is termed *substance analysis*. The simplest case is where an element's rate of

TABLE 23.3   Types of Resource Analyses

Designation	Acronym
Elemental flow analysis	EFA
Elemental stock and flow analysis	ESFA
Molecular flow analysis	MoFA
Molecular stock and flow analysis	MoSFA
Substance flow analysis	SFA
Substance stock and flow analysis	SSFA
Material flow analysis	MFA
Material stock and flow analysis	MSFA

transition from reservoir to reservoir is the subject of analysis. The next most complex approach deals with molecules or alloys, that is, with well defined chemical entities. Finally, if a resource is followed in various chemical states through a series of natural and anthropogenic collective reservoirs such as ores or automobiles, the exercise is titled *material analysis*. For any of these, the approach may involve determination of flows only, or of both flows and stocks. Table 23.3 summarizes the different types of resource analyses. A resource analysis can be performed for natural stocks and flows, for anthropogenic stocks and flows, or for a combination of the two.

23.3.1 Elemental Analyses

In an elemental analysis, the emphasis is on the atom. This may be because the supply of the atom is restricted (gold, for instance), or because it is a biotoxicant (cadmium, for instance). It does not mean that the subject of the analysis is necessarily present in atomic form, but that the chemical form in which the atom exists is not addressed in the analysis. The advantage of this approach is that its data are unambiguous—a flow of sulfur is expressed as mass of sulfur per unit time, for instance, regardless of whether the actual entity being transferred is flowers of sulfur, sulfur dioxide, or sulfuric acid.

An EFA for lead in 1998 is shown in Figure 23.4. It is quickly seen that most lead was used in batteries, that two-thirds of the lead in batteries was recycled, and that pigments were the largest of the dissipative uses. The scale is global, but need not have been: Resource analyses may be performed at any spatial scale.

23.3.2 Molecular Analyses

One can make the case that the first molecular resource analysis in industrial ecology won the Nobel Prize; this was the study of the atmospheric cycle of chlorofluorocarbons by Mario Molina and F. Sherwood Rowland of the University of California–Irvine. These compounds, created by industry as refrigerants, had been detected in the atmosphere. Molina and Rowland deduced that their eventual fate was to be broken apart by high-energy solar radiation in the upper atmosphere, there to react with and remove ozone. The discovery of the ozone hole over Antarctica several years later demonstrated the accuracy of their analysis.

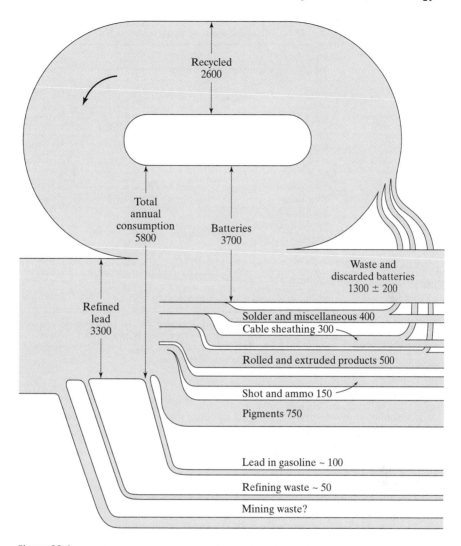

Figure 23.4

An elemental flow analysis for lead in the world economy. The data are in units of Gg / yr for 1988. (Reprinted by permission from V. Thomas and T. Spiro, Emission and exposure to metals: Cadmium and lead, in *Industrial Ecology and Global Change*, R. Socolow, C. Andrews, F. Berkhout, and V. Thomas, eds., Cambridge, U.K.: Cambridge University Press, 297–318, 1994.)

Molecular approaches can be powerful tools to study species that have both natural and anthropogenic sources. Atmospheric methane is a good example; its budget is given in Table 23.4. The analysis shows that the flux from anthropogenic sources pretty clearly exceeds that from natural sources, and that no single anthropogenic source is dominant.

TABLE 23.4   Molecular Flow Analysis for Atmospheric Methane

Budget item	Flux (Tg C / yr)
Sources	
Natural	
Wetlands	120 (100–200)
Termites	20 (10–50)
Ocean	10 (5–20)
Hydrates	0 (0–5)
Anthropogenic	
Coal, gas mining	100 (70–120)
Rice fields	60 (20–100)
Ruminants	80 (65–100)
Waste treatment	80 (60–100)
Biomass burning	40 (20–80)
Sinks	
Reaction with HO ·	430 (350–510)
Removal by soils	30 (15–45)

Data are for 1990 and are from T.E. Graedel and P.J. Crutzen, *Atmospheric Change: An Earth System Perspective*, New York: W. H. Freeman, 1993. For each item, a selected best estimate is provided, together with the total range within which the true value is thought to occur.

23.3.3 Substance Analyses

Molecular analyses are performed because some property of the molecule—its radiation absorption spectrum, for example—is of overriding importance. If one wishes rather to consider recycling potential or some other property in which several chemical forms of a resource are potentially of interest, it may be desirable to perform a substance analysis. An example is the global zinc cycle shown in Figure 23.5. In this analysis, some of the zinc is evaluated in bulk, some as an alloy component, and some as a coating on another metal.

The lead analysis of Figure 23.4 could be regarded as an SFA if the chemical state of the lead in each flow stream were specified. Even without that level of detail, it is apparent that the lead that undergoes recycling tends to be from uses in which the lead is in elemental form (batteries, cable sheeting, rolled lead), while that lost to the environment tends to be from uses where the chemical form of lead is more complex (pigments, leaded gasoline, refining waste). The diagram thus provides immediate advice to the product designer: If possible, avoid using alloys, composites, or other mixtures unless efficient techniques are available for recovery, separation, and recycling.

23.3.4 Material Analyses

In a material analysis, the resource of interest is evaluated within and/or between several reservoirs in which the resource may be either a major or minor constituent. This is the industrial analog of Figure 23.2, in which the stocks and flows of nitrogen, a minor but vital constituent of soil and trees, are determined. In the industrial ecology

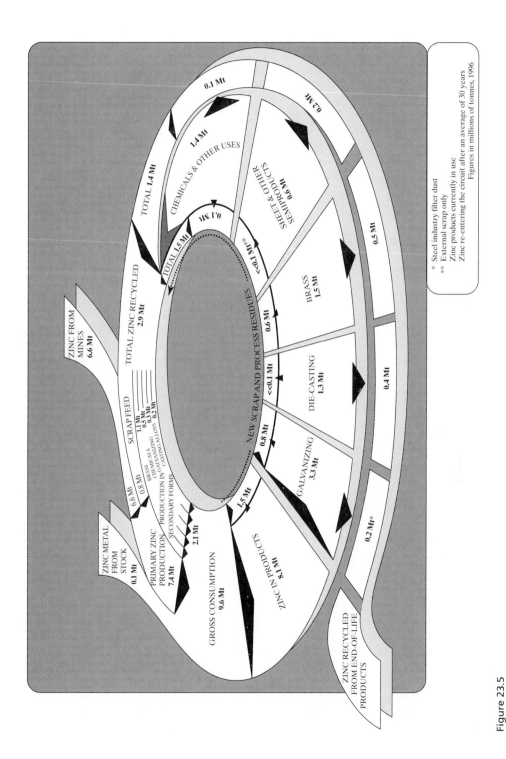

Figure 23.5

A substance flow analysis for zinc in the world economy. The data are in units of Tg / yr for 1996. (Reprinted by permission from *Zinc Recycling: The General Picture*, Brussels: International Zinc Association: Europe, 14 pp., 2000.)

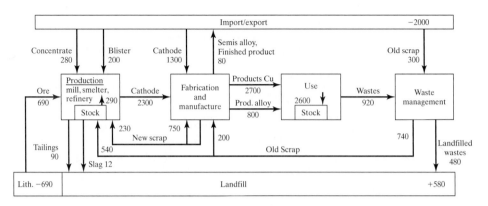

Figure 23.6

A material flow analysis for copper in Europe. The data are in units of Gg/yr for flows and changes in stock; the epoch is the average of 10 years centered on 1995. (Reprinted from S. Spatari et al., The contemporary European copper cycle: One-year stocks and flows, *Ecological Economics, 41*, 27–42, 2002.)

application, the initial reservoir is a natural one—a mineral or oil deposit, for example—while the subsequent reservoirs generally involve human uses such as photocopy machines or airplanes. Such a budget, for copper in Europe in 1995, is presented in Figure 23.6. Here the flow of virgin copper is much higher than that of recycled copper, the major in-use reservoirs are in electrical wire and copper tubing, and several percent of the copper is lost to the environment, largely as tailings and slag in production.

23.4   THE BALANCE BETWEEN NATURAL AND ANTHROPOGENIC MOBILIZATION OF RESOURCES

Only a few of the elements in the periodic table have been the subject of extensive budget and cycle analyses. Nonetheless, it is readily apparent that natural flows are dominant for some elements, while anthropogenic flows are dominant for others. Figure 23.7 presents a periodic table in which the controlling dominance, as best it can be determined, is indicated. It can be seen that natural flows are generally the more important for elements in groups Ia, IIa, and VIa–VIIIa, anthropogenic flows for elements in groups IIIa–Va and Ib–VIIIb.

What is the cause of the patterns in Figure 23.7? It appears that the principal explanation relates to the solubility in water of the principal compound in which an element occurs in nature. Those that are highly soluble—sodium in sea salt, for example—are efficiently mobilized and redeposited by natural processes. These elements are used as building blocks by living organisms—calcium in shells, chlorine in cells, and so forth. Elements in the center of the periodic table, however, tend to occur as highly insoluble oxides or sulfides. Natural processes have no effective way to isolate these elements, nor do they know how to deal with them when they are isolated by human industrial processes. As a result, it is these elements that form corrosion-resistant structures, that provide high-strength materials, that are often biotoxic. For the major-

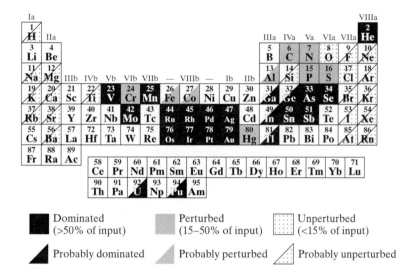

Figure 23.7

An assessment of the degree to which the global cycles of the elements are dominated by nature or by human activity. (Reproduced from T.E. Graedel and R.J. Klee, Elemental cycles: A status report on human or natural dominance, submitted for publication, 2002.)

ity of the elements, though not all, aqueous solubility determines whether nature or humans will control their mobilization and flow.

23.5 THE UTILITY OF METABOLIC AND RESOURCE ANALYSES

The types of analyses discussed in this chapter have proven themselves extremely useful as industrial ecology tools. Organismal metabolic analyses permit us to examine resource use by individuals and by factories, and provide insights into the cultural and technological patterns that are revealed. For geographical regions, such analyses and the predictions that arise from them suggest a variety of public policy instruments to help improve performance—traffic planning, recycling incentives, and waste treatment plant construction, to name a few possibilities.

Resource analyses have substantial utility as well. The results help us gain perspective on resource sustainability, potentials for increased resource cycling, and evaluations of existing and potential environmental hazards. Public policy as well as corporate policy can obviously be brought to bear on these topics once the budgets are determined with satisfactory accuracy.

Environmental impacts result from actions taken to stimulate flows of resources, or from the flows themselves. When we have evaluated these flows, over time and space and with the appropriate temporal and spatial reduction, we will have acquired the information needed to determine the sustainability of our industrial enterprise.

FURTHER READING

Ayres, R.U., Industrial metabolism, in *Technology and Environment*, J.H. Ausubel and H.E. Sladovich, eds., Washington, National Academy Press, 23–49, 1989.

Bringezu, S., and Y. Moriguchi, Material flow analysis, in *A Handbook of Industrial Ecology*, R.U. Ayres and L.W. Ayres, eds., Cheltenham, U.K.: Edward Elgar Publishers, 79–90, 2002.

Graedel, T.E., and W.C. Keene, The budget and cycle of Earth's natural chlorine, *Pure and Applied Chemistry, 68*, 1689–1697, 1996.

Matthews, E., et al., *The Weight of Nations: Material Outflows from Industrial Economies*, Washington, DC: World Resources Institute, 2000.

Molina, M.J., and F.S. Rowland, Stratospheric sink for chlorofluoromethanes: chlorine atom catalyzed destruction of ozone, *Nature, 249*, 810–812, 1974.

Van der Voet, E., J.B. Guinée, and H.A. Udo de Haas, eds., *Heavy Metals: A Problem Solved?*, Dordrecht, The Netherlands: Kluwer Academic Publishers, 2000.

Wernick, I.K., and J.H. Ausubel, National material flows and the environment, *Annual Review of Energy and the Environment, 20*, 463–492, 1995.

EXERCISES

23.1 You are assigned to analyze a reservoir generally similar to that of Figure 23.1. The reservoir has three input flows ($I_1 = 6$ l/s, $I_2 = 13$ l/s, $I_3 = 9.5$ l/s) and two output flows ($O_1 = 16$ l/s and $O_2 = 10$ l/s). Is the system in steady state? If not, at what rate is the reservoir contents changing?

23.2 The reservoir of Exercise 23.1 contains 1500 liters of water. A third output flow, $O_3 = 2.5$ l/s, has been added. Compute the turnover time for the water in the reservoir.

23.3 Table 23.3 provides information on environmental outflows in five countries. If you were the Minister of the Environment in Germany, what policy initiatives might these data suggest? Would the approach differ in Japan? In the Netherlands? What other information would be helpful in evaluating policy initiatives?

23.4 The zinc cycle of Figure 23.5 is not at steady state. What does this imply for global in-use reservoirs of zinc? Which reservoirs are changing most rapidly? What will be the change in the contents of each of the in-use reservoirs after 10 years, given constant flow rates?

Systems Analysis, Models, and Scenario Development

24.1   THINKING AT THE SYSTEMS LEVEL

24.1.1 The Systems Concept

Perhaps the most important operational feature of the industrial ecology approach is its ability to focus not solely on the product, but also on a product's related systems and their behavior. Identifying the appropriate system and properly relating it to its technological context is frequently a critical step in any successful industrial ecology assessment. Accordingly, it is useful to introduce and discuss the general concept of technological systems.

A system may be thought of as a group of interacting, interdependent parts linked by exchanges of energy, matter, and/or information. Defining a system is almost always somewhat arbitrary, requiring the analyst to identify those exchanges that are relevant to the purpose of the definition, to understand the linkages within the system as defined, and to determine the external context within which the system is embedded. For example, one would not want to try to define the environmental aspects of a technology with relation to the solar system—the system is too large and many of the linkages to the technology are too indirect. Rather, one might choose a limited system—for example, what are the environmental effects in a province or region of the inputs, outputs, and processes related to a particular technology, and can the effects be reduced if desired? Even in such a limited system, data uncertainties and complexities may make it appropriate to identify and evaluate only the major impacts involved.

The most important distinction between classes of systems for the industrial ecologist is that between simple and complex. In many cases, the management approaches, regulatory structure, and analytical techniques traditionally applied to issues involving technology, economic efficiency, and environmental efficiency explicitly or implicitly assume the system involved is simple. In virtually all cases, however, the systems involved—whether economic, environmental, or technological—are complex. The terms are used here not in the sense of classical physics (e.g., the "simple" harmonic oscillator), but of the dictionary definitions, "complex" meaning consisting of interconnected parts so as to make the whole difficult to understand; simple is thus not combined or compound. Some of the implications of this distinction are:

1. Simple systems tend to behave in a linear fashion: The output of the system is linearly related to the input. Complex systems, on the other hand, are characterized by strong interactions among the parts and by nonlinear responses. For example, a salt marsh may be relatively resistant to chemical pollution until a threshold is exceeded, after which even a small increment of additional insult will cause it to suddenly degrade precipitously. It is a complex system.

2. Simple systems can generally be evaluated in terms of cause and effect: An action is easily traceable through the system to its predictable effect. With complex systems, on the other hand, one finds feedback loops that often make the linkage between cause and effect difficult to establish.

3. Complex systems, unlike simple systems, are characterized by significant time and space discontinuities. A problem for most people in comprehending the issues surrounding global climate change is that there are large time lags between the forcing function (driving automobiles or using electricity, for example), and the resulting shifts in global climate patterns. Compounding the time lags are spatial lags—the links between local activities, such as driving cars and the global effects that may be produced in distant regions (e.g., coastal flooding in Bangladesh).

4. Simple systems tend to be characterized by a stable and known equilibrium point, to which they return in a predictable fashion if perturbed. Many complex systems, on the other hand, operate far from equilibrium, in a state of constant adaptation to changing conditions. Complex systems often evolve; simple systems generally stay more or less as they are.

5. With simple systems, change is assumed to be an additive function of subsystem characteristics; that is, a system with stressors A and B will have an impact Y given by

$$Y = A + B \tag{24.1}$$

or, to give a common example, if you combine two quarters, you get $0.50, no more and no less. With complex systems, on the other hand, emergent behaviors may not be additive: We cannot predict the characteristics of an anthill by summing up the observed behavior of individual ants, because we have omitted interactive effects. In a complex system with stressors A and B, impact Y may be dictated by these interactive AB terms, the magnitude and behavior of which may be far from obvious.

$$Y = A + B + f(AB) \tag{24.2}$$

An example of interactive effects is the influence of water vapor (A) and carbon dioxide (B) on planetary temperature (Y). A and B in this system are not independent, because if carbon dioxide concentrations increase and the planet warms, additional water will evaporate into the atmosphere, increasing the absorption of outgoing infrared ratiation and warming the planet still more. This is a *positive feedback* effect.

24.1.2 The Automotive Technology System

The relevance of complex systems approaches in industrial ecology relates to the fact that the products of technology do not stand in isolation, but are embedded within and are components of a larger technological system. Just as there is an important place for product-level industrial ecology, there is a role for those employing industrial ecology at the systems level. This latter approach can be best demonstrated by an example familiar to everyone: the automotive technology system.

Figure 6.1 showed a simple schematic diagram of the system. It includes the lowest, relatively technology-rich, levels of an automobile's mechanical subsystems and the manufacturing processes by which they are made. These subsystems and processes have been a predominant focus of environmental regulation.

The next systemic level, automobile use, is more important. There are two major dimensions to this system level: technical and cultural. On the technical side, great progress in reducing environmental impacts has been achieved. An example of the results is the cycle of reductions in emissions of nonmethane hydrocarbons (largely exhaust residues of fuel molecules for which combustion was incomplete), as shown in Figure 24.1. In 1970, emissions were about 2 g/km. In 1975, the introduction of exhaust catalysts reduced this rate by a factor of four. 1980 saw the first electronic sensors and exhaust system controls, and emissions dropped by an additional factor of two. The comprehensive use of precision technology in the 1990s—increasingly uniform machining of combustion chambers, electronic control of fuel injection and ignition, and advanced exhaust gas sensors—has further reduced emissions, so that the vehicle of the late 1990s has hydrocarbon emission rates nearly 100 times lower than those of 1970 vehicles.

From the cultural standpoint, in contrast, failure to address the environmental impacts of automobile use is virtually total. The mix of cars purchased in more developed countries—and soon to be available in rapidly developing countries such as China—is increasingly inefficient. In the United States, families routinely purchase vehicles with four-wheel-drive systems and high gasoline consumption. The simplest measure of lack of popular linkage between automobile use and environmental impact is the increase over time in vehicle distance traveled each year. For the United States, this is reflected in a 440 percent growth between 1950 and 1992, during a period when population increased only 68 percent. This growth has increased for drivers of all ages, and for both men and women (Figure 24.2). Much of this driving is not for business or shopping, but for community or leisure activities: an evening's entertainment, the children's sporting event, a vacation. Furthermore, the gas-saving federal 55 mile per hour speed limit was repealed in 1995 on the grounds that it was an imposition on personal liberty and states' rights; depending on the state, legal speed is now 65, 75, and, in one case, unlimited. In Germany, attempts to impose any speed limits on the Autobahn routinely fail, even though the German environmental party (the Greens) is widely popular.

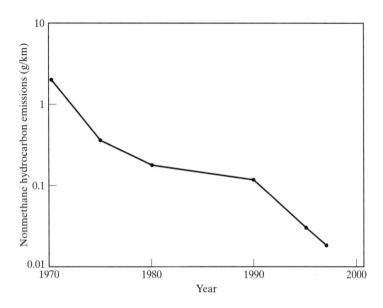

Figure 24.1

Nonmethane hydrocarbon emissions from U.S. vehicles over the past three decades. The data are from the Low Emission Partnership, USCAR Consortium.

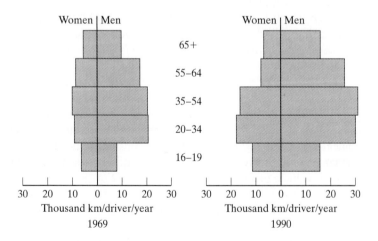

Figure 24.2

Vehicle miles traveled per year as a function of driver age, for U.S. drivers in 1969 and 1990. (Adapted from L. Schipper, Life-styles and the environment: The case of energy, *Daedalus, Journal of the American Academy of Arts and Sciences*, from the issue entitled "The Liberation of the Environment," Summer 1996, Vol. 125, No. 3.)

Even a cursory evaluation of the automotive system indicates that attention is being focused on the wrong subsystem, and illustrates the fundamental truth that a strictly technological solution is unlikely to fully mitigate a culturally influenced problem. Contrary to the usual understanding, the most significant environmental impacts of the automobile technology system should probably be allocated to the highest levels of the system, the infrastructure technologies and the social structure. Consider the energy and environmental impacts that result from just two of the major system components required by the use of automobiles. The construction and maintenance of the "built" infrastructure—the roads and highways, the bridges and tunnels, the garages and parking lots—involve huge environmental impacts. The energy required to build and maintain that infrastructure, the natural areas that are perturbed or destroyed in the process, the amount of materials, from aggregate to fill to asphalt, demanded—all of this is required by the automobile culture, and attributable to it. In addition, the primary customer for the petroleum sector and its refining, blending, and distribution components—and, therefore, causative agent for much of its environmental impacts—is the automobile. Some feeling for the complexity that the systems view introduces to our perception of the automobile and its environmental interactions is provided by the systemic mass flow analysis of Figure 24.3—a lot more than just automotive emissions! Efforts are being made by a few leading infrastructure and energy production firms to reduce their environmental impacts, but these technological and management advances, desirable as they are, cannot in themselves begin to compensate for the increased demand generated by the cultural patterns of automobile use.

The final and most fundamental effect of the automobile may be in the geographical patterns of population distribution to which it has been a primary impetus. Particularly in lightly populated and highly developed countries such as Canada and Australia, the automobile has resulted in a diffuse pattern of residential and business development which is otherwise unsustainable. Lack of sufficient population density along potential mass transit corridors makes public transportation uneconomic within many such areas, even where absolute population density would seem to augur otherwise (e.g., in densely populated suburban New Jersey). This transportation infrastructure pattern, once established, is highly resistant to change in the short term, if for no other reason than residences and commercial buildings last for decades. Moreover, these trends do not seem to be improving. Data indicate that the number of commuters using public transportation in most urban centers is small, and that automobile commuting continues to increase. Thus, high demand for personal transportation (i.e., the automobile) is firmly embedded in the physical structure of countries around the world.

This discussion illustrates the value of studying systems as well as components, and of appreciating the full scope of cultural and economic factors on the environment. This is not to say that technological evolution cannot serve an important role in mitigating environmental impacts even where cultural patterns are crucial components. Indeed, DfE and other methodologies explored in this text attempt precisely that by taking a life cycle approach to automotive performance and design. We are unquestionably better off because automobiles are far more environmentally friendly today than they were in the past. However, we have not yet achieved control over the systems that link technology, society, and the environment.

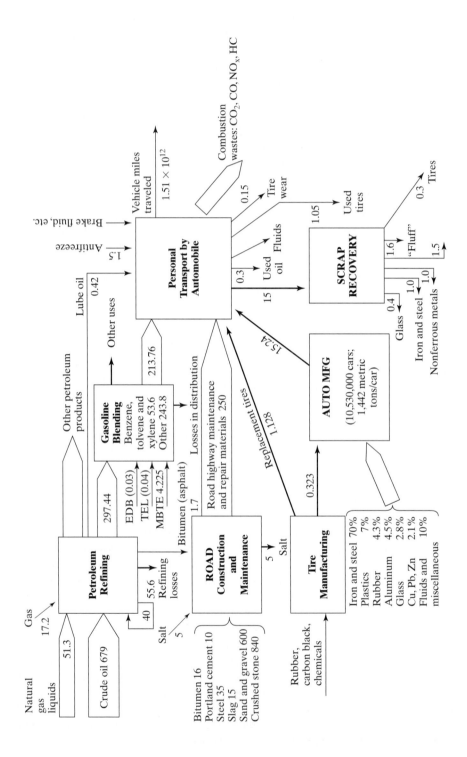

Figure 24.3

1998 mass flows for the U.S. automotive system, in million metric tons. (Adapted from R.U. Ayres and L.W. Ayres, Use of materials balances to estimate aggregate wastes, in P.C. Schulze, ed., *Measures of Environmental Performance and Ecosystem Condition*, Washington, DC: National Academy Press, 96–156, 1999.)

24.2   MODELS OF TECHNOLOGICAL SYSTEMS

24.2.1 The Concept of a Model

Once a system of interest has been defined, the logical next step is to want to determine how the system works. In the case of the automotive system, for example, how is automotive use related to the characteristics of infrastructure or culture within which it must operate? The intellectual construct that attempts to describe those interrelationships is termed a model.

In biological ecology, a common goal of models is to evaluate how the individual actions of plants or animals in the acquisition and use of resources and their interactions with each other produce the observed large-scale and temporal distributions of individuals. If successful, the results provide insights into both species biology and resource consumption. Similarly, in industrial ecology a visionary goal of models is to describe dynamic industrial–environmental systems that are aggregates of heterogeneous units. As in biology, there is a scale-crossing problem here: How do design choices made at the microscale (e.g., the factory) aggregate to determine rates of resource utilization or environmental impact at meso- and macroscales, and how might changes in technological or societal systems be reflected in large-scale effects? These are daunting but important questions, and the construction of models that can address them is a grand challenge indeed.

Models are formulations, generally mathematical, that seek to represent a portion of the real world. In industrial ecology, models relate in some way to the extraction and use of resources and to their environmental consequences. The models can be developed for any or all of the following purposes:

- To see if we understand the behavior of a system of interest
- To predict the behavior of a system of interest in response to a change in one or more of its driving forces or constraints
- To provide guidance to policy makers

Models can be characterized in a number of ways. One of the more important is the temporal approach taken. The most common is the development of a *static model*, which attempts to reproduce events at one point in time—to take a snapshot, as it were. More ambitious is the *dynamic model*, which attempts to follow a system as it evolves over time. A dynamic model requires much more detailed knowledge of interactions and thresholds than does a static model. In systems terms, static models are appropriate for simple systems, while complex systems are best explored through the use of dynamic models.

A second important characteristic is a model's spatial approach. Many models have no spatial resolution at all—they define the geographical boundaries of the system and then compute interactions as though they operated at average rates at all points within the system. This approach is suitable for many problems, especially where data to support the model are scarce. Spatially discrete models are considerably more challenging: They normally require supporting data at the same spatial resolution as in the model, together with representations of processes operating over a range of parametric values rather than averages. Especially for environmental impact studies where a highly sensitive ecosystem may be located near a less sensitive one, however, spatial resolution is often crucial.

The steps involved in building and using a model are as follows:

- Define the relevant characteristics of the system you wish to understand.
- Define the scope of the model and its temporal and spatial approaches.
- Identify primary actors or primary processes.
- Define interactive relationships among actors and/or processes.
- Describe the relationships mathematically.
- Determine the initial conditions.
- Solve the model equations for situations of interest.
- Interpret the results.

There is no one correct model for a system, because models may be constructed for many different purposes and uses. In the automotive case, one might wish to investigate the relationship of automobile use to air quality, in which case social structure components would not be modeled. Alternatively, one might wish to study how social structure relates to automobile recycling, in which case a representation of road networks might not be needed. The scope of a model, like the scope of an LCA, dictates much of the subsequent implementation.

24.2.2 Iron and Steel in the U.K.: A Model Example

An example of an industrial ecology model was constructed in 2000 by the University of Surrey for the U.K. iron and steel industry. The system diagram is shown in Figure 24.4. The model has no spatial resolution, but it is dynamic, computing a number of parameters on an annual basis for a period of several decades. A number of simplifications have been made for tractability purposes, such as adopting an average lifetime for all steel goods in use.

In addition to deriving several flow magnitudes, such as the amount of new scrap or the post-use recycling rate, the model uses average values of exergy rates (available energy rates) from iron and steel production, transport, generation of waste steel, trade, and recycling to compute exergy consumption. An example is shown in Figure 24.5 for the iron and steel production phase. Over the 40-year period of the analysis, the exergy consumption dropped by more than 50%. This had three causes: (1) reduction in the exergy consumption per metric ton of output for each process, (2) substitution of the open hearth process by the more efficient oxygen furnace and electric arc furnace processes, and (3) a lower overall rate of iron and steel production.

24.2.3 Model Validation

How do we know whether model results are valid? The typical output of a model consists of computed values for a number of dependent variables. Ideally, these results are compared with actual data for conditions similar to those simulated by the model. In the U.K. steel model, for example, one can see whether the flows of scrap are consistent with the scrap input to production and with scrap trading figures. If so, an initial

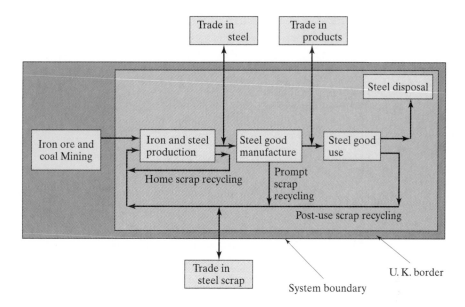

Figure 24.4

A model of the steel sector in the United Kingdom. (Reproduced with permission from P. Michaelis and T. Johnson, Material and energy flow through the U.K. iron and steel sector. Part 1: 1954–1994, *Resources, Conservation, and Recycling, 29*, 131–156, 2000.)

measure of confidence is provided. If the model fails to reproduce the data, then the model is faulty in some way, and studying those discrepancies often provides clues to increased understanding. Finally, the model can be used to make a prediction, say for steel disposed of in 1995, and the prediction evaluated by actual measurement. When a large number of such steps have been carried out, more or less successfully, the model is said to have been *validated*. The model results for unmeasurable parameters, such as exergy consumption in U.K. steel production in 1965, can then be regarded as likely to be good representations of the actual situation.

A model that is validated is one that does not contain known or detectable flaws and is internally consistent. That is not to say that the model results necessarily reflect the behavior of the real world. The results depend in part on the quality and quantity of the inputs, the accuracy with which processes are represented, and the extensiveness of model testing. A model might generate results that are reasonable and useful on one spatial or temporal scale, but not at another. Depending on the purpose of the model, this may be perfectly satisfactory.

Naomi Oreskes of Dartmouth College and her colleagues have written on the uses to which models can legitimately be put, even if they cannot be regarded as "perfectly correct." They say

Models can corroborate a hypothesis by offering evidence to strengthen what may be already partly established through other means. Models can elucidate discrepancies in other models. Models can be also be used for sensitivity analysis—for exploring "what if" questions—thereby illuminating which aspects of the system are

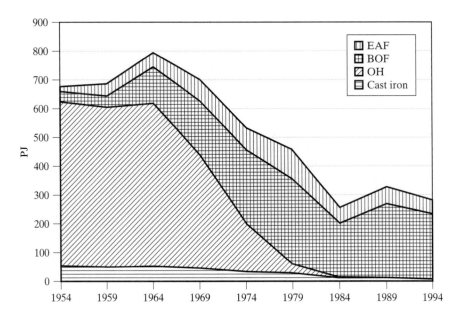

Figure 24.5

The annual exergy consumption from different methods of iron and steel production in the United Kingdom. EAF = electric arc furnace, BOF = basic oxygen furnace, OH = open hearth process. (Adapted with permission from P. Michaelis and T. Johnson, Material and energy flow through the U.K. iron and steel sector. Part 1: 1954–1994, *Resources, Conservation, and Recycling, 29*, 131–156, 2000.)

most in need of further study, and where more empirical data are most needed. Thus, the primary value of models is heuristic: Models are representations, useful for guiding further study but not susceptible to proof.

24.3   DESCRIBING POSSIBLE FUTURES

24.3.1   The Utility of Scenarios

A successful model means that a system is well enough understood so that the factors that influence it are known and their effects can be specified with at least reasonable accuracy. The model can then be used in the predictive mode: Assumptions can be made concerning the magnitudes of the forcing functions in the future and the model can be exercised to generate results that, it is hoped, are realistic projections. Models are generally most useful with "determined" systems: that is, systems that evolve according to well established natural laws (although a deterministic system may still be very complex, such as climate). Human systems, including the economic and industrial, add an additional element of complexity: the contingency introduced by human choice. This means that, in practical terms we not only do not, but cannot, know what course industrial development, the employment of materials, and the evolution of culture and society may take. Accordingly, people such as business planners who try to predict and understand possible future industrial systems often use techniques that are less formal and rigorous than modeling: A common approach is to develop a series of plausible futures, or *scenarios*, and explore the consequences of each of them.

An industrial ecology scenario may take any of a number of forms, and may adopt any appropriate spatial or temporal scale. A global scale is often the easiest, as it avoids complications of resource transfer among countries, for example, and can study overall questions of extraction rate, use, recycling, and loss. If the scenario were based on the assumption of a continuation of current flow rates and the rates of change of those rates, the projection of the driving forces is straightforward. If we wish to assume that some sectors or regions will grow faster than others, or that environmental technology will change over time, more complex scenarios will be required.

Scenarios are not predictions, but mental explorations. They ask "Suppose the world (region, industrial sector, etc.) develops in such and such a way. What does that imply for dependent characteristics in which I have an interest?" Those who are in the business of constructing scenarios often recommend three scenarios as a workable number—one describing what appears to be the most likely path, one describing a very positive path, and one a very negative path. The attempt is to cover the range of possibilities without having so many options that the exercise is more confusing than illuminating. If the scenario analysts find that they can live with any of three varied scenarios, they can feel more confident as the world moves forward into the future.

24.3.2 The IMAGE Model for Climate Change

One of the most ambitious efforts to date at model building and scenario development related to technology and environment is the IMAGE model of the Institute for Public Health and the Environment in The Netherlands (RIVM). The RIVM team and its international collaborators have developed the model and its "conventional wisdom" scenario with the following goals in mind:

- Link scientific and policy aspects of global change in a geographically specific manner
- Provide a dynamic and long-term perspective (50–100 yr) on climate change
- Provide insight into cross-linkages
- Provide a basis for costs and benefits of various policy alternatives

To address this ambitious agenda, the IMAGE team constructed a model composed of three interacting systems, as shown in Figure 24.6. Because the focus was on greenhouse gases, the emphasis of the energy–industry system was on fossil fuel use, that of the terrestrial environment system on carbon exchange with vegetation and the emission of methane from rice growing and livestock production. The Earth system model for the atmosphere–ocean system takes input from the two scenario models in order to calculate future climate conditions.

The spatial approach of IMAGE is interesting in that it is different for each of the three models:

- The energy-industry model is divided into 13 world regions. For each of these, it uses economic, technological, and demographic assumptions to calculate possible future energy use and the consequent emissions.

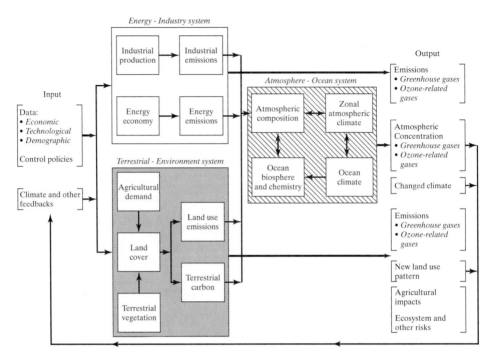

Figure 24.6

A flow diagram for the IMAGE 2 global change model. (Reproduced with permission from J. Alcamo et al., Global modelling of environmental change: An overview of IMAGE 2.1, in *Global Change Scenarios of the 21st Century*, J. Alcamo, R. Leemans, and E. Kreileman, eds., Kidlington, U.K.: Elsevier Science, 3–48, 1998.)

- The terrestrial environment model is divided into $0.5° \times 0.5°$ grid squares across Earth's land surface. Assumptions regarding land use, agricultural demand, and vegetative growth are used to calculate possible greenhouse gas emissions with high spatial resolution.
- The atmosphere–ocean model is divided into $10°$ latitude belts, for which climate change parameters are computed.

A substantial strength of IMAGE is that it explicitly incorporates feedback. If a warmer climate emerges from the atmosphere–ocean model, that changed climate will influence vegetative growth at later simulated times, for example. This characteristic, however, requires that the computation simultaneously deal with the three model systems, straining the computing power that can be brought to bear on the problem.

An example of the results of the IMAGE energy–industry system is shown in Figure 24.7, which illustrates energy consumption by sector for India plus South Asia from 1970 to 1995 (actual) and 1995–2100 (projected). The consumption under this scenario increases about 13 times over that period, with industry and the residential sector being the largest users.

Given the results of the energy–industry model and the terrestrial environment model, the atmosphere–ocean model then computes the scenario-driven climate. An

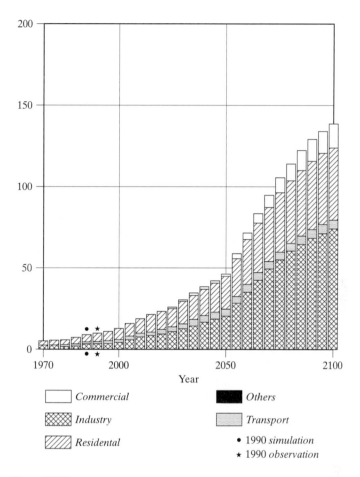

Figure 24.7

The total energy consumption by sector for the region India plus South Asia for the period 1970–2100, as computed by the IMAGE model. (Reproduced with permission from H.J.M. de Vries, J.G.J. Olivier, R.A. van den Wijngaart, G.J.J. Kreileman, and A.M.C. Toet, Model for calculating regional energy use, industrial production and greenhouse gas emissions for evaluating global climate scenarios, *Water, Air, and Soil Pollution, 76*, 79–131, 1994.)

example of the results for the "conventional wisdom" scenario is shown in Figure 24.8. The global average precipitation under this scenario is predicted to increase by about 7% over the century. That increase will be quite spatially nonuniform, and would be expected to lead to substantial flooding and other local climate perturbations in the most highly affected areas.

24.3.3 IPCC 2000 Scenarios

The interest in industrial ecology scenarios is related in part to the potential use of the results of the scenario models in revealing possible environmental conditions in the future, and in studying the policy implications that flow from that understanding. The

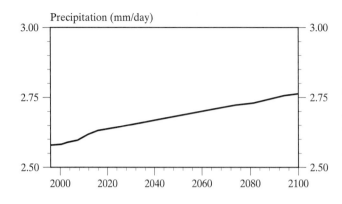

Precipitation (mm/day)

Figure 24.8

Global average precipitation for the 21st century as calculated by IMAGE for a "conventional wisdom" scenario. (Reproduced with permission from K. Klein Goldewijk, J.G. van Minnen, G.J.J. Kreileman, M. Vloedbeld, and R. Leemans, Simulating the carbon flux between the terrestrial environment and the atmosphere, *Water, Air, and Soil Pollution*, *76*, 199–230, 1994.)

conceptual approach is perhaps most familiar from the scenario construction developed by the Intergovernmental Panel on Climate Change (IPCC) to analyze potential climate evolution, impacts, and mitigation options.

In the fall of 2000, the IPCC issued a new set of several scenarios. As with its earlier 1992 versions, these scenarios are directed at greenhouse gas emissions, and thus pay special attention to fossil fuel combustion, agriculture, and other GHG-related activities. The four families of storylines and scenarios are as follows:

1. The *A1* storyline and scenario family describes a future world of very rapid economic growth, global population that peaks in mid-century and declines thereafter, and the rapid introduction of more efficient technologies.
2. The *A2* storyline and scenario family describes a regionally oriented world in which local identities are of high importance, and economic development and technological change occur relatively slowly.
3. The *B1* storyline and scenario family describes a world similar to that of the *A1* storyline but with rapid changes toward a service and information economy, lower material intensities, and a high degree of pollution prevention.
4. The *B2* storyline and scenario family describes a world with an *A2* theme of self-reliance, but a *B1* goal of local and regional sustainability.

Many characteristics of development and its consequences arise from the IPCC scenarios, such as the global carbon dioxide emissions implied by the scenarios and shown in Figure 24.9. Most of the scenarios generate CO_2 emissions two to three times the 1990 level, but in the most extreme case the flux is ten times as much. Those emissions scenarios are then provided to developers of global climate models so that they can calculate possible future climatic conditions. An example of the results from the year 2001 assessment of the IPCC is shown in Figure 24.10. It shows that there is considerable range in projections of sea level rise, as the projections are made from the framework of the large and diverse set of scenarios for 21st century human activity. The model average is for sea level rise over the century of about 35 cm. Should that occur, there is general agreement that low-lying countries such as the Maldive Islands or Bangladesh will suffer severely, and that damaging and expensive consequences will

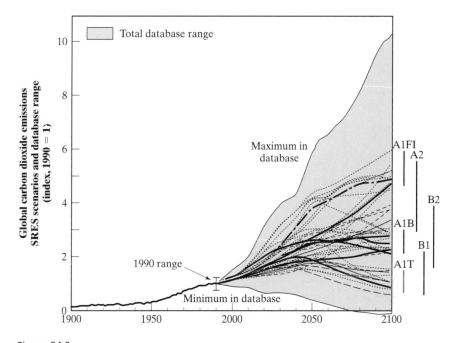

Figure 24.9

Emissions of carbon dioxide in the 21st century for the family of IPCC 2000 scenarios. (Reproduced with permission from Intergovernmental Panel on Climate Change, *IPCC Special Report: Emissions Scenarios*, Cambridge, U.K.: Cambridge University Press, 2000.)

be common along the coastlines of countries worldwide. Both enhanced precipitation and sea level changes are thus among the anticipated effects of global climate change, and models are the tools used to estimate the possible magnitudes of these changes.

24.4 DEVELOPING A PREDICTIVE INDUSTRIAL ECOLOGY

The goal of systems analysis, models, and scenarios is to understand the technology–environment system well enough to make predictions about how that relationship may evolve in the future, and what possible policy initiatives or interventions might be indicated. Predictions cannot be taken seriously until we understand the contemporary system well enough to model it. With the exception of greenhouse gas forcing of global change, these activities in industrial ecology are at an embryonic stage.

A potential goal for industrial ecology is therefore to develop its own set of scenarios for the 21st century. Those scenarios will need to incorporate considerably more breadth, multiscaling, and interdisciplinarity than those, for example, of the IPCC. Such scenarios might begin with the IPCC framework and build upon it to provide more detail concerning social, political, and technological aspects. Were industrial development to increase at a rate of 1% per year, for example, what technological progress is implied? Is such progress realistic? What would be the probable use of scarce resources? What does such use imply, environmentally, sociologically, and eco-

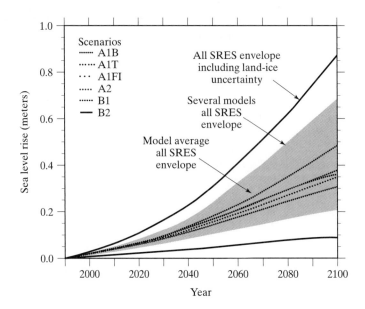

Figure 24.10

Global average sea level rise as computed by several global climate models, based on the family of IPCC 2000 scenarios. SRES = Special Report on Emissions Scenarios. (Reproduced with permission from *Climate Change 2001: The Scientific Basis*, Third Assessment Report, Working Group I, Intergovernmental Panel on Climate Change, *www.ipcc.ch*, 2001.)

nomically, for those countries that supply the resources? Additional scenarios could examine how the world might be transformed by the injection of a high level of technology into the most rapidly developing, populous countries, or how a major regional war might influence resource supplies and environmental impacts. Suppose that political entities in several important countries rejected technological transfer. What then would be the regional and global consequences? The process of devising such scenarios would require substantial interdisciplinary and international efforts, and much would be learned in the process.

As industrial ecology continues to develop, it will become more and more predictive, at increasingly better spatial resolution, and will therefore become compelling at the policy level. At that point, industrial ecology will function to change the way things operate not only at the microscale (the individual product or corporate facility), but at all scales up to those of the planet itself.

FURTHER READING

Hammond, A., *Which World? Scenarios for the 21st Century*, Washington, DC: Island Press, 1998.

Intergovernmental Panel on Climate Change, *IPCC Special Report: Emissions Scenarios*, Cambridge, U.K.: Cambridge University Press, 2000.

Levin, S.A., B. Grenfell, A. Hastings, and A.S. Perelson, Mathematical and computational challenges in population biology and ecosystems science, *Science, 275*, 334–343, 1997.

Low, B., R. Costanza, E. Ostrom, J. Wilson, and C.P. Simon, Human–ecosystem interactions: A dynamic integrated model, *Ecological Economics, 31*, 227–242, 1999.

Oreskes, N., K. Shrader-Frechette, and K. Belitz, Verification, validation, and confirmation of numerical models in the Earth sciences, *Science, 263*, 641–646, 1994.

Schneider, S.H., Integrated assessment modeling of global climate change: Transparent rational tool for policy making or opaque screen hiding value-laden assumptions? *Environmental Modeling and Assessment, 2*, 229–249, 1997.

Schneider, S.H., What is "dangerous" climate change?, *Nature, 411*, 17–19, 2001.

Strogatz, S.H., Exploring complex networks, *Nature, 410*, 268–276, 2001.

EXERCISES

24.1 The automobile technology system is illustrated in Figure 6.1. Construct a similar diagram for the pharmaceutical technology system. Describe a technological change that would noticeably perturb the system. Repeat for a cultural change and for a government policy change.

24.2 Develop three scenarios for the next three decades for the geographical region of which you are a part. (You may choose the scale, but no larger than national.) What are the characteristics of your scenarios? What governmental policies might be suggested, and when should they be considered?

24.3 Develop the framework for a model of the life cycle of rigid plastic (i.e., not bags and wrappings) within a city. The framework should resemble Figure 24.4, but provide more detail in the use stage. What information will be easy to acquire, what difficult? If such a model is constructed, what useful results might be anticipated?

CHAPTER 25

Earth Systems Engineering
and Management

25.1   INTRODUCING THE CONCEPT

Traditional engineering is the art of transforming materials so as to satisfy a human need or desire. In so doing, the relationship of technology to the natural systems of the planet has not been a topic for consideration. The consequences, not surprisingly, have often been that natural systems were degraded, either instantaneously or over time, by industrial activity.

Most of industrial ecology, and indeed most of this book, is directed toward altering our technological society so that we reduce or eliminate impacts on the environment. This is essentially a proactive approach. Other engineering activities that some might term industrial ecology relate to trying to manage Earth systems that are already being affected by human activity: we term this Earth system engineering and management (ESEM).

ESEM is a new area of study arising from a confluence of trends in different fields. It springs from the same insight that underlies industrial ecology: that, as a result of the Industrial Revolution and associated changes in culture, technology, economic systems, human population levels, and economic activity, the dynamics of many fundamental natural systems are increasingly dominated by the activities of a single species—ours. Examples of perturbed natural systems include the carbon, nitrogen, sulfur, phosphorous, and hydrologic cycles; atmospheric and oceanic systems; and the biosphere at scales from the genetic to the species and community levels.

The observation that many of nature's systems are now human-dominated is not new. It was powerfully expressed over a hundred years ago in the classic *Man and Nature* by George Perkins Marsh. William Clark of Harvard University's Kennedy School noted in a 1989 special issue of *Scientific American* entitled "Managing Planet Earth," "[S]elf-conscious, intelligent management of Earth is one of the great challenges facing humanity as it approaches the 21st century." What is unique at present is the scale of human impact, and the increasing coupling between human activities and different natural systems. This has been especially evident in the concern for global climate change, and how it derives from, and perhaps can be reduced through, technological innovation.

A formal definition of ESEM is the engineering and management of Earth systems (including human systems) so as to provide desired human-related functionality in an ethical manner. Important elements of this definition include treating human and natural systems as coherent complexes, to be addressed from a unified perspective, as well as an understanding that requisite functionality includes not just the desired output of the technological system, but also respect for, and protection of, the relevant aspects of coupled natural systems. This can include things valued by humans, such as aesthetics, or ecosystem services such as flood control, as well as respecting biodiversity or the global water cycle as independent values in themselves.

One important difference between ESEM, traditional engineering disciplines, and much of industrial ecology practice to date, is worth emphasis. While virtually all engineering activities reflect the cultures within which they are embedded, ESEM activities require that special attention be paid to cultural, religious, political, and institutional dynamics.

25.2   EXAMPLES OF ESEM, IMPLEMENTED AND PROPOSED

What scale of complexity, or what spatial level, qualifies as an Earth system? To say it another way, what level of disturbance of the natural world qualifies as Earth system engineering and management? The answer becomes ambiguous once we go below the planetary scale to increasingly modest disturbances and eventually to the family garden, but complexity certainly exists no matter how large or small the system may be. Perhaps the appropriate distinction is that ESEM deals with transformations rather than perturbations. For convenience, we adopt here the approach that ESEM applies to any natural system that is significantly perturbed by human action, realizing that a highly specific definition is not crucial to our purpose, and is hard to come by in any case.

25.2.1 Brownfields Restoration

The most common example of small-scale ESEM is the restoration of "brownfields"— locations that historically have been used for industrial purposes and have been contaminated by dumping or leaking of solvents, heavy metals, or other hazardous chemicals. These sites are customarily in urban areas, and thus are near concentrations of population, public transportation and public utilities, characteristics that generally appear to make restoration far preferable to abandonment. The most common restoration technique is to remove the contaminated soil, transport it to a low-population area and bury it deeply. Alternatively, the site may be restored only to the degree that allows specific new uses that are compatible with the state of restoration.

25.2.2 Dredging the Waters

The aquatic analog of brownfields restoration is the dredging of rivers, lakes, and harbors that have been contaminated by the leakage or dumping of pollutants. Sediment is removed from beneath the water body and treated in the same manner as is contaminated soil from brownfield sites. Dredging is often contentious, because the remediation activity may liberate contaminants into the water rather than leaving them sequestered in the sediment. Decisions are made on a case-by-case basis depending on the characteristics of the contaminant and the sediment.

25.2.3 Restoring Regional-Scale Wetlands

Wetlands restoration seeks to recreate historic water flows and associated terrestrial and aquatic communities in severely disrupted wetlands. Done at small scale, it is a difficult and not well understood example of restoration ecology. Done at large scale, such as the South Florida Everglades, it rises to the level of a highly controversial ESEM activity. The Everglades ecosystem has been transformed over the past century by restricting and channeling water flows to deal with increasing human use for agriculture and residential and commercial consumption. Some 3000 km of canals now divert seven billion liters of water per day. The natural cycles that once defined the system have been profoundly affected by human settlement patterns, agriculture, tourism, industry, and transportation systems. In response to clear signs of ecosystem collapse—the nesting success of birds has declined 95% since the mid 1930s, for example—a major restoration project has been proposed, features of which are shown in Figure 25.1. It is obvious, however, that the Everglades has long been, is now, and will continue to be a product of human design. The Everglades of the future will be an engineered system that will display those design objectives (water flow patterns, floral and faunal ecosystems, etc.) that humans choose. The challenge faced is not to restore some hypothetical past state, but to exercise ethical and rational choice in addressing the system as it now exists.

25.2.4 Combating Global Warming

Global warming is the phenomenon in which human-associated emissions of gases to the atmosphere result in increased trapping of outgoing infrared radiation. The principal anthropogenic greenhouse gas is carbon dioxide, though contributions from methane, CFCs, nitrous oxide, and other gases also contribute. Because of the potential severity of global warming, a number of mitigative approaches have been proposed. We discuss them below.

Capturing Carbon Dioxide.   Human emissions of carbon dioxide, the main anthropogenic greenhouse gas, are not just a phenomenon of the Industrial Revolution. In fact, initial perturbations to atmospheric CO_2 concentrations arose from the deforestation of Europe and North Africa between the 10th and 14th centuries. Much later, the development of the internal combustion engine and, as a result, the automotive industry, greatly accelerated emissions of carbon dioxide. As fossil fuel use increases in our modern world, atmospheric CO_2 concentrations continue to increase as well. A possible ESEM alternative is to capture carbon dioxide from stack gases, liquefy it, and in-

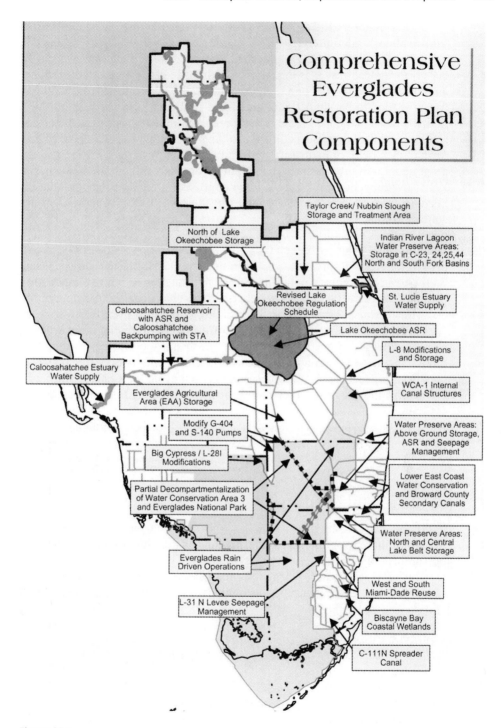

Figure 25.1

The 2001 restoration plan for the Florida Everglades. (Reprinted from *www.evergladesplan.org*)

ject it into underground or undersea aquifers and geologic formations. There, it may remain almost indefinitely.

Strictly speaking, CO_2 sequestration from power plant stacks is not ESEM, but pollution control, just as is the capture of volatile organic gases before they can leave an industrial facility. In common usage, however, all proposals for dealing with global warming tend to be lumped into an ESEM framework. Regardless of how it is classified, if a fossil fuel power plant is designed to combust carbon-based feedstocks, transfer the combustion products directly to long life reservoirs, and produce energy in the form of electricity or hydrogen, the perspective on power generation and the environment undergoes fundamental change. In Figure 25.2, rather than being part of a significant environmental problem—a large emitter of greenhouse gases—the power plant becomes part of the solution—a factor in the control of greenhouse gas atmospheric concentrations.

Sequestering Carbon in Vegetation. CO_2 sequestration, as described above, is an approach that aims to prevent CO_2 from being emitted into the atmosphere. Once it is there, however, another potential ESEM approach is to remove a portion of it. One technique that has been fairly widely embraced has been the planting of fast-growing trees, since atmospheric CO_2 is the building block for the cellulose from which trees are made. It is clear that reforestation of previously forested areas will indeed store CO_2 for at least a period of time, though the absorption rate slows as the trees age. Although the degree of long-term gain is imperfectly understood, tree planting has many beneficial aspects besides carbon storage, and may be increasingly adopted. Increased biomass proposals raise other issues, however: Can significant increases in biomass

Figure 25.2

The Sleipner natural gas/condensate complex in the North Sea off Norway. The natural gas in this region typically contains about 9% CO_2, which must be reduced to less than 2.5% before it can be sold. The larger platform comprises the natural gas extraction and condensation equipment, while the smaller platform to the left houses the CO_2 extraction and sequestration equipment. (Photo courtesy of Statoil.)

production be done in such a way as to avoid destabilizing the global nitrogen cycle? How critical are genetically engineered forms of biomass (e.g., trees designed to fix their own nitrogen) to the implementation of this plan? Once again we are challenged to think of any action from a very broad systems perspective.

Sequestering Carbon in Marine Organisms. CO_2 is a building block for phytoplankton, the tiny marine organisms that carry out nearly half of the photosynthesis on Earth. The reproduction and growth of those organisms is limited in most parts of the oceans by the availability of nutrients, particularly iron. It was thus proposed in the early 1990s that if the oceans were fertilized with iron, the resulting growth of organisms would remove considerable CO_2 from the atmosphere.

Spreading fertilizer on the ocean surface on a regular and widespread basis is an enormously ambitious project, but a few tests have been made to assess the feasibility of the idea. In the most extensive of these, a seeding experiment in the Southern Ocean south of Tasmania, increased phytoplankton growth was stimulated and maintained for over a month. It was not clear that the CO_2 that was incorporated was then transferred to the deep ocean, as would be required for the approach to be effective. In addition, there is concern for unintended side effects such as deoxygenating the deep ocean and disrupting the structure of marine food webs. Given the current scientific uncertainty, it remains unclear whether, and at what scale, this approach should be employed.

Scattering Solar Radiation with Sulfur Particles. Another potential ESEM approach to the mitigation of global warming avoids dealing with CO_2, but rather with preventing incoming solar radiation from reaching the planet's surface. This idea was proposed some years ago by Russian climatologist Mikhail Budyko, who envisioned injecting some 35 Gg of sulfur dioxide annually, about 25% of the amount presently released by fossil fuel burning, directly into the stratosphere. He calculated that such an amount, once converted there to sulfate aerosol particles, should significantly enhance the backscattering of solar radiation to space. The success of the technique depends on an accurate assessment of stratospheric sulfur dioxide to sulfate transition rates at all latitudes and seasons, and it is uncertain whether this information can be precisely derived. Still more challenging, however, are the logistical difficulties of delivering many gigagrams of gases or particles to an altitude near the limit of modern aircraft by fleets of thousands of planes. A possible alternative is to load sulfur particles into ballistic shells and shoot them into the stratosphere using the guns of the world's large naval vessels (several thousand rounds per day, day after day, year after year). By either method, the cost would be in the tens of billions of U.S. dollars annually, and the potential environmental impacts appear highly problematic.

Reflecting Solar Radiation with Mirrors in Space. An alternative to injecting scattering particles into Earth's upper atmosphere is to send mirrors or other reflective devices by space satellite to the Lagrangian L1 point (a point along the Earth-Sun line where no net forces act on a small object), so that the objects might remain at that location and reflect radiation away from Earth indefinitely (Figure 25.3). Unfortunately for the concept, gravitational displacement forces from planets and other celestial ob-

Sun   Earth

Figure 25.3

Conceptual diagram for solar reflectors placed at Lagrangian point L1 of the Earth-Sun system to decrease the amount of solar radiation incident on Earth. There are five Lagrangian points, three of which are shown (the other two are at the points of equilateral triangles from the Earth-Sun line). Objects located at these points are stable, or nearly so, because gravitational acceleration and centrifugal acceleration are exactly in balance. L1 is approximately 1% of the distance from Earth to the sun.

jects are to be expected at the L1 point, so the objects would require an active positioning system in order to remain in place. Active positioning systems imply such things as mechanical components and compressed gases, and preliminary assessments of this idea suggest that spacecraft stabilization over long periods of time may be impractical. In addition, of course, the enterprise would be extremely expensive, not to mention beset with political intricacies, and the size and reflectivity of the devices would have to be determined very precisely in order not to heat or cool Earth more than desired.

Global Warming ESEM.   As we have seen, there are many ways in which technologies, economic practices, and cultures can be used to modify human impacts on the carbon cycle and climate systems. Figure 25.4 illustrates how these options might be gathered together into a portfolio of options to enable management of the carbon cycle and climate. From an ESEM perspective, however, it is also important to recognize that there are high levels of interconnection among fundamental natural and human cycles that are not addressed by these targeted approaches. Imperfectly understood couplings and complexities raise caution flags for most or all of these ideas.

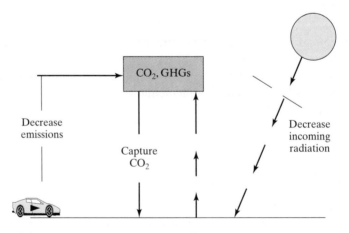

Figure 25.4

The system of potential ESEM approaches to combat global warming.

25.3   THE PRINCIPLES OF ESEM

As the examples in the previous section suggest, the institutional, cultural, and knowledge capabilities to carry on ESEM do not yet exist. The ethical structure necessary to support integrated ESEM activities is lacking as well. However, Earth systems that are so severely degraded that they are or threaten to become nonfunctional may nonetheless require that we take some action. In these cases, one can draw on experience to date with complex systems engineering projects to generate a basic set of ESEM principles that, although still illustrative, create an operational foundation. These principles can be sorted into three categories: theory, governance, and design and engineering.

25.3.1   Theoretical Principles of ESEM

The theoretical underpinnings of ESEM reflect the complexity of the systems involved and our current levels of ignorance. The overriding dictum in ESEM is thus the precautionary principle, from which those below are derived.

- Only intervene when necessary, and then only to the extent required, because minimal interventions reduce the probability and potential scale of unanticipated and undesirable system responses.
- ESEM projects and programs are not just scientific and technical in nature, but unavoidably have powerful economic, political, cultural, ethical and religious dimensions. An ESEM approach should integrate all these factors.
- Unnecessary conflict surrounding ESEM projects and programs can be reduced by recognizing the difference between social engineering—efforts to change cultures, values, or existing behavior—and technical engineering. Both need to be part of ESEM projects, but they are different disciplines and involve different issues and world views.
- ESEM requires a focus on the characteristics and dynamics of the relevant systems as systems, rather than as sums of individual components. The components will, of course, also have to be considered: ESEM augments, rather than replaces, traditional engineering activities.
- Boundaries around ESEM initiatives should reflect real world linkages through time, rather than disciplinary simplicity.
- Major shifts in technologies and technological systems should be evaluated before, rather than after, implementation of policies and initiatives designed to encourage them. For example, encouraging reliance on biomass plantations as a global climate change mitigation effort should not become policy until predictable implications such as further disruption of nitrogen, phosphorus, and hydrologic cycles are explored.

25.3.2   Governance Principles of ESEM

As discussed in Chapter 6, the global governance system is rapidly evolving and becoming more complex. These changes, especially when combined with the inherent complexity of human and natural systems, give rise to a second category of principles involving ESEM governance.

- ESEM initiatives by definition raise important scientific, technical, economic, political, ethical, theological and cultural issues in the context of global polity. Given the need for consensus and long-term commitment, the only workable governance model is one that is inclusive, transparent, and accountable.

- ESEM governance models that deal with complex, unpredictable systems must accept high levels of uncertainty as inherent in the process. ESEM policy development and deployment must be understood as a continuing dialog with the relevant systems rather than a definitive endpoint, and should thus emphasize flexibility. Moreover, the policymaker must be understood as part of an evolving ESEM system, rather than an agent outside the system guiding or defining it.

- Continual learning at the personal and institutional level must be built into the process.

- There must be adequate resources available to support both the project, and the science and technology research and development that are necessary to ensure that the responses of the relevant systems are understood.

25.3.3 Design and Engineering Principles of ESEM

Finally, there is a set of principles that informs the design and engineering of ESEM systems:

- Know from the beginning what the desired and reasonably anticipated outcomes of any intervention are, and establish quantitative metrics by which progress may be tracked.

- Unlike simple, well known systems, the complex, information-dense and unpredictable systems that are the subject of ESEM cannot be centrally or explicitly controlled. Rather than being outside the system, the Earth systems engineer will have to see herself or himself as an integral component of the system itself, closely coupled with its evolution and subject to many of its dynamics.

- Whenever possible, engineered changes should be incremental and reversible, rather than fundamental and irreversible. Any scale-up should allow for the fact that in complex systems, discontinuities and emergent characteristics are the rule, not the exception.

- An important goal in Earth systems engineering projects should be to support the evolution of resiliency, not just redundancy, in the system. Thus, inherently safe systems are to be preferred to engineered safe systems.

25.4   FACING THE ESEM QUESTION

While the global climate change discussions provide, perhaps, the most dramatic example of prospective ESEM, one might also cite the ongoing efforts to manage the Baltic Sea, managing regional forests to be sustainable, restricting exploitation of local and regional fisheries, and meeting continued challenges from invasive species. This last example is a product in large part of previous patterns of human migration, and illustrates that ESEM is not something that humans should now begin to do, but that we have been overtly influencing natural systems for centuries. Similarly, it is not unrea-

TABLE 25.1 Characteristics of Implemented or Proposed ESEM Projects

ESEM activity	Environmental challenge	Potential severity	Scale of impact	Understanding of challenge	Scale of activity	Public visibility	State of implementation
Brownfield redevelopment	Human toxicity	Moderate	20–100 m	Good	20–100 m	Very high	Extensive
Dredging the waters	Pollutants in sediment	Low	100–1000 m	Good	100–1000 m	Modest	Several projects
Wetlands restoration	Ecosystem degradation	High	100–10,000 km	Modest	100–10,000 km	High	Several projects
CO_2 sequestration	Global warming	Very high	Global	Modest	1–10 km	None	Sleipner project
Tree planting	Global warming	Very high	Global	Modest	1–100 km	None	Several projects
Ocean fertilization	Global warming	Very high	Global	Modest	10–100 km	None	Preliminary experiments
Stratospheric sulfate	Global warming	Very high	Global	Poor	Global	None	None
Mirror in space	Global warming	Very high	Global	Poor	Extraplanetary	None	None

sonable to view global agricultural systems, tightly linked as they now are by trade and commodity markets, as an ESEM process—and, obviously, another one that has been going on for centuries.

For the most part, the specialty of environmental engineering is distinct from industrial ecology, as it is largely centered on the proper treatment of residues and the remediation of relatively small-scale contaminated sites. To the extent that the systems aspects of ESEM may be regarded as industrial ecology, however, ESEM potentially brings these two specialties together. It is less important to debate whether ESEM belongs to one or the other than to realize that ESEM at the smaller scales, at least, is an ongoing activity that joins the technological society and the environment in intricate and problematic ways.

In Table 25.1, we collect characteristics of implemented or proposed ESEM projects. Some important lessons are revealed by the table. It is interesting to ask why some ESEM proposals are being implemented and others are not. The table shows that implementation is not related to the potential severity of the environmental challenge, or to the spatial scale of the proposed ESEM activity, or to the spatial scale of the impact. It is related, however, to the public visibility of the environmental challenge and (to a lesser extent) to the degree of scientific understanding. The implication is that if we can readily witness a problem and know how to attack it, ESEM implementation will occur. Conversely, with more complex but less visible perturbations such as climate change, it is likely to be difficult to engage the public in addressing them.

ESEM raises a fundamental issue: What level of ESEM is appropriate? ESEM by its nature is a means to an end which can only be defined in ethical terms. Simply put, the question "To what end are humans engineering, or *should* humans engineer, the planet Earth?" is a moral as well as a technical one. It is also not just hypothetical: Human institutions are implicitly answering that question every day by invoking ESEM activities on a variety of spatial and temporal scales.

ESEM will assume increased relevance as discussions move from a goal of environmental improvement to one of sustainability. The latter implies some sort of targets for technology–environment interactions, together with policies designed to meet those targets, monitoring to evaluate progress toward those targets, and periodic review to assess whether mid-course corrections are needed. This is essentially the approach that was taken to the ozone hole problem in the Montreal Protocol, which ultimately involved industrial ecology in devising alternatives to CFCs, engineering in capturing and destroying existing CFC stocks, atmospheric science in monitoring progress, and international political action to make it all happen. If we as a society are serious about sustainable development, we will need ESEM approaches to approach the societal, economic, and environmental goals that sustainable development implies.

FURTHER READING

Allenby, B.R., Earth systems engineering and management, *Technology and Society, 19*(4), 10–24, 2000/2001.

Clark, W., Managing planet earth, *Scientific American, 251 (3)*:47–54, 1989.

Diamond, J., 1997, *Guns, Germs and Steel: The Fates of Human Societies* (New York: W. W. Norton & Company).

Marsh, G., 1864, *Man and Nature* (Cambridge, MA: Cambridge Belknap Press).

Schneider, S.H., Earth systems engineering and management, *Nature, 409*, 417–421, 2001.

Science, 1997, Special report: Human-dominated ecosystems, *277*:485–525.

Turner, B. L., W. C. Clark, R. W. Kates, J. F. Richards, J. T. Mathews, and W. B. Meyer, eds., *The Earth as Transformed by Human Action*, Cambridge, U.K.: Cambridge University Press, 1990.

EXERCISES

25.1 An area frequently identified as an ESEM case study is genetic engineering of human and nonhuman species. Create a conceptual framework analogous to Figure 25.4 for biotechnology.

25.2 What are some of the moral or ethical issues inherent in the global climate change negotiations? Are they implicit or explicit?

25.3 What is the relationship among industrial ecology, ESEM, and sustainable development?

CHAPTER 26

The Future
of Industrial Ecology

26.1   INDUSTRIAL ECOLOGY IN THE MIDST OF CHANGE

As the new millennium begins, society, the environment, and governance structures are all undergoing rapid and transitional changes. The environment is clearly a system under substantial stress. Ecosystems are in trouble, biodiversity is diminishing, and planetary processes that have evolved over very long periods of time show signs of dysfunction. These changes will have dramatic effects on technology and its interactions with the natural and human worlds. It seems likely that the period 2010–2050 will encompass extremely rapid societal transformations and newly dramatic examples of environmental degradation. Human governance systems will have no choice but to respond vigorously to environmental challenges.

We are not completely unaware of these dangers. Our scope for thinking and analysis is broadening and our temporal and spatial assessment scales are increasing. Among the environmentally related approaches of note are the following:

- Not looking to the past, but to the future
- Not fragmented, but systemic
- Emphasizing not gross insults, but microtoxicity
- From focusing on environmental improvement to focusing on sustainability

This new vision needs to play out in the midst of a dramatic increase in global population and an equally vigorous reliance on technology. The demographic transition from a high birth rate/high death rate society to a low birth rate/low death rate society will be essentially complete by about 2050, with a global population of nine billion or so. The majority of these people will live in urban areas, probably megacities. They are likely to have increased income, and will demand increasing amounts of resources (or the services thereof). They will desire buffering from surprises. Five transitions capture most of this societal evolution.

- Not local, but global
- Not pastoral, but urban
- Not isolated, but connected
- Not more technology, but better technology
- Not emphasizing the developed world, but the developing world

Important transitions are occurring in the governmental and corporate policy arenas as well. Although precise regulatory instruments have done well at minimizing some of the most damaging environmental impacts of technology, it is increasingly clear that such approaches will take us only part of the way to a more sustainable world. As a result, several policy transformations are increasingly apparent.

- An emphasis on cooperation, not regulation
- The gradual incorporation of environmental costs into price structures
- Corporations rather than governments making key decisions on technology and the environment
- A more important role for nongovernmental organizations, with greater public accountability for their activities

Many of the environmental, societal, and policy changes have implications for industrial ecology, where we expect the following transitions:

- From a product focus to a system focus
- From a manufacturing plant concern to a life-cycle and service-oriented focus
- From greening current products to designing inherently green products
- From environmental thinking as peripheral to environmental thinking as strategic

In the midst of these changes, many of them dramatic and unprecedented, industrial ecology becomes a field attempting to achieve major improvements in a technology–environment system that is itself in rapid evolution. Many of the general directions to be taken seem well identified, but it is realistic to anticipate that lots of mid-course correction will be needed.

26.2   THE INDUSTRIAL ECOLOGY HARDWARE STORE

The transformation of technology to enable a substantially more sustainable world is a most challenging project. Nonetheless, as we have seen in this book, a number of useful tools are available for deployment in such a project, even as they see continual development. We might think of these tools as the shelf stock in an industrial ecology hardware store, a store with sections designed to have particular appeal to product and process designers, to managers, to technological analysts (systematists), and to policy makers. A number of these tools are primitive; it has been said that we often apply flint-age tools to space-age problems, especially in the DfE area. One obvious deficiency is that many of our tools are static, while the problems they address are dynamic. However, they are what we have to work with at present, and they have proven generally useful. We summarize the shelf stock of each set of tools below.

26.2.1   Tools for the Product and Process Designer

Designers of products and processes make small-scale decisions with enormous multiplicative potential. A more energy-efficient motor design is extremely influential when hundreds of thousands of those motors are produced, as is the use of environmentally benign solvents when millions of doses of a pharmaceutical product are being manufactured. The toolkit that aids in such designs is described in more detail in chapters 8–18; it includes:

- "Inviolates" lists
- Design guidance handbooks
- Life-cycle assessment
- Streamlined life-cycle assessment

26.2.2   Tools for the Corporate Manager

The corporate manager integrates environmental responsibility into corporate philosophy and practice. In this role, environmental thinking and action become strategic parts of corporate operations. The toolkit available to the manager includes those integrated with other corporate assessment approaches, and those that are distinct:

- The Green Product Realization Process, in which environmental considerations are integrated into existing corporate management tools
- ISO 14000, in which a corporation's environmental goals and implementation approaches are established and then evaluated by external auditors
- Environmental Management Systems, in which management plans for all aspects of a corporation's environmental performance are carried out
- The Triple Bottom Line approach, where economic, environmental, and social performance are all valued and measured

26.2.3   Tools for the Service Provider

The importance of the service sector in modern society is indisputable, as is the sector's potential for environmentally beneficial action. The sector's industrial ecology toolkit overlaps to some degree those of the designer and the corporate manager, but surely includes the following:

- Regulatory Compliance audits, to ensure a base level of environmental performance

- Resource Use audits (energy, water, waste), to identify opportunities to save money and improve environmental performance
- Comet diagrams, to evaluate alternative approaches to environmentally related husbandry of equipment
- Reverse Fishbone diagrams, to optimize reuse of obsolete equipment, its components, and its materials

26.2.4 Tools for the Systematist

Technological analysts evaluate industry–environment interactions at meso- and macroscales. Their goal is the understanding of the interactions at the level of systems, encompassing the influences of the range of industrial sectors, products, and emissions, as well as the stimuli or constraints imposed by patterns in space, time, or state of economic development. Among the tools available to the industrial ecology systematist are:

- Food web analysis
- Assessments of flows of resources (materials flow analysis, substance flow analysis, elemental flow analysis)
- Industrial symbiosis analysis
- Models and scenarios

26.2.5 Tools for the Policy Maker

Policy makers look at industrial ecology from the perspective of social and cultural systems enabled by technology. Their tools include:

- The Grand Objectives
- Earth Systems Engineering and Management
- Quantitative sustainability evaluations

26.3   INDUSTRIAL ECOLOGY AS AN EVOLVING SCIENCE

Industrial ecology is a science, if a very young one. What perspective should industrial ecologists take? One consideration is that of scale, spatial and temporal. In most scientific fields, there are people working at very small scales and very large scales, as shown in Figure 26.1. Industrial ecology has components that operate in very different scale regions as well, and specialists are also common. The loci of these specialists in industrial ecology space–time reflect not only their preferences, but also the state of maturation of the field.

On the smallest scale, the activity is in pollution prevention (Figure 26.2), such as efforts of manufacturers to decrease water use. Those activities have been very successful, but the achievements are ultimately limited by the amount of this "low-hanging fruit" available to be harvested and by the limitation of pollution prevention activities to contemporary processes and to early life-cycle stages.

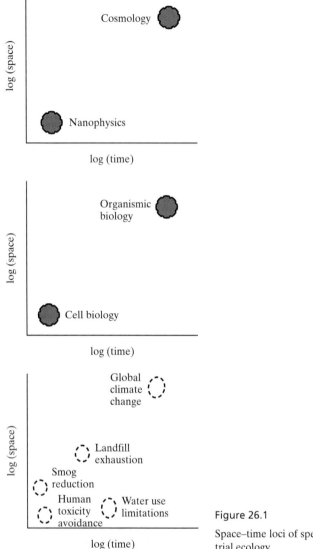

Figure 26.1

Space–time loci of specialists in physics, biology, and industrial ecology.

The next largest IE regime can be termed the technological phase. The field is mostly in that phase at present, and is concentrated on developing new approaches to product and process design and implementation.

The more forward-thinking organizations, corporations, and governments are beginning to move into the transition phase where the fruit is more succulent, but harder to reach, as it involves consideration of systems and interactions. The final stage, well into the future, is the metadisciplinary phase, when IE will combine forces with the social and policy sciences to optimize the technology–society relationship. This knowledge integration can be pictured schematically as shown in Figure 26.3.

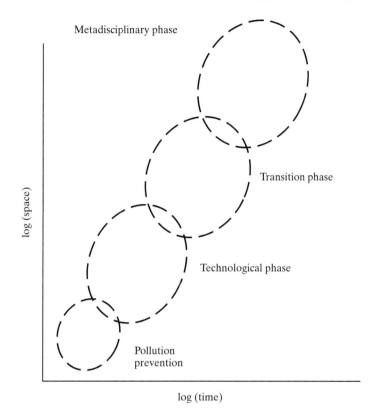

Figure 26.2

A graphical representation of industrial ecology space–time suggests the positional relationship of various system approaches. The issues that are addressed in the metadisciplinary approach are orders of magnitude larger in space and time than those addressed in pollution prevention.

As we move forward, the challenge is to be rational about the concept of sustainability. John Ehrenfeld of the Massachusetts Institute of Technology uses these words:

> Sustainability is a possible way of living or being in which individuals, firms, governments, and other institutions act responsibly in taking care of the future as if it belonged to them today, in equitably sharing the ecological resources on which the survival of all life depends, and in assuring that all who live today and in the future will be able to satisfy their needs and human aspirations.

Sustainability, including sustainable development, does *not* mean doing a little bit better job of what we are doing now. Rather, as William McDonough says, it is about "changing the story" to *transform* the relationship between technology and the environment. The path to optimum change is not completely clear, of course. Nonetheless, we can probably identify a path leading in the direction of environmental superiority, and to have reasonable confidence that mid-course corrections to that path will lead in the direction of sustainability.

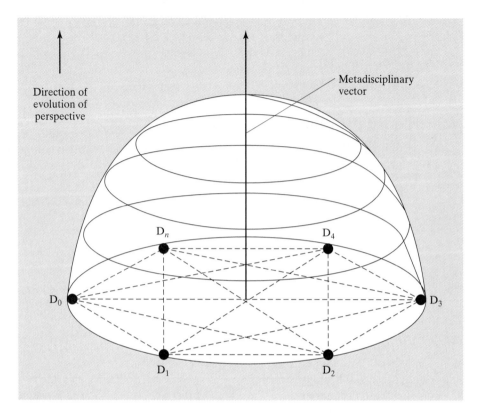

Figure 26.3

The metadisciplinary approach framework, in which knowledge from scientific, technological, social, and economic fields is synthesized in a focus on global sustainability and the human prospect. Locations D indicate specific intellectual disciplines; the dashed lines between them represent interactions between the knowledge bases of two disciplines. As one moves further up the metadisciplinary vector, the reduction in circumference of the usable knowledge circle suggests improved integration of information. (Based on a diagram by H. Koizumi, Hitachi Corporation.)

No matter what the phase of its evolution as a field, industrial ecology must work to attain a high level of intellectual rigor. In its first decade or so, industrial ecology has addressed important but relatively straightforward and highly visible challenges, largely at short time scales and small spatial scales. These advances have required engineering diligence, but limited amounts of innovation, detailed data acquisition, or in-depth analysis. Too often we have tried to simplify our field for our own ease or that of others—"reduce dissipation to zero," or "extract nothing" are examples. This is not intellectual rigor and it is not complex systems analysis. If we are dissipating a resource at a rate assimilable by the ecosystem, and the energy cost to recover the resource is high, we might well wish to preserve energy and accept dissipation. We have, however, no satisfactory tradeoff analysis tools.

Science and technology reach the highest level when the scientific method is fully employed: ask an interesting and relevant question, devise an experiment or analysis to address the question, acquire the necessary high-quality data, analyze the data from the perspective of the question asked, and answer or reform the original question. Intellectual rigor in science, engineering, economics, and policy studies takes different forms, but if industrial ecology achieves rigor its credibility and influence will markedly increase. We need to accept nothing less.

26.4   AN INDUSTRIAL ECOLOGY RESEARCH ROADMAP

Research and development activity is customarily considered to have three stages:

1. *Basic research*: The study of fundamental aspects of phenomena and of observable facts, without specific applications in mind
2. *Applied research*: The development of the knowledge or understanding necessary for determining the means by which a recognized need may be met
3. *Development*: Systematic use of the knowledge or understanding gained from research

Intellectually, these stages may be perceived as sequential. In practice, they often occur more or less in inverse order, at least early in the development of a field. Industrial ecology is pretty clearly still in this early development phase.

What then are the research frontiers for industrial ecology? Or, as Valerie Thomas of Princeton University asks, "If we are successful over the next few years in doing industrial ecology, what will we have done?" Every field of science and technology has a small number of key research questions, and in the basic sciences these can be precisely stated. In fields where closed systems can be established, scientists such as molecular biologists and nuclear physicists ask how cells divide or what the structure of the atom might be. In open-system sciences like biological ecology, the key questions are more complex and tend to deal with interactions rather than with basic properties:

- Why are ecosystems the way they are?
- How do ecosystems function?
- What establishes species populations?
- Why are there so many species?

In these open-system fields, one can also ask questions related to external, uncontrollable systems conditions. For biological ecology, for example, one might be interested in the effects on ecosystems of changes in external drivers (e.g., climate), or in the relationships between biogeochemical cycles and related systems at all scales (habitat alteration, etc.).

Drawing on these questions for guidance, we can attempt to define analogous key questions for industrial ecology. The list includes the following:

- What are the types of anthropogenic resource use?
- What are the relationships between technology and environment?

- What are the trends in anthropogenic resource use?
- What are the trends in technology-related environmental impact?
- How do the resource-related aspects of human cultural systems operate?
- What is an operational and quantitative definition of sustainability?
- What scenarios describe possible technology–environment futures?

These key questions suggest a number of research goals for industrial ecology, which we can divide into three categories: (1) theoretical, (2) experimental, and (3) applied.

26.4.1 Theoretical Industrial Ecology Goals

- *Develop a theory of industrial ecosystems.* Can we construct a robust theory of technological change? As industrial systems develop and change over the next few decades, can we predict the types and magnitudes of environmental interactions that will occur? Do evolutionary patterns of technology–environment interaction differ with industrial sector? With spatial pattern? With type of material input? With different cultures and societies?
- *Develop a multiscale energy budget for technology.* Technology involves the transformation of materials, and the transformations are accomplished by using energy. Existing energy budgets tend to be at national or global scales, but smaller scales could well reveal systematic patterns that could optimize the energy–technology interface. Many studies at different scales and degrees of detail will be needed in order to derive generalized conclusions.
- *Model interactions between human and natural systems.* Human and natural systems, traditionally studied as separate entities, are obviously strongly interactive. The interactions need to be studied and understood well enough to model the impacts on human technological systems of transformative natural systems change, and on natural systems of transformational human technological systems change.
- *Develop a theory of quantitative sustainability.* If sustainability is to be more than a concept, it must be rendered in quantitative form, allocating resource quantities and permissible environmental impacts among individuals and corporations. Defining allowable rates of use and loss and achieving agreement on allocation will be difficult, and will depend on the particular form of sustainability our species chooses. Without those steps, however, it will not be possible to address the challenge of achieving a sustainable world.

26.4.2 Experimental Industrial Ecology Goals

- *Develop budgets for the materials of technology.* In recent years, natural scientists have constructed cycles for the "grand nutrients": carbon, nitrogen, sulfur, and phosphorus. The results have provided scientific understanding and a basis for policy initiatives. In a similar way, it will be valuable to construct cycles for the materials of modern technology: metals, plastics, renewables, etc. The results should provide insight on resource use rates and losses to the environment, and guidance for policy decisions related to the materials cycles.

- *Analyze the design and development of ecoindustrial parks.* It appears probable that the reuse of industrial residue streams can be enhanced by carefully planned ecoindustrial parks, but we have as yet few data to support this conjecture. The input, output, potential flows, and spatial and temporal evolution of existing ecoindustrial parks of all types need to be collected, grouped, analyzed in detail, and integrated into the municipal planning and development process.

- *Gather and analyze data on industrial food webs.* Biological ecologists use food web analysis to understand the linkages among species, and the interspecies resource transfers that occur. In perhaps similar ways, industrial species share metals, plastics, paper, and other resources in systems where linkages might be revealed by extensive industrial food web analysis.

- *Analyze the metabolism of cities.* Cities are known to be strong attractors of resources, and weak dispersers of them. If we think of cities as organisms, however, we have little quantitative information on their metabolisms. Research is needed to establish metabolic differences between cities in different parts of the world, of different sizes, and in different stages of evolution. The results will help us better understand the relationship among urban regions, resources, and the environment.

26.4.3   Applied Industrial Ecology Goals

- *Continue development of Design for Environment and Manufacturing for Environment.* As with all engineering design approaches, DfE and MfE will always be works in progress. New materials, new analytical capacity, and new product concepts will require appropriate responses from the DfE and MfE communities. Increasingly, the use of these tools must become second nature for the designer as they become fully integrated into design systems.

- *Link industrial ecology to land use.* Technology–environment interactions are ultimately place-based, and the relationship of industrial operations to the use and health of the land is a topic demanding increasing attention. There has been no detailed exploration of this topic, but it deserves study because of its central role in the planet's future.

- *Develop policy instruments to incentivize industrial ecology.* Industrial ecology activities are sometimes frustrated by governmental policies, as in the operational difficulty of reusing material labeled hazardous waste. Conversely, government policies such as buying energy-efficient equipment can encourage industrial ecology practice. The development of new policy instruments to more fully stimulate industrial ecology should be investigated and subsequently implemented.

- *Diffuse industrial ecology into developing countries.* As populations and affluence increase in the developing countries, environmental opportunities and challenges related to industrial ecology will move there as well. It will be important to devise educational and outreach methodologies to provide these evolving geographical regions with the industrial ecology tools they will need.

26.5   REDEFINING THE CHALLENGE

Edward O. Wilson of Harvard University says that we are in the "Environmental Century," and offers the following prediction: "We are entering the century of the environment, when science and politics will give the highest priority to settling humanity down before we wreck the planet." His perspective is not very encouraging. As industrial ecologists, however, we can perhaps afford to be cautiously optimistic. Many corporations are implementing industrial ecology, citizens everywhere are talking seriously about global warming, nearly everybody realizes that we are living unsustainably. Thus, an initial vision of sustainability is beginning to evolve, the toolkits are getting better and better, and environmental sensitivity is being incorporated into educational activities at all levels. Industrial ecology is thus evolving into a discipline that can be given a new, enhanced definition:

> Industrial ecology is the science of multiscale planetary stewardship, involving the practice of intelligent oversight of the planet as it undergoes natural and anthropogenically driven variability.

Were we to actively implement this vision, we could change the story. This implementation is far from an easy task, of course. It will take the efforts of all those who read this book, and many others. The job is large, but increasingly it appears possible. Good luck to all of us!

FURTHER READING

Ostrom, E., J. Burger, C.B. Field, R.B. Norgaard, and D. Policansky, Revisiting the commons: Local lessons, global challenges, *Science, 284*, 278–282, 1999.

Rejeski, D., Metrics, systems, and technological choices, in *The Industrial Green Game*, D.J. Richards, ed., Washington, DC: National Academy Press, 48–72, 1997.

Vandenburg, W.H., *The Labyrinth of Technology*, Toronto: University of Toronto Press, 2000.

Wallner, H.P., M. Narodoslawsky, and F. Moser, Islands of sustainability: A bottom-up approach towards sustainable development, *Environment and Planning A, 28*, 1763–1778, 1996.

Wilson, E.O., Integrated science and the coming century of the environment, *Science, 279*, 27 March 1998.

Electronic Solder Alternatives: A Detailed Case Study

A.1   INTRODUCTION

An alternative to the SLCA 5 × 5 matrix approach described in Chapter 17 is to use multiple assessment matrices. The goal is to capture the full scope of potential attributes—manufacturing, social and political, toxicity and exposure, and environmental. In this way, we treat all aspects of business decision making while retaining the relative efficiency of a streamlined assessment approach.

Comparative DfE analyses can be quite complex, and it is useful to provide an example to illustrate some of the considerations involved. We choose an important one for modern industry: Whether the use of lead-containing solder is the preferable way to fasten electronic components to circuit boards, given that lead is a toxic substance. (This study complements one described in Chapter 10 for a five-component solder, focused solely on the sustainable supply of solder materials.) The example here uses the steps laid out in Chapters 15 and 16: (1) scoping of options and depth of analysis, (2) capture of relevant data in a group of DfE matrices, and (3) the translation of results into design directions. Because the purpose of the exercise is to illustrate the procedure, not make a soldering expert of the reader, much of the detail of the analysis has been omitted.

A.2   DFE ANALYSIS OF LEAD SOLDER IN PRINTED WIRING BOARDS

The principal use of lead solder in electronics manufacturing is in the assembly of printed wiring boards, which form the heart of every electronics device. This use of lead solder is a concern to industry for two reasons. First, lead is unquestionably a toxic heavy metal, so any consumption that might result in human environmental exposure raises health and safety concerns. Second, government regulation of lead use is increasing, and in some countries regulatory constraints have been placed on the use of lead. Nonetheless, the comparative volume of lead involved in this particular use is quite small (less than one % of overall consumption—see Figure 23.3).

Indium and bismuth, two metals frequently suggested as substitutes for lead in soldering applications, are little used. As a consequence, full replacement of lead solders with indium/tin alloy in the U.S. (for example) would result in a jump in demand for indium from approximately 28 metric tons to some 11,000 metric tons; the comparable figures for bismuth are 1400 metric tons of current consumption compared to approximately 12,000 metric tons for replacement of lead in solder. A third alternative, using silver as the metallic filler for conductive epoxies, is intermediate in comparative resource use depending on the technologies chosen.

Both indium and bismuth occur in extremely low concentrations in their ores and are produced almost entirely as byproducts of the mining of other metals: indium from zinc and bismuth from lead. Again silver occupies an intermediate position: Some silver deposits are relatively rich, but approximately two-thirds of silver is produced as a byproduct from mining other ores.

A.2.1   The Scoping Function

Because electronics manufacturing technologies and techniques have coevolved with lead solder, any significant shift to an alternative material, especially one with a substantially different operating temperature, could involve substantial redesign of existing processes and considerable capital investment in new equipment. Redesign of existing products and components would also be a likely consequence. An unsuccessful shift to an alternate technology could result in significant competitive penalties and quality degradation. Accordingly, a "cradle-to-reincarnation" analysis of potential alternatives is clearly warranted.

As a rule, options chosen for comparative analysis should be limited in number. This restriction is obviously critical if the analysis is to be manageable. In the present case, there are myriad potential substitutes for tin/lead solder, many of which can be immediately rejected because they also contain lead or cadmium. Of those that are usable in the same general temperature range as lead solders, alloys based on indium and bismuth are predominant. Metal-containing polymer matrices are also potentially attractive. Accordingly, a generic tin/indium solder, a tin/bismuth solder, and a nonalloy alternative consisting of an epoxy matrix filled with conducting silver beads are selected for analysis.

A.2.2   Manufacturing Matrices

The first of the assessment matrices is the manufacturing matrix. From a manufacturing standpoint, potential problems exist with substitution for lead as a result of the scarcity and byproduct status of indium and bismuth, and, to a lesser extent, silver. These problems are reflected in the cost and availability categories and, for indium and bismuth, the energy consumption category (these metals occur in such low concentrations that significant energy must be expended to extract and purify them). A matrix evaluation has been performed for each alternative; the indium manufacturing matrix is presented in Figure A.1 as a sample.

Among the questions raised by the manufacturing matrix analysis is that of inappropriate resource consumption for indium, bismuth, and, to a lesser extent, silver. These relatively rare metals are being taken from ores, where they are recoverable, and put into product where they will probably be unrecoverable. The current price structure does not reflect this potential cost to future generations—indeed, cannot, given the inherent uncertainty of the future value of existing materials, which in turn depends on future technological developments.

A.2.3   Social/Political Matrices

The indium social/political matrix is illustrated in Figure A.2. This matrix is the most difficult to evaluate because of data and methodological deficiencies. The unspoken but standard assumption is that the price structure for various options captures these effects: This is usually not the case. Especially when operations involve developing world countries, most social costs simply remain externalities. Take, for example, the increases in mining and ore processing activity implied by substantial substitution of indium, bismuth, or silver (epoxy) options for lead solder. Not only will there be greatly increased environmental disruptions (captured in the environmental matrices), but the impact of such activities on local communities can be profound. Mining by definition draws down the natural capital of a region. Mining operations also create temporary needs for labor, resulting in temporary communities. This negatively affects not only the communities from which the labor will be drawn, but the mining community itself, which will be abandoned in due course. Such effects are more likely to be associated with byproduct metals than with lead, because lead used in printed wiring board assembly is such a small fraction of existing overall lead demand.

An additional conundrum is raised where an option presupposes substantial demand increases, as here with indium, bismuth and, to some extent, silver. The demand restructuring will raise prices and reduce materials availability for competing uses. Bismuth is heavily used in pharmaceuticals, as a substitute for more toxic lead in a number of metallurgical applications, and in fire control systems. Indium is used for, among other things, infrared reflecting window coatings, windshield defrosters for airplanes, as a surface coating on engine bearings, and in new generation solar cells. While some concern is indicated in the significant externalities category for this reason, the more general point is that methodologies for evaluating such social welfare externalities do not exist, and will be difficult to develop because ethical and value judgments are required.

A.2.4   Toxicity/Exposure Matrices

The indium toxicity/exposure matrix is shown in Figure A.3. This section of the analysis covers topics that have traditionally been part of environmental, health, and safety practices in government and private industry. Accordingly, they are relatively self-explanatory. The only caution that must be exercised is not to let the available data distort the analysis. This is difficult because even for a substance used as long and as extensively as lead, there will be far more data available on mammalian toxicology

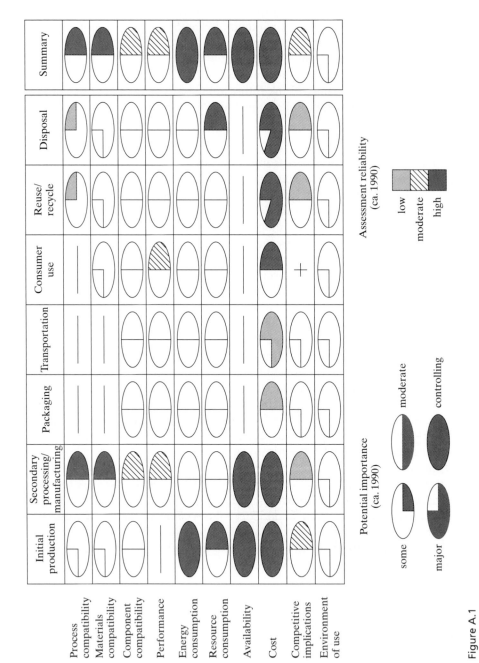

Figure A.1

The indium manufacturing matrix for printed wiring board assembly.

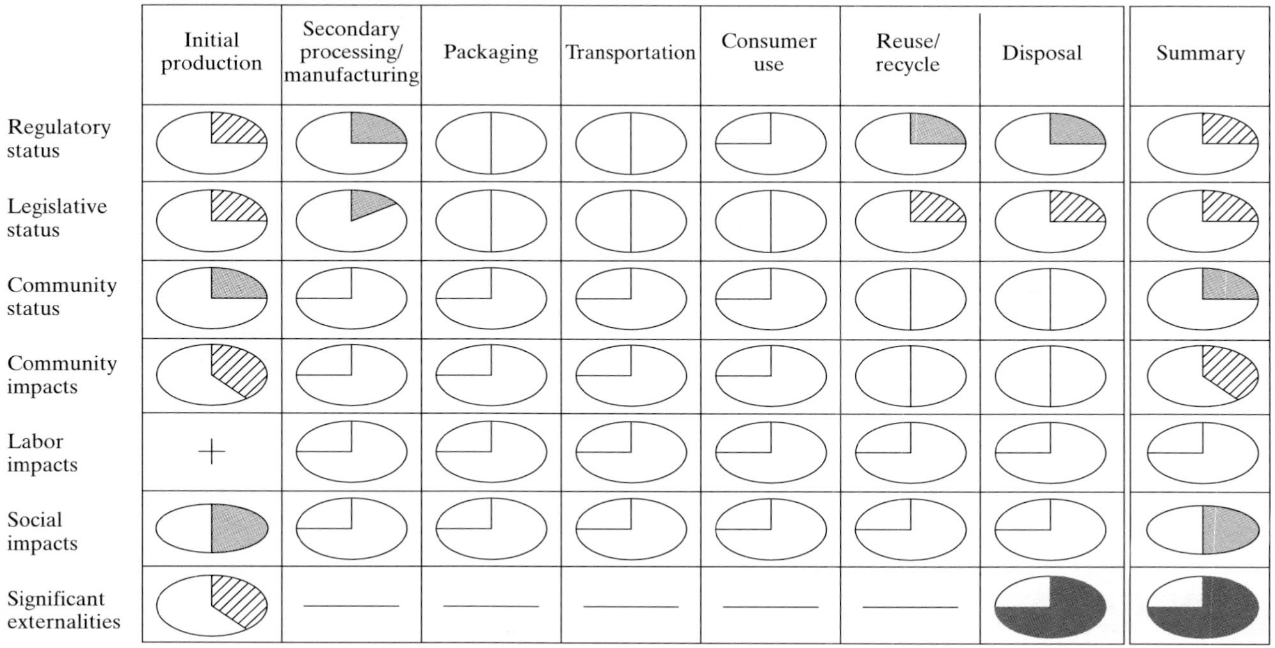

Figure A.2

The indium social/political matrix for printed wiring board assembly. (The key to the symbols is given in Figure A.1.)

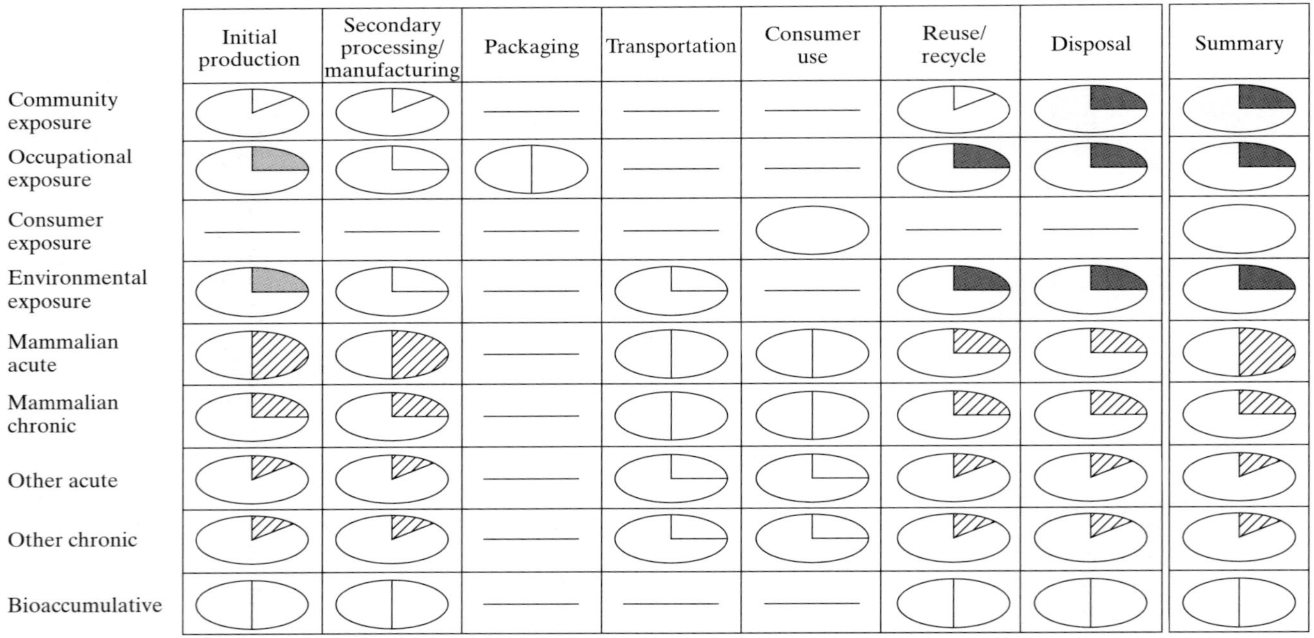

Figure A.3

The indium toxicity/exposure matrix for printed wiring board assembly. (The key to the symbols is given in Figure A.1.)

than on nonmammalian toxicology. For less well-studied materials, such as indium and bismuth, the nonmammalian toxicology is essentially unknown, and projections must be made from few data indeed.

In the same sense, reliably predicting exposures may be difficult, especially where exposures would change dramatically as a result of the implementation of an option. In the present assessment, one recognizes that there are currently no indium or bismuth recycling operations, so it is difficult to evaluate what exposures would result if substantial substitution of these materials for lead in solder were to occur and recycling operations were to be developed. Projections from current recycling operations can be made, of course, but they must be treated as approximations at best.

The point to be made is that even in an area where substantial attention has been paid to date, a systems-based approach reveals considerable data and methodological deficiencies. Nonetheless, qualitative assessments turn out to be both possible and defendable.

A.2.5   Environmental Matrices

The indium environmental matrix is presented in Figure A.4. What is most notable is that almost all severe impacts occur at life-cycle stages not under the control of the firm using the lead solder to manufacture printed wiring boards: The initial production life-cycle stage is controlled by the mining/processing firm and the reuse/recycle and disposal stages occur after consumer use. Accordingly, were the firm to evaluate only the environmental impacts of its operations in choosing among options, it would be ignoring virtually all of the major environmental impacts associated with its choice. The necessity of a life-cycle, systems-based approach is thus clearly demonstrated by these matrices.

In the secondary processing/manufacturing life-cycle stage, several additional considerations arise. First, replacement of any existing technology before its useful life is over in some sense wastes existing capital stock. All other things equal this is undesirable, as the embodied energy and resources in that stock are therefore not used as extensively as they otherwise might be. This potential loss is particularly likely for the conductive epoxy technology, which would render existing soldering equipment obsolete.

A.2.6   Summary

The four matrices have been presented as though they were aspects of the same stage of analysis. In actuality, they perform (in a qualitative way) two stages of life-cycle assessment. The first stage, inventory analysis, is effectively captured by the manufacturing matrix. The other three matrices deal not with emissions, consumption, or other inventory parameters, but with the effects of those activities; they constitute a qualitative impact assessment, stage 2 of LCA. Thus, although the analysis does not involve the level of computational detail of, for example, the Swedish EPS technique, it goes well beyond one-stage LCA approaches.

The results of the detailed analyses can be captured in a summary matrix (Figure A.5), which is constructed by recording the most serious levels of concern obtained for the bismuth, indium, lead and epoxy options in each detailed matrix. For the toxic-

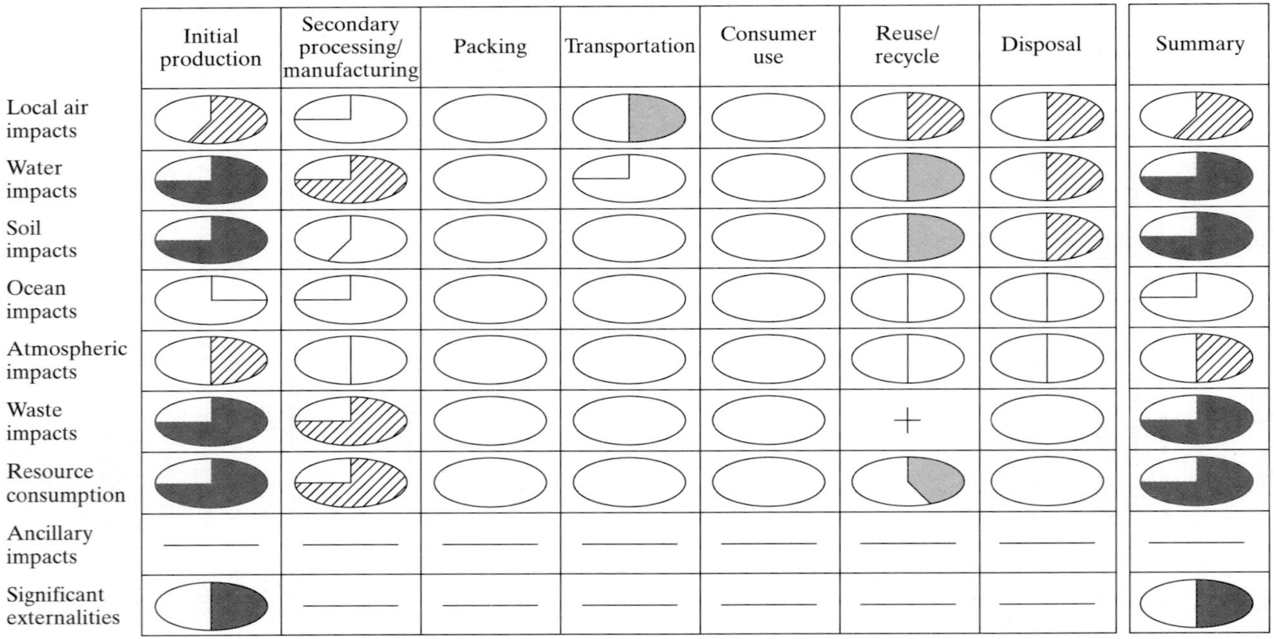

Figure A.4

The indium environmental matrix for printed wiring board assembly. (The key to the symbols is given in Figure A.1.)

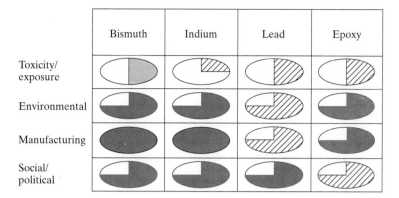

Figure A.5

The summary matrix for printed wiring board assembly. (The key to the symbols is given in Figure A.1.)

ity/exposure matrix, however, the exposure and toxicity rankings for each substance are combined into one ranking in the summary matrix. This ranking reflects the overall hazard risk posed by the option, and some additional explanation is appropriate. For bismuth, the ranking reflects a low level of concern because exposure potentials are limited and existing data on bismuth toxicity do not raise serious issues. Indium receives a moderate concern ranking for overall risk based on moderate levels of toxicity but limited exposures across the lifecycle. Lead poses clear and severe toxicity issues, especially in the mammalian chronic category. However, the amount of exposure associated with this particular use of lead is so low that it is not appropriate to rank that risk as serious. Finally, some of the organic constituents of epoxy pose significant toxicity risks and the silver component exhibits substantial aquatic toxicity for some organisms. Moreover, there is the possibility that some exposure to epoxy precursors may occur at the initial production and secondary processing/manufacturing life-cycle stages. Accordingly, a moderate risk ranking for silver epoxy is appropriate.

The results of the impact assessment can be stated simply: The status quo, lead solder, is preferable to substantial substitution of alloys containing significant amounts of bismuth or indium, or by epoxies containing significant amounts of silver. When the relatively minor component of overall lead demand attributable to printed wiring board assembly applications is contrasted with the significantly expanded mining and processing activities that the other options would entail, lead-based solders are the least environmentally harmful choice. Thus, a systematic analysis has led to what is a counterintuitive result.

A.3   FROM ASSESSMENT TO PLANS

The third stage of LCA, improvement analysis, follows directly from the summary matrix. It is clearly appropriate to reduce the use of lead in printed wiring board assembly where feasible. First, it makes sense to investigate low volume, niche applications for indium, bismuth and conductive epoxy alternatives. Second, it is advisable to investigate new fastening techniques and new electronic circuit designs requiring fewer interconnections.

The analysis also identified important areas for future research, including the possibility of niche applications of indium and bismuth alloys and the development of epoxy and polymer systems containing minimal amounts of conductive metal. This research and development effort would be guided by the analysis: Most resources would flow toward the generally environmentally preferable alternatives of low metal content polymeric systems, with relatively less expended to evaluate indium and bismuth alternatives.

There are two other noteworthy facets of the analysis. The first is that the environmental impacts that drive the analytical conclusions occur in life-cycle stages that a traditional environmental analysis by the manufacturer, focused on its operations and its customers, would miss. The second is that this exercise raises value and ethical issues that have not yet been addressed in a comprehensive manner. Such value questions require cultural and social responses, not responses by private firms responding to short-term profit motives. They can be identified, but not resolved, by DfE analyses.

APPENDIX B

Units of Measurement in Industrial Ecology

The basic unit of energy is the joule (1×10^7 erg). One will often see the use of the British thermal unit (Btu), which is 1.55×10^3 J. For very large energy use, a unit named the *quad* is common; it is shorthand for one quadrillion British thermal units. Thus, 1 quad $= 1 \times 10^{15}$ Btu $= 1.55 \times 10^{18}$ J.

The units of mass in the environmental sciences and in this book are given in the metric system. Since many of the quantities are large, the prefixes given in Table B.1 are common. Hence, we have such figures as 2 Pg $= 2 \times 10^{15}$ g. Where the word tonne is used, it refers to the metric ton $= 1 \times 10^6$ g.

The most common way of expressing the abundance of a gas phase atmospheric species is as a fraction of the number of molecules in a sample of air. The units in common use are *parts per million* (ppm), *parts per billion* (thousand million; ppb), and *parts per trillion* (million million; ppt), all expressed as volume fractions and therefore abbreviated ppmv, ppbv, and pptv to make it clear that one is not speaking of fractions in mass. Any of these units may be called the *volume mixing ratio* or *mole fraction*. Mass mixing ratios can be used as well (hence, ppmm, ppbm, pptm), a common example being that meteorologists use mass mixing ratios for water vapor. Since the pressure of the atmosphere changes with altitude and the partial pressures of all the gaseous constituents in a moving air parcel change in the same proportions, mixing ratios are preserved as long as mixing between air parcels can be neglected.

Particles can be mixtures of solid and liquid, so a measure based on mass replaces that based on volume, the usual units for atmospheric particles being micrograms per cubic meter ($\mu g\ m^{-3}$) or nanograms per cubic meter (ng m^{-3}). For particles in liquids, micrograms per cubic centimeter ($\mu g\ cm^3$) is common. It is sometimes convenient to compare quantities of an element or compound present in more than one phase, say as

TABLE B.1   Prefixes for Large and Small Numbers

Power of 10	Prefix	Symbol
+24	yotta	Y
+21	zetta	Z
+18	exa	E
+15	peta	P
+12	tera	T
+9	giga	G
+6	mega	M
+3	kilo	k
−3	milli	m
−6	micro	mu
−9	nano	n
−12	pico	p
−15	femto	f
−18	atto	a
−21	zepto	z
−24	yocto	y

both a gas and as a particle constituent. In that case, the gas concentration in volume units is converted to mass units prior to making the comparison.

For constituents present in aqueous solution, as in seawater, the convention is to express concentration in volume units of moles per liter (designated M) or some derivative thereof (one mole [abbreviated mol] is 6.02×10^{23} molecules). Common concentration expressions in environmental chemistry are millimoles per liter (mM), micromoles per liter (μM), and nanomoles per liter (nM). Sometimes one is concerned with the combining concentration of a species rather than the absolute concentration. A combining concentration, termed an *equivalent*, is that concentration which will react with 8 grams of oxygen or its equivalent. For example, one mole of hydrogen ions is one equivalent of H^+, but one mole of calcium ions is two equivalents of Ca^{2+}. Combining concentrations have typical units of equivalents, milliequivalents, or microequivalents per liter, abbreviated eq/l, meq/l, and μeq/l. A third approach is to express concentration by weight, as mg/l or ppmw, for example. Concentration by weight can be converted to concentration by volume using the molecular weight as a conversion factor.

Acidity in solution is expressed in pH units, pH being defined as the negative of the logarithm of the hydrogen ion concentration in moles per liter. In aqueous solutions, pH = 7 is neutral at 25°C. Lower pH values are characteristic of acidic solutions; higher values are characteristic of basic solutions.

Glossary

Acid deposition—The deposition of acidic constituents to a surface. This occurs not only by precipitation but also by the deposition of atmospheric particulate matter and the incorporation of soluble gases.

Acquirotroph—An industrial organism that extracts ores or other useful materials to inaugurate industrial food webs.

Acute—In toxicology, an effect that is manifested rapidly (e.g., minutes, hours, or even a few days) after exposure to a hazard; either the exposure that generates the response, or the response itself, may be termed acute (compare with *Chronic*, below). For example, an oral poison such as cyanide would be acutely toxic.

Anthropogenic—Derived from human activities.

Aquifer—Any water-bearing rock formation or group of formations, especially one that supplies ground water, wells, or springs.

Autotroph—An organism capable of assimilating energy from inorganic compounds.

Bioaccumulation—The concentration of a substance by an organism above the levels at which that substance is present in the ambient environment. Some forms of heavy metals, and chorinated pesticides such as DDT, are bioaccumulated.

Biomagnification—The increasing concentration of a substance as it passes into higher trophic levels of a food web. Many substances which are bioaccumulated are also biomagnified.

Biosphere—That spherical shell encompassing all forms of life on Earth. The biosphere extends from the ocean depths to a few thousand meters of altitude in the atmosphere, and includes the surface of land masses. Alternatively, the life forms within that shell.

This glossary was compiled from various sources, including: T.E. Graedel and P.J. Crutzen, *Atmospheric Change: An Earth System Perspective*, W.H. Freeman, New York, 446 pp., 1993; and B.W. Vigon, D.A. Tolle, B.W. Cornaby, H.C. Latham, C.L. Harrison, T.L. Boguski, R.G. Hunt, and J.D. Sellers, *Life Cycle Assessment: Inventory Guidelines and Practices*, Report EPA/600/R - 92/245, U.S. Environmental Protection Agency, Washington, DC, 1993.

Budget—A balance sheet of the magnitudes of all of the sources and sinks for a particular species or group of species in a single reservoir.

By-product—A useful product that is not the primary product being produced. In life-cycle analysis, by-products are treated as coproducts.

Carcinogen—A material that causes cancer.

Cascade recycling—See *Open-loop recycling*.

CFCs—Chlorofluorocarbon compounds, i.e., organic compounds that contain chlorine and/or fluorine atoms. CFCs are widely recognized as hazardous to stratospheric ozone.

Chronic—In toxicology, an exposure or effect of an exposure which becomes manifest only after a significant amount of time—weeks, months, or even years—has passed. Many carcinogens (substances causing cancer) are chronic toxins, and low-level exposure to many heavy metals, such as lead, produces chronic, rather than acute, effects.

Closed-loop recycling—A recycling system in which a particular mass of material is remanufactured into the same product (e.g., glass bottle into glass bottle). Also known as *Horizontal recycling*.

Comet diagram—A diagram that details the stages of product lifetime and the opportunities for end-of-life reuse of products, components, and materials.

Commons—A geographical region or entity under common ownership or without ownership (e.g., a village green, the oceans, the atmosphere).

Coproduct—A marketable by-product from a process. This includes materials that may be traditionally defined as waste, such as industrial scrap that is subsequently used as a raw material in a different manufacturing process.

Cultural construct—A social idea or approach that has become absolute and unquestioned.

Cycle—A system consisting of two or more connected *Reservoirs*, where a large part of the material of interest is transferred through the system in a cyclic manner.

Depletion time—The time required to exhaust a resource if the present rate of use remains unchanged.

Design for environment—An engineering perspective in which the environmentally related characteristics of a product, process, or facility design are optimized.

Discount rate—A rate applied to future financial returns to reflect the time value of money and inflation.

Disposal—Discarding of materials or products at the end of their useful life without making provision for *Recycling* or *Reuse*.

Dissipative product—A product that is irretrievably dispersed when it is used (e.g., paint, fertilizer).

Dose-response curve—A curve plotting the known dose of a material administered to organisms against the percentage response of the test population. If the material is not directly administered, but is present in the environment surrounding the organism (e.g., water, air, sediment), the resulting curve is know as a concentration-response curve.

Ecology (biological)—The study of the distribution and abundance of organisms and their interactions with the physical world.

Ecoprofile—A quick overview of a product or process design to ensure that no disastrous features are included or to determine whether additional assessment is needed.

Elemental flow analysis—An analysis of the flows of a specific element within and across the boundaries of a particular geographical region.

Embodied energy—The energy employed to bring a particular material or product from its initial physical reservoir or reservoirs to a specific physical state.

Emissions inventory—An assessment of the types, magnitudes, and receptors of losses from a manufacturing process, a facility, a corporation, or a geographical entity.

Employotroph—An industrial organism that employs products for various purposes.

Energy audit—An accounting of input flows, output flows, and losses of energy within an industrial process, a facility, a corporation, or a geographical entity.

Exergy—The available energy in a system. Strictly speaking, the work that could be extracted by taking a system with the same temperature and pressure as the environment reversibly to the same chemical composition as in the environment.

Expert system—A computer-based system combining knowledge in the form of facts and rules with a reasoning strategy specifying how the facts and rules are to be used to reach conclusions. The system is designed to emulate the performance of a human expert or a group of experts in arriving at solutions to complex and not completely specified problems.

Exposure—Contact between a *Hazard* and the target of concern, which may be an organ, an individual, a population, a biological community, or some other system. The confluence of exposure and hazard give rise to risk.

Externality—A cost that is not captured within the economic system and thus is not reflected in prices.

Flux—The rate of emission, absorption, or deposition of a substance from one *Reservoir* to another. Often expressed as the rate per unit area of surface.

Food chain—A sequence in which resources flow in linear fashion from one trophic level to the next.

Food web—A pattern in which resources flow largely from one trophic level to the next but may also flow across trophic levels in nonlinear fashion.

Fossil fuel—A general term for combustible geological deposits of carbon in reduced (organic) form and of biological origin, including coal, oil, natural gas, oil shales, and tar sands.

Fugitive emissions—Emissions from valves or leaks in process equipment or material storage areas that are difficult to measure and do not flow through pollution control devices.

Full-cost accounting—An accounting system in which environmental costs are built directly into the prices of products and services.

Gangue—The nonmetalliferous or worthless materials contained in ore.

Global warming—The theory that elevated concentrations of certain anthropogenic atmospheric constituents are causing or will cause an increase in Earth's average temperature.

Green accounting—An informal term referring to management accounting systems that specifically delineate the environmental costs of business activities rather than to include those costs in overhead accounts.

Green chemistry—Employing chemical techniques and methodologies that reduce or eliminate the use or generation of feedstocks, products, by-products, solvents, and reagents that are hazardous to the environment.

Greenhouse effect—The trapping by atmospheric gases of outgoing infrared energy emitted by Earth. Part of the radiation absorbed by the atmosphere is returned to Earth's surface, causing it to warm.

Greenhouse gas—A gas with absorption bands in the infrared portion of the spectrum. The principal greenhouse gases in Earth's atmosphere are H_2O, CO_2, O_3, CH_4, and N_2O.

Hazard—(as used in risk assessment) A material or condition that may cause damage, injury, or other harm, frequently established through standardized assays performed on biological systems or organisms. The confluence of hazard and exposure create a risk.

Heterotroph—An organism that derives its energy and nutrients from organisms at lower trophic levels.

Heuristics—The intuitive rules used by experts in arriving at decisions. In expert systems, representations of heuristics are combined with algorithms (precise definitions and procedures) to reach conclusions concerning problems whose characteristics cannot be completely specified.

Hitchhiker resource—A resource acquired principally as a byproduct of the mining of other resources.

Home scrap—The waste produced within a fabricating plant, such as rejected material, trimmings, and shavings. Home scrap is recirculated within the fabricating plant and does not become external waste. (Also called *Prompt scrap.*)

Horizontal recycling—See *Closed-loop recycling.*

Impact analysis—The second stage of life cycle assessment, in which the environmental impacts of a process, product, or facility are determined.

Improvement analysis—The final stage of life cycle assessment, in which design for environment techniques are used in combination with the results of the first and second LCA stages to improve the environmental plan of a process, product, or facility.

Indicator—A nonquantitative measure of the status of a chosen parameter, environmental or otherwise.

Industrial ecology—An approach to the design of industrial products and processes that evaluates such activities through the dual perspectives of product competitiveness and environmental interactions.

Industrial symbiosis—See *Symbiosis.*

Inventory analysis—The first stage of life cycle assessment, in which the inputs and outputs of materials and energy are determined for a process, product, or facility.

Inviolates list—A list of design decisions never allowed to be taken by product or process designers.

Leachate—The solution that is produced by the action of percolating water through a permeable solid, as in a landfill.

Life cycle—The stages of a product, process, or package's life, beginning with raw materials acquisition, continuing through processing, materials manufacture, product fabrication, and use, and concluding with any of a variety of waste management options.

Life cycle assessment—A concept and a methodology to evaluate the environmental effects of a product or activity holistically, by analyzing the entire life cycle of a particular material, process, product, technology, service, or activity. The life cycle assessment consists of three complementary components: (1) goal and scope definition, (2) inventory analysis, and (3) impact analysis, together with an integrative procedure known as improvement analysis.

Material flow analysis—An analysis of the flows of materials within and across the boundaries of a particular geographical region.

Metabolic analysis—The analysis of the aggregate of physical and chemical processes taking place in an organism, biological or industrial.

Metric—A quantitative measure of the status of a chosen parameter, environmental or otherwise.

Mineral—A distinguishable solid phase that has a specific chemical composition, e.g., quartz (SiO_2) or magnetite (Fe_3O_4).

Molecular flow analysis—An analysis of the flows of a specific molecule within and across the boundaries of a particular geographical region.

Mutagen—A hazard that can cause inheritable changes in DNA.

Neurotoxin—A hazard that can cause damage to nerve cells or the nervous system.

New scrap—See *Prompt scrap.*

Nonpoint source—See *Source.*

NO_x—The sum of the common pollutant gases NO and NO_2.

Old scrap—See *Postconsumer solid waste.*

Omnivory—The acquisition of resources from organisms at several different trophic levels.

Open-loop recycling—A recycling system in which a product from one type of material is recycled into a different type of product (e.g., plastic bottles into fence posts). The product receiving recycled material itself may or may not be recycled. Also known as *Cascade recycling.*

Ore—A natural rock assemblage containing an economically valuable resource.

Overburden—The material to be removed or displaced that is overlying the ore or material to be mined.

Ozone depletion—The reduction in concentration of stratospheric ozone as a consequence of efficient chemical reactions with molecular fragments derived from anthropogenic compounds, especially CFCs and other halocarbons.

Packaging, primary—The level of packaging that is in contact with the product. For certain beverages, an example is the aluminum can.

Packaging, secondary—The second level of packaging for a product that contains one or more primary packages. An example is the plastic rings that hold several beverage cans together.

Packaging, tertiary—The third level of packaging for a product that contains one or more secondary packages. An example is the stretch wrap over the pallet used to transport packs of beverage cans.

Plating—The act of coating a surface with a thin layer of metal.

Point source—See *Source*.

Pollution prevention—The design or operation of a process or item of equipment so as to minimize environmental impacts.

Postconsumer scrap—See *Postconsumer solid waste*.

Postconsumer solid waste—A material that has served its intended use and has become a part of the waste stream. (Also called *Old scrap* and *Postconsumer scrap*.)

Prompt scrap—Waste produced by users of semifinished products (turnings, trimmings, etc.). This scrap must generally be returned to the materials processor if it is to be recycled. (Also called *Howe scrap* and *New scrap*.)

Recycling—The reclamation and reuse of output or discard material streams useful for application in products.

Remanufacture—The process of bringing large amounts of similar products together for purposes of disassembly, evaluation, renovation, and reuse.

Reserve—The total known amount of a resource that can be mined with today's technology at today's market prices.

Reserve base—The total known amount of a resource that can be mined, without regard for technology or market prices.

Reservoir—A receptacle defined by characteristic physical, chemical, or biological properties that are relatively uniformly distributed.

Reuse—Reemploying materials and products in the same use without the necessity for *Recycling* or *Remanufacture*.

Reverse fishbone diagram—A diagram detailing the steps involved in the disassembly of a product.

Risk—The confluence of exposure and hazard; a statistical concept reflecting the probability that an undesirable outcome will result from specified conditions (such as exposure to a certain substance for a certain time at a certain concentration).

Risk assessment—An evaluation of potential consequences to humans, wildlife, or the environment caused by a process, product, or activity, and including both the likelihood and the effects of an event.

Scenario—An alternative vision of how the future might unfold.

Servicizing—Satisfying a need by providing a service rather than by providing products that meet that need.

Slag—The fused residue that results from the separation of metals from their ores.

Smog—Classically, a mixture of smoke plus fog. Today the term has the more general meaning of any anthropogenic haze. Photochemical smog involves the production, in stagnant, sunlit atmospheres, of oxidants such as O_3 by the photolysis of NO_2 and other substances, generally in combination with haze-causing particles.

Solute—The substance dissolved in a *Solvent*.

Solution—A mixture in which the components are uniformly distributed on an atomic or molecular scale. Although liquid, solid, and gaseous solutions exist, common nomenclature implies the liquid phase unless otherwise specified.

Solvent—A medium, usually liquid, in which other substances can be dissolved.

Source—In environmental chemistry, the process or origin from which a substance is injected into a reservoir. *Point sources* are those where an identifiable source, such as a smokestack, can be identified. *Nonpoint sources* are those resulting from diffuse emissions over a large geographical area, such as pesticides entering a river as runoff from agricultural lands.

Stratosphere—The atmospheric shell lying just above the *Troposphere* and characterized by a stable lapse rate. The temperature is approximately constant in the lower part of the stratosphere and increases from about 20 km to the top of the stratosphere at about 50 km.

Streamlined life cycle assessment (SLCA)—A simplified methodology to evaluate the environmental effects of a product or activity holistically, by analyzing the most significant environmental impacts in the life cycle of a particular product, process, or activity. The streamlined life cycle assessment consists of three complementary components—(1) restricted inventory analysis, (2) abridged impact assessment, and (3) improvement analysis—together with an integrative procedure known as scoping.

Stressor—A set of conditions that may lead to an undesirable environmental impact.

Substance flow analysis—An analysis of the flow of a specific chemically identifiable substance within and across the boundaries of a particular geographical region. The word substance typically incorporates atoms and molecules of the entity of interest without regard for chemical form.

Sustainability—In the context of industrial ecology, the state in which humans living on Earth are able to meet their needs over time while nurturing planetary life-support systems.

Symbiosis—A relationship within which at least two willing participants exchange materials, energy, or information in a mutually beneficial manner.

Tailings—The residue remaining after ore has been ground and the target metallic minerals separated and retained.

Teratogen—A hazard that can cause birth defects.

Transformotroph—An industrial organism that transforms materials from one physical or chemical form to another.

Trophic level—A group of organisms that perform similar resource exchanges as part of natural food chains or food webs.

Troposphere—The lowest layer of the atmosphere, ranging from the ground to the base of the stratosphere at 10–15 km altitude, depending on latitude and weather conditions. About 85% of the mass of the atmosphere is in the troposphere, where most weather features occur. Because its temperature decreases with altitude, the troposphere is dynamically unstable.

Visibility—The degree to which the atmosphere is transparent to light in the visible spectrum, or the degree to which the form, color, and texture of objects can be perceived. In the sense of visual range, visibility is the distance at which a large black object just disappears from view as a recognizable entity.

Waste—Material thought to be of no practical value. One of the goals of industrial ecology is the reuse of resources, and hence the minimization of material regarded as waste.

Waste audit—An accounting of output flows and losses of wastes within an industrial process, a facility, a corporation, or a geographical entity.

Water audit—An accounting of input flows, output flows, and losses of water within an industrial process, a facility, a corporation, or a geographical entity.

Index

3M, 116
Acid deposition, 9
Acidification, 247
Acquirotroph, 46
Actinium resource status, 130, 135
Airbus Corporation, 261
Allen, D., 108
Aluminum:
　automotive, 127, 169, 220
　can production, 142
　energy consumption in production, 140–142
　recycling, 126
　resource status, 129, 134
American Electronic Association, 180
American Fiber Manufacturers Association, 163
Ametek Corporation, 153
Antarctica, 20
Antimony, resource status, 130
Argon, resource status, 129
Arsenic:
　resource status, 129, 135
　toxicity, 175
Arsine, 121
Asbestos, 119
Assimilation efficiency, 42
Astatine, resource status, 130, 135
AT&T Bell Laboratories, 121, 166

AT&T, 152, 217, 265
Ausubel, J., 289
Automobile:
　EPS case study, 199–202
　generic, 218–224
　material, 127
　recycling, 169–171
　resources in, 48
Automotive shredder residue, 169
Automotive technology system, 69, 301
Autotroph, 46

Barium, resource status, 129, 135
Basel Convention, 80
Bavarian Motor Works (BMW), 178
Benefit-cost analysis, 87–88
Bennett, E., 236
Beryllium, resource status, 129, 135
Biodiversity, 9
Biological ecology, 39–52, 268
Biological organism, 40–41
Biological system, 49–51
Biomass, 60, 139
Bismuth:
　in cartriges, 128
　in solder, 125, 339–248
　resource status, 130